# Essentials of Numerical Analysis
# with Pocket Calculator Demonstrations

Peter Henrici

1807 1982

175 YEARS OF PUBLISHING

**John Wiley & Sons**

New York    Chichester    Brisbane    Toronto    Singapore

*To Garrett Birkhoff*

## Other Books by Peter Henrici

*Elements of Numerical Analysis* (1964)

*Discrete Variable Methods in Ordinary Differential Equations* (1962)

*Computational Analysis with the HP-25 Pocket Calculator* (1977)

*Applied and Computational Complex Analysis Vol. 1* (1974)

*Applied and Computational Complex Analysis Vol. 2* (1977)

QA
297
.H42
1982

**Library of Congress Cataloging in Publication Data:**

Henrici, Peter, 1923–
    Essentials of numerical analysis, with pocket
calculator demonstrations.

    Bibliography: p.
    Includes index.
    1. Numerical analysis—Data processing.  2. Programmable
calculators.  I. Title.
QA297.H42        519.4        81-10468
ISBN 0-471-05904-8        AACR2

Printed in the United States of America

10 9 8 7 6 5 4 3 2 1

# Preface

This text grew out of an attempt to modernize my *Elements of Numerical Analysis* which first appeared in 1964. However, a new book has resulted. The treatment of the difference calculus has been eliminated. There now are substantial chapters on numerical linear algebra and on numerical Fourier analysis. The chapter on approximation has been enriched by sections on Hermite interpolation and on spline functions. A chapter on discrete computation, emphasizing such concepts as the numerical stability of an algorithm and the condition of a problem, now forms the basis of the whole work.

In addition, the number of numerical examples has been greatly increased. All examples, or *demonstrations* as they are called here, have been calculated on programmable pocket calculators. The numerous computational problems also are designed to be solved by means of such calculators. The majority of the demonstrations and problems can be performed on a calculator of the size of the HP-33E; a few (marked by the symbol $\oplus$) require the capacity of the HP-34C. Specific calculators are rarely mentioned in the text; however, in order to make the demonstrations more easily reproducible for the student and to encourage further experimentation, companion booklets will be available for some current calculators which contain fully documented programs for virtually all demonstrations and for selected computational problems. In this pragmatic manner, a student working with this book will not only acquire firsthand experience with the joys and pitfalls of numerical computation, but will also learn some rudiments of scientific computer programming.

The methods of numerical analysis are, of course, largely independent of the computing equipment on hand. In this sense, the book may serve as a general introduction to numerical analysis and may be used in conjunction with any computing system. The emphasis on pocket calculators serves to free the teaching of numerical analysis from the organizational constraints of a computing center, and to enable the student to do experimental work in a relaxed atmosphere without having the next student who is waiting for a turn at the terminal breathing down his neck.

The prerequisites for this book are modest. They include calculus (including functions of several variables), plus a smattering of linear algebra and of differential equations. Courses covering most of the material of the Chapters 1 through 6 are regularly taught at the Swiss Federal Institute of Technology to first-year students of electrical engineering, geodesy, earth sciences, physics, and mathematics. I am grateful for having been allowed in the fall of 1978 to teach this course to a similar audience of junior-level students at Stanford University in California.

A glance at the table of contents will show that I am not trying to teach, or to allude to, everything that is known in numerical analysis. Something must be left to more advanced courses. Thus, the important topics of eigenvalue problems (differential and algebraic), iterative methods in linear algebra, and partial differential equations are totally absent. Even within the compass of topics that I am trying to do justice to, I prefer to treat a few methods and algorithms in reasonable depth rather than to superficially mention a large number of methods without adequate discussion. My judgement of what is most useful and pedagogically feasible has usually dictated my choice of topics. In a few instances, when such considerations did not point strongly in any particular direction, I have given preference to methods that have small memory requirements, and thus are suitable for pocket calculators.

While restricting the number of algorithms treated, I have taken pains, as I did in my *Elements*, to emphasize the connection of numerical analysis with certain topics of theoretical analysis that are in danger of being slighted in the structure-oriented treatment of mathematics which has become fashionable. Some examples are the emphasis on the asymptotic behavior of sequences throughout the book; the allusion to the calculus of variations in the treatment of splines in Chapter 5; the Euler-Maclaurin sum formula in Chapter 6; and the introduction to Fourier series from a least squares point of view in Chapter 7. In all these instances, an effort has been made to preserve an elementary and relaxed level of presentation.

In the Introduction I try to make clear that I regard numerical analysis as an essential tool of *applied* mathematics. To stress this fact I have often attempted, especially in the problem sections, to embed the numerical problem under study in an applied environment. I realize that even much more should be done in this direction. As anyone involved in the teaching of elementary courses knows, however, there is the difficulty that, at the level for which this book is intended, the acquaintance with mathematical models of applied situations is still very limited, especially considering the diversity of the backgrounds of my potential readers.

It is my pleasure to express my thanks to a number of individuals who helped me to shape this book. W. Robert Mann and David R. Kincaid have read the entire manuscript and have suggested many valuable improvements. Several of my assistants, notably Peter Geiger, have contributed to the exercises and have ironed out inconsistencies and outright errors. My wife, Marie-Louise Henrici, has examined the numerical work with great care. I have profited from conversations with Walter Gander who has generously shared his expertise in practical numerical analysis. Finally, anybody familiar with the work of Heinz Rutishauser will recognize my indebtedness to this giant of modern numerical analysis; my way of presenting the material (especially in Chapter 4) often is rooted in his teachings, as preserved in his *Vorlesungen über numerische Mathematik* (2 vols., Birkhäuser, Basel 1976).

I dedicate this volume to Garrett Birkhoff who in his lectures, in his teaching, and in his written work, has set a standard for doing justice to both dialectic and algorithmic mathematics, and for dealing with genuine applications while maintaining a high level of mathematical polish and intellectual purity. May his example prolong the traditional unity of mathematics.

Peter Henrici
*Zurich*

# Contents

# INTRODUCTION

C. G. J. Jacobi (b. 1804, d. 1851), the famous mathematician from Königsberg (the Jacobian matrix mentioned in §1.7 is named after him) once declared: "Mathematics solely and exclusively serves the honor of the human spirit." Nevertheless, mathematics time and again has been *applied*; that is, used to control man's relation to his physical environment.

The old Babylonians used special cases of the theorem of Pythagoras to generate right angles. Eratosthenes of Kyrene (b. −290, d. −214) knew that the planet earth is a sphere and used simple trigonometry to estimate its radius. Archimedes (b. −285, d. −215), in addition to numerous mathematical achievements, discovered the basic laws of statics and used them to construct engines of war which enabled Syracuse, his home, to withstand the Roman siege for two years. Johannes Kepler (b. 1571, d. 1630) explained the motions of the planets by means of conic sections. Isaac Newton (b. 1643, d. 1727) formulated the basic laws of dynamics and reduced all of celestial mechanics to a single mathematically formulated law, the law of gravitation, which even today serves to compute the trajectories of satellites and space ships. The Swiss mathematician Leonhard Euler (b. 1707, d. 1783), while still a student, won a Paris Academy prize with a paper on where to place masts on sailing vessels. His collected works abound with numerical applications of mathematics to all branches of physics that were then known. Carl Friedrich Gauss (b. 1777, d. 1855) was active as a surveyor and astronomer and created the method of least squares, a basic tool of applied mathematics. Bernhard Riemann (b. 1826, d. 1866), considered by many the greatest mathematician of the nineteenth century, laid the groundstone for what was to become a standard textbook of mathematical physics. Henri Poincaré (b. 1854, d. 1912), the founder of modern topology, wrote a *Méchanique Céleste* in three volumes.

Since World War II, the application of mathematics has been facilitated by the electronic computer with its enormous speed of calculation and its automatically executed programs. The atom bomb would have been impossible without large-scale computations, and the same holds for the conquest of space or for the peaceful uses of atomic energy. Today, mathematical methods are even used in fields, such as medicine and economics, that formerly seemed rather removed from quantification.

Most applications of mathematics exhibit a common pattern. First, an intellectual *model* is constructed of the phenomenon to be investigated. Usually one will try to keep this model as simple as possible; however, all relevant effects should be taken into account in an unambiguous manner. To keep the model simple it is quite customary to make simplifying assumptions. In studies of planetary motion, the planets are assumed to be points of mass. In models of economic science, functional dependencies are frequently assumed to be *linear*.

If complex phenomena are modelled mathematically, one first will have to ask whether the model used is *meaningful*. For instance, in the modelling of phenomena in classical physics the model frequently consists of a *differential equation* or of a system of such equations. Here, one will have to ask whether a solution of the equation exists, whether it depends continuously on the data (which usually are not precisely known), and whether the solution (as in nature) is determined uniquely by the data. Similar questions will be asked where the model consists of a system of algebraic equations. Questions of this kind, which relate to the meaningfulness of a mathematical model, are frequently resolved in a completely satisfactory manner by theoretical, "dialectic" mathematics, worked solely with pencil and paper.

The methods of dialectic mathematics may fail, however, if the model is to be used to draw quantitative conclusions. Consider space travel. Theoretical mathematics is very well able to assure the *existence* of a solution of the system of differential equations describing the trajectory of a space vehicle, but if the problem is to land on the moon, a knowledge of the mere existence of the correct trajectory is not sufficient. This trajectory must also be *constructed*; that is, the differential equations have to be solved.

This is the point where numerical mathematics springs into action. Its task is to develop methods for extracting quantitative answers from mathematical models. In this sense, numerical mathematics is a secondary science. Its points of reference are not the axioms of pure mathematics, but models and concrete problems which usually originate elsewhere. If one considers the ideal mathematical theory to be one that is purely abstract, axiomatic, and independent of extraneous considerations, then the preoccupation of numerical mathematics with *ad hoc* problems is clearly unsatisfactory. The inflexible concern with a given problem, which cannot be changed at will, somehow seems to stifle the free flight of imagination and at times gives numerical mathematics an aspect more akin to engineering than to pure mathematics. Whatever feelings of constraint may result from this should be compensated for by an awareness of the indispensable role which numerical mathematics has played time and again in solving man's problems in a modern world.

As far as certain standardized models of applied mathematics are concerned (ordinary differential equations, linear equations, approximation), numerical mathematics has created standardized methods of solution, called **algorithms,** which solve the problem in a routine fashion if the data do not exhibit pathological properties. Such standard algorithms are incorporated in the program libraries of most large computing centers and may be used without a detailed knowledge of numerical mathematics.

In addition to these standard models, in practice, models of a non-standard type frequently occur which require the development of special methods of computation, taking into account the pecularities of the model. In such instances, a knowledge of some principles of numerical mathematics is usually helpful. The art of *computer programming*, in addition to developing useful methodologies, serves to codify solution algorithms so that by means of a compiler they can be translated directly into a machine program.

In Chapters 2 through 7 of this book, solution algorithms for some typical problems of applied mathematics will be presented and, to the extent made possible by the modest prerequisites required, studied mathematically. In Chapter 1, we address the basic problem of *stability*, which must always be faced when the mathematical model involves a continuum, regardless of the special problem on hand or the algorithm used for solving it.

# CHAPTER 1

# Computation

In this chapter, we call attention to the fact that the number system of any computer is *discrete*, which implies that the results of even the simplest arithmetic operations and function evaluations are inaccurate. Unless appropriate measures are taken, these inaccuracies through such effects as cancellation and smearing can significantly diminish, and sometimes destroy, the accuracy of the result of an extended computation.

Using examples we then proceed to a discussion of the notions of *numerical* and *mathematical* stability. The former is a property of an algorithm; the latter (also called *condition*) is a property of the problem to be solved. These notions are basic to all of numerical analysis.

## §1.1 THE DISCRETE NUMBER SYSTEM OF THE COMPUTER

In principle, numerical mathematics is concerned with all kinds of computation, including computations involving Boolean expressions, algebraic formulas, or formulas from the predicate calculus. In practice, however, numerical mathematics is concerned mainly with computations involving *numbers* or systems of numbers, such as vectors or matrices.

Frequently, the models of applied mathematics are based on the idea of a *continuum* in space or time. The variables of the models then are (systems of) *real numbers.* Many models in electrical engineering and in mathematical physics work with *complex numbers.* The models of economics frequently involve numbers of units, that is, *integers.* In mechanical engineering, for instance in the design of gearboxes, it is sometimes appropriate to assume that the variables of the model are *rational numbers.*

In principle, the representation of *integers* in a computer does not present any difficulties. Any computer internally works with a fixed **base** $b$. Here $b$ is an integer $\geq 2$; frequently $b$ is chosen as a power of 2; less frequently $b = 10$. (However, the input and output of numbers is nearly always performed in the base $b = 10$.) As is well known, a given integer $n \neq 0$ possesses a unique representation,

$$n = \pm (n_0 + n_{-1}b + n_{-2}b^2 + \cdots + n_{-k}b^k),$$

where the $n_i$ are integers satisfying $0 \leq n_i < b$ $(i = 0, -1, \ldots, -k)$, and where the representation is made unique by requiring that $n_{-k} \neq 0$. In ordinary notation, $n$ would be shown as

$$n_{-k}n_{-k+1} \cdots n_{-2}n_{-1}n_0 .$$

(The digits $n_i$ are numbered so that the indices increase from left to right.) Any integer $n$ thus is completely specified by its sign and by the sequence of its digits. The number 0 requires a special representation; among other things, it has no sign. (Regrettably, however, $-0 \neq +0$ on some calculators.) A difficulty arises from the fact that $k + 1$, the number of digits, is bounded. If, for instance, $b = 2$, and if 48 digits can be accommodated, then the largest integer that can be represented in the computer equals

$$2^{48} - 1 = 281, 474, 976, 710, 655.$$

Some applications, such as certain investigations in number theory, require the accommodation of larger integers. These would be represented in the "super" base $b = 2^{48}$, with digits which themselves are 48-digit integers in the base $b = 2$.

   *Rational numbers* are represented in the computer as they are conceived in pure mathematics, namely, as pairs of integers.

   The problems of numerical computation begin with the representation of *real numbers*. (Complex numbers, as in pure mathematics, are treated as ordered pairs of real numbers, subject to certain rules of computation.) It is clear that a computer cannot represent every real number. As is well known, the set of real numbers is not denumerable, whereas in the computer, a number must be characterized by denumerably many (in fact by finitely many) yes-or-no states. Two methods of representing real numbers are in use:

(A)  **Fixed point representation.**  This was the system used by the very early computers. The computer here works exclusively with numbers of the form,

$$x = \pm \sum_{i=m}^{n} x_i b^{-i},$$

where $m$ and $n$ are fixed integers (depending only on the computer) satisfying $m < n$ and (usually) $m \leq 0$, $n > 0$, and where the $x_i$ are integers satisfying $0 \leq x_i < b$. Every number that can be represented in the computer thus is determined by its sign and by the sequence of its digits,

$$\left(x_m, \, x_{m+1}, \, \ldots, \, x_n\right).$$

In ordinary notation this would be read as,

$$x_m x_{m+1} \cdots x_{-1} x_0 \cdot x_1 x_2 \cdots x_n.$$

In fixed point representation, the decimal (or binary, or "$b$-ary") point is always at the same place. For instance, the SEAC, the pioneering computer completed in 1951 at the National Bureau of Standards, worked with $b = 2$, $m = -1$, $n = 48$, which enabled it at least to accommodate approximations to such mathematically important numbers as $e$ and $\pi$.

If fixed point representation is used, the largest number that can be represented in the computer is $b^{1-m} - b^{-n}$; thus, for instance, on SEAC $4 - 2^{48}$. Before running a problem on a fixed point computer, it had to be prepared so that no final or intermediate result of the computation exceeded the maximum number that could be handled. This advance preparation, called **scaling**, could be extremely tedious. It had the advantage, however, that the details of the computation had to be looked into in advance with much greater care than is customary today.

**(B)** **Floating point representation.** This representation, which is much more flexible than fixed point representation, is almost universally used today. The computer works with numbers of the form

$$x = \pm y \cdot b^z. \tag{1}$$

Here,

> $b$ is the **base** of the number system used,
> $y$ is the **mantissa**,
> $z$ is the **exponent**

of the number $x$. The *mantissa* is a fixed point number,

$$y = \sum_{i=m}^{n} y_i b^{-i},$$

made unique (if $x \neq 0$) by the condition that $y_m \neq 0$. Frequently, in large computers $m = 1$, so that (if $x \neq 0$), the $b$-ary point is to the extreme left, and

$$b^{-1} \leq y < 1.$$

The *number of digits* of the mantissa in any case is $n - m + 1$. The *exponent z* is an integer, usually also represented in the base $b$.

**EXAMPLE 1.1-1:**  On the CDC computer in use at the ETH Zurich, $y$ is a 48-digit binary number with $m = 1$. The range of the exponent $z$ is approximately indicated by

$$10^{-293} \le 2^z \le 10^{322}.$$

**EXAMPLE 1.1-2:**  On the HP-33E pocket calculator, the mantissa is externally shown as a 7-digit number with $m = 0$. Thus, the mantissa shown satisfies

$$1 \le y \le 9.999999.$$

Internally, however, this calculator works with a 10-digit mantissa. The whole mantissa can be made visible by a special instruction. The range of the exponent is

$$-99 \le z \le 99.$$

We see that rather small and rather large numbers can be handled in floating point representation. This does not change the fundamental fact that the set of numbers which the computer has at its disposal is finite.

A number of interesting problems pose themselves to the engineer concerned with the design of digital computers, especially if the floating point representation is used. How should the basic arithmetic operations be performed? How should input and output be handled? How should the number 0 be treated? When should one round up, when down? We do not concern ourselves with these questions here. We simply recognize that the results of almost all operations have to be *rounded*.

Somewhat formalizing this fact, let $\mathbb{M} = \mathbb{M}_C$ be the set of numbers representable in the computer $C$. (This set generally depends on the computer; however, the index $C$ is omitted in the following.) If

$$x \in \mathbb{M}, \qquad y \in \mathbb{M},$$

the numbers

$$x + y, \qquad x - y, \qquad xy, \qquad \frac{x}{y}$$

are in general not in $\mathbb{M}$. The results of the basic arithmetical operations can be represented only approximately in the computer. Thus, for instance, the exact value of the product $xy$ must be approximated by some number $(xy)^*$ which is in $\mathbb{M}$. Assuming the engineers have worked well, the quantity

$$|(xy)^* - xy|$$

will be as small as possible; that is, the exact value $xy$ is represented by the (or a) machine number $(xy)^*$ closest to $xy$.

## PROBLEM

---

1  Find out the base, the number of digits in the mantissa, and the exponent range of your calculator. What is the smallest positive number representable in your calculator? What does your calculator do in the case of exponent underflow or overflow?

---

## §1.2  M IS NOT A FIELD

As a consequence of the discreteness of the number system of the computer, the simplest laws of algebra are no longer valid on the computer. We substantiate this claim, which at first sight may seem astonishing, by some simple examples. These examples were calculated on a programmable calculator of type HP-33E; however, similar examples could be found for any computer, large or small.

**DEMONSTRATION 1.2-1:**  In any mathematical system known as a group (for instance, in the additive group of real numbers), there holds the associative law

$$(a + b) + c = a + (b + c),$$

where the parentheses indicate the operation that is to be executed first. However, on our calculator, the data

$$a = 1, \qquad b = 3 * 10^{-10}, \qquad c = 3 * 10^{-10}$$

yield the different results:

$$(a + b) + c = 1.000000000,$$

$$a + (b + c) = 1.000000001.$$

**DEMONSTRATION 1.2-2:**  This is a more complicated example of the same kind. We consider the sum

$$s_n := 1 + \sum_{k=1}^{n} \frac{1}{k^2 + k}.$$

By writing this in the form,

$$s_n = 1 + \sum_{k=1}^{n} \left( \frac{1}{k} - \frac{1}{k+1} \right)$$

$$= 1 + (1 - \tfrac{1}{2}) + (\tfrac{1}{2} - \tfrac{1}{3}) + \cdots + \left( \frac{1}{n} - \frac{1}{n+1} \right)$$

it is clear that, mathematically,

$$s_n = 2 - \frac{1}{n+1},$$

so that, for instance,

$$s_{9999} = 1.9999.$$

Computing the sum numerically by the usual summation algorithm,

$$s_0 := 1, \qquad s_k := s_{k-1} + \frac{1}{k(k+1)}, \qquad k = 1, 2, \ldots,$$

we get

| $n$ | $s_n$ |
|---|---|
| 9 | 1.900000000 |
| 99 | 1.990000003 |
| 999 | 1.999000003 |
| 9999 | 1.999899972 |

However, if the summation is started at the tail end,

$$s_n = \frac{1}{(n+1)n} + \frac{1}{n(n+1)} + \cdots + \frac{1}{3 \cdot 2} + \frac{1}{2 \cdot 1} + 1,$$

then the results are as follows:

| $n$ | $s_n$ |
|---|---|
| 9 | 1.900000000 |
| 99 | 1.990000000 |
| 999 | 1.999000000 |
| 9999 | 1.999900000 |

The values obtained in the two modes of summation are different. Thus, the associative law also is violated here. Why does summing backwards yield the more accurate values? When the small terms are summed first, the exponents of the partial sums remain small until almost the very end. Accordingly, the rounding errors in the individual additions remain small until the exponent of the running sum reaches its final value, which in our example happens only at the very end. We conclude that, in forming a sum with a large number of terms, the terms with a small absolute value should be summed first.

**DEMONSTRATION 1.2-3:** In an additive group, an equation of the form $a + x = b$ always has a uniquely determined solution $x$. On our calculator the equation,

$$1 + x = 1$$

in addition to the mathematical solution $x = 0$ also has the solution $x = 10^{-10}$. More generally, every machine number $x$, satisfying $|x| < 5 * 10^{-10}$, is a solution of the equation. We conclude that on a computer equations need not have uniquely determined solutions.

**DEMONSTRATION 1.2-4:** It is well known that for arbitrary non-negative numbers $x$ and $y$, there holds the inequality of the arithmetic and the geometric mean,

$$\tfrac{1}{2}(x + y) \geq \sqrt{xy}$$

with equality holding only for $x = y$. On our calculator, however, the values

$$x = 5.000000001, \qquad y = 5.000000002$$

yield

$$\tfrac{1}{2}(x + y) = 5.000000000, \qquad \sqrt{xy} = 5.000000002.$$

Not only the basic laws of arithmetic, but also some fundamental facts of analysis lose their validity on the computer.

**DEMONSTRATION 1.2-5:** Let $a < b$, let $f$ be a real function continuous on the interval $[a, b]$, and let $f(a) < 0, f(b) > 0$. Then, according to the intermediate value theorem, there exists $x$ so that

$$f(x) = 0.$$

Consider however

$$f(x) := x^3 - 3.$$

We find

$$f(1.442249570) = -0.000000002,$$

whereas at the next greater number that can be handled by the computer the evaluation yields

$$f(1.442249571) = 0.000000004.$$

Thus, the set of machine numbers does not contain a number $x$ for which $f(x) = 0$. The equation $f(x) = 0$ on our calculator has no solution.

On all larger calculators the so-called elementary functions ($\sin x$, $\arcsin x$, $e^x$, $\log x$, and so on) are preprogrammed. (Some methods for the computation of such functions will be discussed in Chapter 5.) Here, we merely wish to call attention to the fact that it would be foolish to expect the function values calculated by the computer to be absolutely accurate. It could be postulated reasonably that for each machine number $x$ the computer should furnish, in place of the exact value $y = f(x)$, which in general is not a machine number, a machine number $y^*$ closest to $y$. The following demonstration shows that not even this postulate is always satisfied.

**DEMONSTRATION 1.2-6:** Many pocket calculators are internally programmed to calculate $y^x$ for arbitrary machine numbers $x$ and $y > 0$. For $x = y = 3$, the HP-25 yields

$$3^3 - 27 = 0.000000030.$$

The explanation lies in the fact that the evaluation of $y^x$ is carried out by means of the formula

$$y^x = e^{x \log y}$$

and thus is reduced to the computation of $e^x$ and of the natural logarithm $\log x$. Neither of these functions is computed accurately. At the time the HP-25 was designed, it was considered too cumbersome to provide for a special treatment of the function $y^x$ for integer values of $x$ and $y$. More recent calculators do have this provision. For instance, on the HP-33E, we get

$$3^3 - 27 = 0.000000000.$$

It could be argued reasonably that, in view of the large number of digits that are usually provided, rounding errors are harmless. In large numerical problems, however, such as the numerical solution of differential equations, the final result may well be the result of some $10^6$ or even $10^9$ individual operations. In unfavorable circumstances, the effects of the individual rounding errors may accumulate to such an extent that the final result is meaningless.

One of the main tasks of the numerical analyst is to design algorithms so that the effects of the discrete arithmetic of the computer remain harmless, even if a large number of operations are performed. In the following sections, we call attention to certain numerical effects, which we call

Cancellation,

Smearing,

Numerical instability,

Ill-conditioning.

These can endanger the performance of even very simple algorithms severely.

## PROBLEMS

1  Devise a program that tests the accuracy of the elementary functions provided by your calculator. For which percentage of machine numbers (approximately) does there hold

(a)  $\sqrt{x^2} = x$?

(b)  $(\sqrt{x})^2 = x$?

(c)  $e^{\log x} = x$?

(d)  $\log(e^x) = x$?

2  Assuming your calculator computes the sine function as well as possible, what is the largest number $x$ so that, on your calculator, $x = \sin x$? Establish your result by theoretical reasoning and verify experimentally.

## §1.3  CANCELLATION

**Cancellation** occurs in the subtraction of two nearly equal numbers. Assuming floating point arithmetic, let $a$ and $b$ both have the exponent $z$. When the difference $a - b$ is formed, it also will first be represented with the exponent $z$. Some leading digits of the mantissa will be zero, however; that is, significant digits will have been lost or "cancelled." Subsequently, the mantissa will be **normalized** by moving the digits to the left to satisfy the condition that the leading digit should be $\neq 0$. The exponent will be diminished accordingly. At the tail end of the mantissa there will appear zeros, which are meaningless.

**DEMONSTRATION 1.3-1:**   In floating point arithmetic,

$$\sqrt{9876} = 9.937806599 * 10^1,$$

$$\sqrt{9875} = 9.937303457 * 10^1.$$

Thus, to begin with

$$\sqrt{9876} - \sqrt{9875} = 0.000503142 * 10^1.$$

Normalization changes this to

$$5.031420000 * 10^{-3}.$$

The four zeros at the end of the mantissa are meaningless. How can we obtain a more accurate result? By using the identity

$$\sqrt{x} - \sqrt{y} = \frac{x - y}{\sqrt{x} + \sqrt{y}},$$

we obtain

$$\sqrt{9876} - \sqrt{9875} = \frac{1}{\sqrt{9876} + \sqrt{9875}} = 5.031418679 * 10^{-3},$$

which is accurate to all digits given.

**DEMONSTRATION 1.3-2:**   Suppose we wish to solve the quadratic equation

$$x^2 - 1634x + 2 = 0.$$

The textbook formula yields

$$x = 817 \pm \sqrt{667,487},$$

thus,

$$x_1 = 817 + 816.9987760 = 1.633998776 * 10^3,$$

$$x_2 = 817 - 816.9987760 = 1.224000000 * 10^{-3}.$$

Four of the six zeros at the end of the mantissa of $x_2$ are the result of cancellation, and thus are meaningless. Another more accurate result can be obtained by means of a simple arithmetical trick. According to the theory of equations,

$$x_1 x_2 = 2,$$

hence,

$$x_2 = \frac{2}{x_1}.$$

Using this formula, we obtain

$$x_2 = 1.223991125 * 10^{-3},$$

where all digits now are significant.

**DEMONSTRATION 1.3-3:**  Here we seek to determine the *length of a polygon*, whose vertices $P_k$ are given in polar coordinates $(r_k, \phi_k)$,

$$P_k = (r_k \cos \phi_k, r_k \sin \phi_k),$$

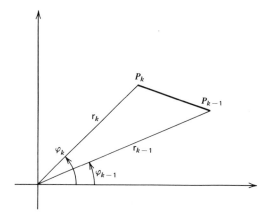

**Fig. 1.3a.**  Length of a polygon.

$k = 0, 1, 2, \ldots, n$. (See Figure 1.3a.) According to trigonometry, the length of the $k$-th segment of the polygon is

$$\overline{P_{k-1} P_k} = \sqrt{r_{k-1}^2 + r_k^2 - 2r_{k-1} r_k \cos(\phi_k - \phi_{k-1})}. \tag{1}$$

Let us now assume that the polygon consists of a large number of short segments. (We may wish to calculate the length of a curve approximately by inscribing a polygon.) Then, in each expression (1)

$$r_k \sim r_{k-1}, \qquad \phi_k \sim \phi_{k-1}, \qquad \cos(\phi_k - \phi_{k-1}) \sim 1,$$

and we see that the expression under the square root is subject to cancellation. Taking the root magnifies the effect, because for small $x > 0$, a small change in $x$ causes a larger change in $\sqrt{x}$. For a partial remedy, we use

$$\cos(\phi_k - \phi_{k-1}) = 1 - 2\{\sin \tfrac{1}{2}(\phi_k - \phi_{k-1})\}^2.$$

This yields

$$\begin{aligned}
\overline{P_{k-1} P_k} &= \sqrt{r_{k-1}^2 + r_k^2 - 2r_{k-1} r_k[1 - 2\{\sin \tfrac{1}{2}(\phi_k - \phi_{k-1})\}^2]} \\
&= \sqrt{(r_k - r_{k-1})^2 + 4r_{k-1} r_k\{\sin \tfrac{1}{2}(\phi_k - \phi_{k-1})\}^2}. \tag{2}
\end{aligned}$$

The two main terms under the root now are positive. Thus, there is no cancellation in forming their sum. There still will be some cancellation when forming $r_k - r_{k-1}$.

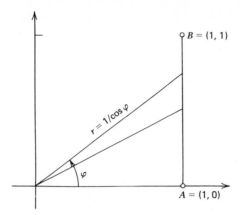

**Fig. 1.3b.**   Length of straight-line segment.

Because the order of magnitude of the terms of the difference now is* $O(r)$ rather than $O(r^2)$, however, the effect of cancellation will be less serious.

For numerical confirmation we (foolishly) compute the length of the straight-line segment joining the points $A = (1, 0)$ and $B = (1, 1)$. (See Figure 1.3b.) We subdivide the segment into $n$ non-congruent subsegments so that each subsegment is seen under the same angle from O. The vertices then are given by

$$\phi_k = \frac{\pi}{4n} k, \qquad r_k = \frac{1}{\cos \phi_k}, \qquad k = 0, 1, \ldots, n.$$

Computing the length of the subsegments by both formulas (1) and (2), we obtain the following results for the length of the whole segment:

| $n$ | (1) | (2) |
|---|---|---|
| 2 | 1.000000001 | 1.000000001 |
| 4 | 0.999999996 | 1.000000001 |
| 8 | 1.000000002 | 1.000000000 |
| 16 | 1.000000079 | 1.000000000 |
| 32 | 1.000000312 | 1.000000000 |
| 64 | 0.999999333 | 1.000000000 |
| 128 | 0.999996236 | 1.000000006 |
| 256 | 0.999986218 | 1.000000005 |
| 512 | 0.999985107 | 1.000000006 |
| 1024 | 0.999817631 | 1.000000006 |

---

* We write $f(r) = O(g(r))$, and say $f(r)$ is of the order of $g(r)$, if the ratio $f(r)/g(r)$ exists and is bounded as $r$ approaches some limit, such as 0, that is obvious from the context.

**DEMONSTRATION 1.3-4:** *Determination of π.* Following Archimedes, we calculate π as the limit of one half of the circumferences of the regular *n*-gons inscribed to the unit circle as *n* tends to ∞. (See Figure 1.3c.) The rule of the game is that the only non-rational operation that is permitted is the extraction of square roots.

Let $f(n)$ denote one half of the circumference of the regular *n*-gon. According to elementary trigonometry,

$$f(n) = n \sin \frac{\pi}{n}.$$

The fact that $\lim f(n) = \pi$ thus is obvious analytically. Our purpose, however, is to calculate this limit numerically. We do not assume that $\sin \alpha$ can be computed for arbitrary $\alpha$. (Otherwise, π could be computed as the smallest positive zero of the function $\sin x$.) If for $k = 0, 1, 2, \ldots$, however, we put

$$y_k := f(2^k) = 2^k \sin(2^{-k}),$$

then $y_1 = 2$, $y_2 = 2\sqrt{2}$; and further $y_k$ may be calculated by the formula

$$\sin \frac{\alpha}{2} = \sqrt{\tfrac{1}{2}(1 - \cos \alpha)}, \qquad 0 \le \alpha \le 2\pi.$$

Indeed, letting $\alpha := 2^{-k}\pi$, we have

$$y_{k+1} = 2^{k+1} \sin \frac{\alpha}{2}$$

$$= 2^{k+1}\sqrt{\tfrac{1}{2}(1 - \cos \alpha)}$$

$$= 2^{k+1}\sqrt{\tfrac{1}{2}(1 - \sqrt{1 - \sin \alpha^2})},$$

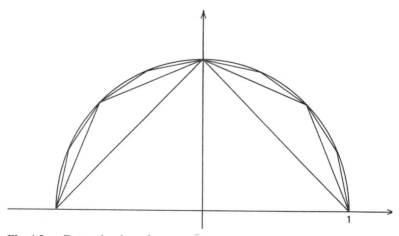

**Fig. 1.3c.** Determination of π.

and, in view of $\sin \alpha = 2^{-k} y_k$, this yields

$$y_{k+1} = 2^{k+1} \sqrt{\tfrac{1}{2}(1 - \sqrt{1 - [2^{-k} y_k]^2})}. \tag{I}$$

The resulting values are given in column (I) of the following table.

| k | (I) | (II) |
|---|-----|------|
| 1 | 2.00000000 | 2.00000000 |
| 2 | 2.82842713 | 2.82842712 |
| 3 | 3.06146746 | 3.06146746 |
| 4 | 3.12144515 | 3.12144515 |
| 5 | 3.13654849 | 3.13654849 |
| 6 | 3.14033115 | 3.14033116 |
| 7 | 3.14127726 | 3.14127725 |
| 8 | 3.14151378 | 3.14151380 |
| 9 | 3.14157167 | 3.14157294 |
| 10 | 3.14158627 | 3.14158772 |
| 11 | 3.14159462 | 3.14159142 |
| 12 | 3.14166137? | 3.14159234 |
| 13 | 3.14192837 | 3.14159257 |
| 14 | 3.14299615 | 3.14159263 |
| 15 | 3.14299615 | 3.14159264 |
| 16 | 3.14299615 | 3.14159264 |
| 17 | 3.21059520 | 3.14159264 |
| 18 | 3.21059520 | 3.14159264 |
| 19 | 3.70727600! | 3.14159264 |

$$\pi = 3.14159265 \ldots$$

Beginning with about $k = 12$, the values computed by means of (I) are quite inaccurate. There is no convergence. The disaster originates in forming the difference

$$1 - \sqrt{1 - [2^{-k} y_k]^2} \, ;$$

since (theoretically) $y_k \to \pi$, the expression under the square root tends to 1; thus, the root itself tends to 1, and on subtraction from 1 experiences violent cancellation.

Once again the situation can be remedied by a simple algebraic transformation. Consider

$$1 - x = \frac{1 - x^2}{1 + x}.$$

18 COMPUTATION

Letting $x := \sqrt{1 - [2^{-k}y_k]^2}$, this yields

$$y_{k+1} = y_k \sqrt{\frac{2}{1 + \sqrt{1 - [2^{-k}y_k]^2}}} . \tag{II}$$

The resulting values are found in column (II) of the table above. The sequence now is stable; it comes to rest at 3.14159264. The rounding error of one unit in the last digit must be expected in a non-self-checking process of this kind.

One still could try to generate the sequence $\{y_k\}$ more efficiently. Setting $\alpha := 2^{-k}\pi$, $y_k = 2^k \sin \alpha$, equation (II) says that

$$y_{k+1} = y_k \sqrt{\frac{2}{1 + \cos \alpha}} .$$

Above we used

$$\cos \alpha = \sqrt{1 - [\sin \alpha]^2} = \sqrt{1 - [2^{-k}y_k]^2}.$$

Remembering, however, the relation

$$\sin 2\alpha = 2 \sin \alpha \cos \alpha,$$

we have

$$\cos \alpha = \frac{\sin 2\alpha}{2 \sin \alpha} = \frac{2^{k-1} \sin(2^{-k+1}\pi)}{2^k \sin(2^{-k}\pi)} = \frac{y_{k-1}}{y_k},$$

and thus get

$$y_{k+1} = y_k \sqrt{\frac{2y_k}{y_{k-1} + y_k}} . \tag{III}$$

Because $y_1$ and $y_2$ are known, this formula may be used recursively to generate the whole sequence $\{y_k\}$. The resulting numerical values are (within the accuracy reported here) identical with the values furnished by (II); however, the computation runs about twice as fast.

This is not the last word about the computation of $\pi$ by the method of Archimedes. It is possible to achieve a spectacular *acceleration* of the convergence of the sequence $\{y_k\}$ by the *Romberg algorithm*. (See §6.3.)

**DEMONSTRATION 1.3-5:** *Running mean and standard deviation of a sequence of data.* Let $x_1, x_2, \ldots, x_n$ be given real numbers. In statistics their **mean** is defined by

$$\mu_n := \frac{1}{n} \sum_{k=1}^{n} x_k, \tag{3}$$

and their **variance** by

$$\sigma_n^2 := \frac{1}{n} \sum_{k=1}^{n} (x_k - \mu_n)^2. \tag{4}$$

The **standard deviation** is

$$\sigma_n := \sqrt{\sigma_n^2}\,; \tag{5}$$

this is a convenient measure of the average deviation of the $x_k$ from their mean. The definition of variance presupposes that the mean is already known; the calculation of the variance (and of the standard deviation) from (4) thus requires two runs through the data $x_k$. Elementary algebra now offers the trump card that mean and variance may be calculated in a single run by accumulating simultaneously the sum of the data

$$s_n := \sum_{k=1}^{n} x_k$$

and the sum of the *squares* of the data,

$$q_n := \sum_{k=1}^{n} x_k^2.$$

Indeed, we have

$$\mu_n = \frac{1}{n} s_n \tag{6}$$

and

$$\sigma_n^2 = \frac{1}{n} \sum_{k=1}^{n} (x_k^2 - 2\mu_n x_k + \mu_n^2)$$

$$= \frac{1}{n} \sum_{k=1}^{n} x_k^2 - 2\mu_n \frac{1}{n} \sum_{k=1}^{n} x_k + \mu_n^2 \frac{1}{n} \sum_{k=1}^{n} 1$$

$$= \frac{1}{n} q_n - 2\mu_n^2 + \mu_n^2\,;$$

thus,

$$\sigma_n^2 = \frac{1}{n} q_n - \mu_n^2. \tag{7}$$

The formulas (6) and (7) would seem to be particularly convenient for dealing with open-ended sets of data, where the number of data points $x_k$ is not known *a priori*, and where data points are added or removed at will. The resulting algorithm would be described by the flow diagram shown in Figure 1.3d. Here, upper signs refer to data being added, lower signs to data being removed.

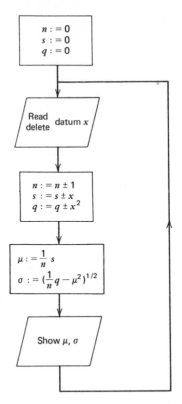

**Fig. 1.3d.** Running mean and standard deviation, classical version.

Many computer programs (including a built-in program of the HP-33E) use Algorithm 1.3–5a to compute means and standard deviations. In certain situations, however, this algorithm is subject to severe cancellation. Suppose the data all are approximately equal to their mean $\mu_n$,

$$x_k = \mu_n + \varepsilon_k, \qquad |\varepsilon_k| < \varepsilon,$$

where $\varepsilon$ is small, and $\mu_n \neq 0$. Then,

$$\frac{1}{n} q_n = \mu_n^2 + O(\varepsilon^2),$$

and the quantity

$$\sigma_n^2 = \frac{1}{n} q_n - \mu_n^2 = \mu_n^2 + O(\varepsilon^2) - \mu_n^2$$

is generated by subtracting two nearly equal numbers, and therefore tends to be inaccurate, as is shown by the following very simple example:

| $x_k$ | $\mu_k$ | $\sigma_k$ computed by (7) | $\sigma_k$ exact |
|---|---|---|---|
| 100,000 | 100,000.0000 | 0.00000000 | 0.00000000 |
| 100,101 | 100,000.5000 | 0.00000000 | 0.50000000 |
| 100,002 | 100,001.0000 | 0.00000000 | 0.81649658 |

It may even turn out that the numerical value of $\sigma_n^2$ as calculated by (7) turns out negative. Therefore, to avoid embarrassment, some programs compute the standard deviation by the hideous formula

$$\sigma_n = \sqrt{\left|\frac{1}{n} q_n - \mu_n^2\right|},$$

tantamount to forcing the computer to lie.

A more stable scheme results by computing the variances recursively from the definition (4). We have

$$(n+1)\sigma_{n+1}^2 = \sum_{k=1}^{n+1} (x_k - \mu_{n+1})^2$$

$$= \sum_{k=1}^{n} (x_k - \mu_{n+1})^2 + (x_{n+1} - \mu_{n+1})^2$$

$$= \sum_{k=1}^{n} [(x_k - \mu_n) - (\mu_{n+1} - \mu_n)]^2 + (x_{n+1} - \mu_{n+1})^2$$

$$= \sum_{k=1}^{n} (x_k - \mu_n)^2 - 2(\mu_{n+1} - \mu_n) \sum_{k=1}^{n} (x_k - \mu_n)$$

$$+ n(\mu_{n+1} - \mu_n)^2 + (x_{n+1} - \mu_{n+1})^2$$

$$= n\sigma_n^2 + 0 + n(\mu_{n+1} - \mu_n)^2 + (x_{n+1} - \mu_{n+1})^2.$$

From (6),

$$(n+1)\mu_{n+1} = n\mu_n + x_{n+1};$$

that is,

$$\mu_n = \frac{n+1}{n} \mu_{n+1} - \frac{1}{n} x_{n+1},$$

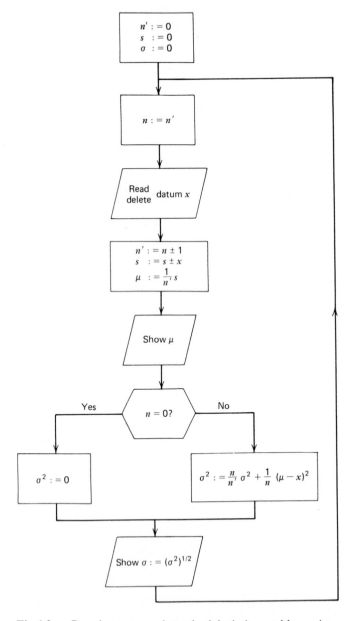

**Fig. 1.3e.** Running mean and standard deviation, stable version.

which implies

$$n(\mu_{n+1} - \mu_n)^2 = n\left(\mu_{n+1} - \frac{n+1}{n}\mu_{n+1} + \frac{1}{n}x_{n+1}\right)^2 = \frac{1}{n}(x_{n+1} - \mu_{n+1})^2.$$

There follows

$$(n+1)\sigma_{n+1}^2 = n\sigma_n^2 + \frac{n+1}{n}(x_{n+1} - \mu_{n+1})^2,$$

and finally

$$\sigma_{n+1}^2 = \frac{n}{n+1}\sigma_n^2 + \frac{1}{n}(x_{n+1} - \mu_{n+1})^2. \tag{8}$$

This yields the stable algorithm for computing standard deviations, shown in the flow diagram in Figure 1.3e.

**NUMERICAL EXAMPLE:** Let the given data be

$$x_k = 1 + \varepsilon\frac{2k - m - 1}{m - 1}, \qquad k = 1, 2, \ldots, m.$$

The exact values of $\mu_m$ and $\sigma_m$ are

$$\mu_m = 1, \qquad \sigma_m = \varepsilon\sqrt{\frac{1}{3}\frac{m+1}{m-1}}.$$

Keeping $m := 10$ fixed, we compute the standard deviations for various values of $\varepsilon$, using the two algorithms indicated above.

| $\varepsilon$ | Classical algorithm | Stable algorithm |
|---|---|---|
| $0.1\,\pi$ | 0.20052306 | 0.20052306 |
| $0.01\,\pi$ | 0.02005231 | 0.02005231 |
| $0.001\,\pi$ | 0.00200524 | 0.00200523 |
| $0.0001\,\pi$ | 0.00020000 | 0.00020052 |
| $0.00001\,\pi$ | Error | 0.00002005 |
| $0.000001\,\pi$ | — | 0.00000201 |
| $0.0000001\,\pi$ | — | 0.00000020 |
| $0.00000001\,\pi$ | — | 0.00000002 |

Digits in error are underlined. The values computed by the stable algorithm are correct to all digits given; whereas the classical algorithm fails to deliver for moderately small values of $\varepsilon$ due to an error halt.

# PROBLEMS

1   If $p$ and $q$ are real, and $p^3 + q^2 > 0$, the cubic equation

$$x^3 + 3px + 2q = 0 \tag{9}$$

has a real root $x_1$, which is given by Cardano's formula

$$x_1 = u - v,$$

where

$$u^3 = \sqrt{p^3 + q^2} - q, \qquad v^3 = \sqrt{p^3 + q^2} + q.$$

(a)  Prove that $x_1$ is a root by substituting into (9).

(b)  By discussing the graph of the function

$$y = x^3 + 3px + 2q,$$

show that under the conditions stated $x_1$ is the only real root of (9).

(c)  Show that Cardano's formula is subject to cancellation when $p^3$ is very much larger than $q^2$.

(d)  Transform Cardano's formula into a stable expression for $x_1$, using the identity

$$(u - v)(u^2 + uv + v^2) = u^3 - v^3.$$

(e*)  Assuming that $q^2/p^3$ is small, express $x_1$ as $-2q/3p$ times a series in powers of $q^2/p^3$ by applying the binomial theorem first to

$$\sqrt{p^3 + q^2} = p^{3/2}\left(1 + \frac{q^2}{p^3}\right)^{1/2},$$

and then, with the result thus obtained, to $u$ and $v$. (Two non-trivial terms of the series suffice.)

(f)  Formulate Newton's method (see §2.2) for equation (9) and show that if Newton's method is started with $x_0 = 0$, then $x_1$ just equals the first term of the series obtained in (e).

---

\* Here and later, a \* is used to indicate problems of above average difficulty.

(g) Implement the algorithms discussed in (c), (d), (e), (f). Apply them to the equation (9) where $p = q = 10^k$, $k = 0, 1, 2, \ldots, 10$.

2 An experiment involves $2^n$ data points $x_k$, where $n$ is a positive integer. The $x_k$ fall into $n + 1$ groups, numbered from $m = 0$ to $m = n$. In the $m$-th group, there are $\binom{n}{m}$ $x_k$, all having the same value

$$x_k = \mu + \frac{2m - n}{\sqrt{n}}. \tag{10}$$

Because

$$\sum_{m=0}^{n} \binom{n}{m} = 2^n, \tag{11}$$

this adds up to just $2^n$ data points. (We are dealing with a form of the *binomial distribution*.)

(a) Show that $\mu$ is the mean of the $x_k$.

(b) Calculate the variance of the $x_k$ from its definition (4), and show that it equals 1.

(c) For $n = 5$ and $\mu = 10^2$, $10^4$, $10^6$, calculate the variance numerically, using (1) the definition, (2) the classical algorithm, (3) the stable algorithm, and compare the results.

3 Instead of working with inscribed polygons as in Demonstration 1.3-4, $\pi$ also may be determined as the limit of $g(n)$ as $n \to \infty$, where $g(n)$ is one half of the circumference of the regular $n$-gon *circumscribed* to the unit circle.

(a) Show that

$$g(n) = n \tan \frac{\pi}{n}.$$

(b) If $z_k := g(2^k)$, show that $z_k$ satisfies either of the recurrence relations,

$$z_{k+1} = 2^{2k+1} \frac{\sqrt{1 + (2^{-k} z_k)^2} - 1}{z_k}$$

and

$$z_{k+1} = \frac{2 z_k}{\sqrt{1 + (2^{-k} z_k)^2} + 1}.$$

Predict which of these relations will be safer to use numerically, and confirm your conclusion by the experiment.

**4\*** The recurrence relation of Demonstration 1.3-4,

$$y_{k+1} = y_k \sqrt{\frac{2y_k}{y_k + y_{k-1}}},$$

can be started with arbitrary positive values of $y_0$ and $y_1$. If $y_1 = y_0$, then all $y_k = y_0$. Thus, we assume that $y_1 \neq y_0$.

**(a)** Show that $y_k$ can be expressed in the form

$$y_k = 2^k \tfrac{1}{2}(a^{2^{-k}} - a^{-2^{-k}})c,$$

where $a$ and $c$ are suitable (real or complex) constants depending on $y_0$ and $y_1$.

**(b)** Determine $\lim y_k$ as a function of $y_0$ and $y_1$.

**(c)** Devise an algorithm, using square roots only, for computing the function $\log x (x > 1)$, and discuss its numerical properties.

**(d)** Same problem for the function $\arcsin x (0 < x \leq 1)$.

---

## §1.4  SMEARING

Cancellation does not occur only when two nearly equal numbers are directly subtracted from each other. The two numbers themselves may be the result of summations. In particular, this situation occurs in the evaluation of a sum which has partial sums that are large compared to the final result.

Let

$$s := \sum_{k=1}^{n} a_k$$

be the sum to be computed, where the $a_k$ may have either sign. We assume that the evaluation is done by forming the sequence of partial sums according to

$$s_1 := a_1, \qquad s_k := s_{k-1} + a_k, \qquad k = 2, 3, \ldots, n,$$

so that $s = s_n$.

If the sum is evaluated in fixed point arithmetic, nothing worse can happen than that each $a_k$ is afflicted with some error, which is bounded by one and the same $\varepsilon$ for all $k$. If no overflow occurs, the error in the final sum $s$ thus will be at most $n\varepsilon$. Actually, because usually not all errors in the $a_k$ have the same sign, the error in $s$ in all likelihood will be much smaller than $n\varepsilon$.

As with cancellation, however, a new phenomenon can occur if the sum $s$ is evaluated in floating arithmetic. Assume that one of the intermediate sums $s_k$ is considerably larger than the final sum $s$, in the sense that the exponent of $s_k$

exceeds the exponent of $s$ by several, say $m$, units. (Obviously this can happen only if not all $a_k$ have the same sign.) If we simulated the accumulation of such a sum in fixed point arithmetic (using for all partial sums the exponent of the final sum $s$), then we would have to replace the last $m$ digits of $s_k$ by zeros. Because these digits influence the last $m$ digits in $s$, these digits will in general be wrong, and it cannot be argued that due to probabilistic effects the actual error will be smaller. The loss of significant digits due to large intermediate sums is called **smearing**. The number of digits lost due to smearing approximately equals

$$\left[ \log_b \left| \frac{s_{max}}{s} \right| \right],$$

where $[x]$ denotes the greatest integer not exceeding $x$, $b$ is the base of the number system, and $s_{max}$ is the partial sum of greatest absolute value. Since $s_{max}$ is usually not known beforehand, one may use the trivial estimate

$$|s_{max}| \geq \tfrac{1}{2} |a_{max}|,$$

where $a_{max}$ is the term of greatest absolute value in the sum, to obtain the approximate lower bound

$$\left[ \log_b \left| \frac{a_{max}}{2s} \right| \right]$$

for the number of digits lost.

**DEMONSTRATION 1.4-1:**  *Evaluation of the exponential series.* The following mathematical result is well known. For all real numbers $x$,

$$e^x = 1 + \frac{x}{1!} + \frac{x^2}{2!} + \cdots = \sum_{k=0}^{\infty} \frac{x^n}{n!}. \tag{1}$$

If $e^x$ is computed numerically using this formula, the series must be truncated. If the last term taken into account is the one with $k = n$, then we neglect the remainder

$$\sum_{k=n+1}^{\infty} \frac{x^k}{k!} = \frac{x^n}{n!} \left[ \frac{x}{n+1} + \frac{x^2}{(n+1)(n+2)} + \cdots \right].$$

If $n > 2[x]$, the absolute value of the expression in brackets is less than 1, and the mathematical truncation error thus will certainly be less than the absolute value of the last term taken into account. Nevertheless, the above series is not well suited for the evaluation of $e^{-x}$ for $x \gg 0$. Writing the series in the form,

$$e^{-x} = \sum_{k=0}^{\infty} (-1)^k a_k,$$

the $a_k$ satisfy the recurrence relation

$$a_0 = 1, \qquad a_k = a_{k-1}\frac{x}{k}, \qquad k = 1, 2, \ldots,$$

which shows that

$$a_{\max} = a_m,$$

where $m$ is the largest integer not exceeding $x$. Using Stirling's formula (see §6.2),

$$m! \sim \sqrt{2\pi m}\left(\frac{m}{e}\right)^m,$$

we have, approximating $m$ by $x$,

$$a_{\max} \sim \frac{x^x}{x!} \sim \frac{x^x}{\sqrt{2\pi x}\left(\dfrac{x}{e}\right)^x} = \frac{1}{\sqrt{2\pi x}}e^x.$$

In view of $s = e^{-x}$, the number of unreliable digits in the mantissa representing the sum $e^{-x}$ equals at least

$$p = \left[\frac{\log\left(\dfrac{e^{2x}}{2\sqrt{2\pi x}}\right)}{\log b}\right]$$

$$= \left[\frac{1}{\log b}\left(2x - \tfrac{1}{2}\log x - 1.612\right)\right]. \tag{2}$$

For numerical confirmation, we evaluate $e^{-x}$ for various $x > 0$ by summing the exponential series. Digits in error are underlined, $b = 10$.

| $x$ | Value of $e^{-x}$ by summing (1) | Exact value | Estimate (2) of number of unreliable digits |
|---|---|---|---|
| 5 | 6.7379<u>57805</u>* $10^{-3}$ | 6.737946999* $10^{-3}$ | 3 |
| 10 | 4.480<u>305719</u>* $10^{-5}$ | 4.539992976* $10^{-5}$ | 7 |
| 15 | <u>5.722010666</u>* $10^{-5}$ | 3.059023205* $10^{-7}$ | 11 |

Of course, the determination of accurate values of $e^{-x}$ for $x \gg 0$ does not present a real problem. By exploiting the relation $e^{-x} = (e^x)^{-1}$, the computation of $e^{-x}$ may be reduced to the computation of $e^x$ for which reliable values may be obtained from the exponential series, even if $x$ is very large. Such simple functional relationships are not always available, however.

**DEMONSTRATION 1.4-2:** Here and in subsequent demonstrations, we consider the problem of computing accurate numerical values of the integrals

$$y_n := \int_0^1 \frac{x^n}{x + a} \, dx \tag{3}$$

for a fixed value of $a \gg 1$ and for $n = 0, 1, \ldots, 10$. Before attempting any numerical computation, it is usually helpful to form an idea about the behavior and the order of magnitude of the result. In this case, the numbers $y_n$ are clearly positive: moreover, since $x^{n+1} < x^n$ for $0 < x < 1$, they form a monotonically decreasing sequence. From

$$\int_0^1 \frac{x^n}{1 + a} \, dx < y_n < \int_0^1 \frac{x^n}{a} \, dx,$$

we even have the fairly accurate estimate

$$\frac{1}{(n + 1)(a + 1)} < y_n < \frac{1}{(n + 1)a}. \tag{4}$$

Thus, for instance, for $a = 10$,

$$0.0090909 < y_9 < 0.0100000.$$

It is not difficult to express the integrals (3) by a closed formula. By the binomial theorem,

$$x^n = \{(x + a) - a\}^n = \sum_{k=0}^n (-1)^k \binom{n}{k}(x + a)^{n-k} a^k.$$

Using this expansion in (3), we get

$$y_n = \int_0^1 \sum_{k=0}^n (-1)^k \binom{n}{k}(x + a)^{n-k-1} a^k \, dx$$

$$= \sum_{k=0}^n (-1)^k a^k \binom{n}{k} \int_0^1 (x + a)^{n-k-1} \, dx$$

$$= \sum_{k=0}^{n-1} (-1)^k a^k \binom{n}{k}\{(1 + a)^{n-k} - a^{n-k}\}\frac{1}{n - k} + (-a)^n \log \frac{1 + a}{a}. \tag{I}$$

Straightforward evaluation of the expression (I) yields for $a = 10$:

| $n$ | $y_n$ from (I) |
|---|---|
| 0 | 0.095310180 |
| 1 | 0.046898202 |
| 2 | 0.031017980 |
| 3 | 0.023153500 |
| 4 | 0.018465000 |
| 5 | 0.015320000 |
| 6 | 0.012800000? |
| 7 | 0.012000000 |
| 8 | −0.020000000 |
| 9 | 0.200000000 |
| 10 | −2.000000000 |

Digits in error are underlined. It is clear that, due to the many trailing zeros, the values of $y_n$ obtained in this manner are unreliable; from $n = 8$ onward they are meaningless. The reason is extreme smearing. For instance, for $n = 10$, the term corresponding to $k = 5$ in the sum (I) equals

$$-252a^5\{(a + 1)^5 - a^5\}\tfrac{1}{5} = -3.13 * 10^{11}.$$

Thus, on a calculator with a 10-digit decimal mantissa, at least 2 digits in front of the decimal point will be unreliable, as well as the digits after the decimal point.

# PROBLEMS

1 The Bessel function, $J_0$, of order zero, which is often required in problems of mathematical physics, is defined by the power series

$$J_0(x) := \sum_{n=0}^{\infty} (-1)^n \left(\frac{x^n}{2^n n!}\right)^2, \tag{5}$$

which mathematically converges for all values of $x$. This function satisfies $|J_0(x)| \le 1$ for all real $x$ and has an infinite number of zeros, which are located near the points $x_n := (n - \tfrac{1}{4})\pi$, $n = 1, 2, \ldots$.

(a) Write a program for the evaluation of the series (5), where the terms of the series are generated recursively. [Letting $J_0 = \sum a_n$, how is $a_n$ related to $a_{n-1}$?]

(b) Denoting by $s_n$ the $n$-th partial sum of the series, why is it impractical or dangerous to use the machine independent termination criterion $s_n = s_{n-1}$?

(c) Devise a termination criterion which on a computer with an infinite number of decimal places would guarantee a truncation error of $< 10^{-15}$.

(d) Assuming $x$ to be an integer, which is the largest term in the series (5), and how large is it approximately? [Use Stirling's formula to approximate factorials.]

(e) Sum the series (5) on your calculator, using the termination criterion which you have found in (c). How many digits in the sum can you guarantee to be correct? Carry out the sum for $x = 5$, 10, 15, and compare with the exact values

$$J_0(5) = -0.17759\ 67713\ 14338$$

$$J_0(10) = -0.24593\ 57644\ 51348$$

$$J_0(15) = -0.01422\ 44728\ 26781.$$

2  The functions

$$f_n(x) := \int_0^x t^n e^{-t}\, dt, \tag{6}$$

where $n = 0, 1, 2, \ldots$ and $x > 0$, are related to the probability integral of the $\chi^2$ distribution (see Fisz [1966], page 340). Here, and in later assignments, we study the computation of numerical values of $f_n(x)$. Because these functions are positive and vary widely in magnitude ($\lim_{x \to \infty} f(x) = n!$), it is reasonable to ask for a small *relative* error. The following algorithms should be tried and compared in their numerical performance for $n = 0, 1, 2, \ldots, 10$ and $x = 0.2, 0.5, 1.0, 2.0, 5.0, 10.0$.

(a)  *Closed form expression.* Using repeated integration by parts, show that

$$f_n(x) = n! \left\{ 1 - e^{-x} \left( 1 + \frac{x}{1!} + \frac{x^2}{2!} + \cdots + \frac{x^n}{n!} \right) \right\}. \tag{7}$$

Write a program for the efficient evaluation of this expression. For which values of $n$ and of $x$ is the expression subject to cancellation?

(b)  *Alternating series.* Obtain a representation of $f_n(x)$ by using the exponential series

$$e^{-t} = \sum_{k=0}^{\infty} \frac{(-t)^k}{k!}$$

in (6) and integrating term by term. Show that the series is subject to smearing for large values of $x$, and estimate the number of correct significant digits that can be expected.

(c) *Series of positive terms.* Obtain a representation of $f_n(x)$ as a series of positive terms by transforming (7) by means of the relation

$$1 + \frac{x}{1!} + \frac{x^2}{2!} + \cdots + \frac{x^n}{n!} = e^x - \sum_{k=n+1}^{\infty} \frac{x^k}{k!}.$$

Write an efficient program for the resulting series, and show that a machine-independent convergence test may be used. Show that the resulting series is reliable for all values of $x$ and $n$.

3   The function

$$F(x) := \frac{2}{\sqrt{\pi}} \int_0^x e^{-t^2} \, dt, \tag{8}$$

called the **error function**, is of fundamental importance in the theory of probability. It is monotonically increasing for $x \geq 0$ and satisfies $\lim_{x \to \infty} F(x) = 1$. The integral (8) cannot be evaluated in closed form. Here we discuss two methods to compute (8) by series.

(a) Expressing $e^{-t^2}$ by the exponential series and integrating term-by-term yields the series

$$F(x) = \frac{2}{\sqrt{\pi}} \sum_{n=0}^{\infty} (-1)^n \frac{x^{2n+1}}{n! \, (2n+1)}, \tag{9}$$

which theoretically converges for all $x$. Show that the series (9) is subject to severe cancellation for $x$ about $> 3$, and estimate the number of digits lost. Experiment, and compare with the exact values

$$F(3) = 0.999977910$$

$$F(4) = 0.999999985$$

$$F(5) = 1.000000000.$$

(b)   Express $F$ in the form

$$F(x) = e^{-x^2} g(x),$$

and show that $g$ satisfies the differential equation

$$g' = 2xg + \frac{2}{\sqrt{\pi}}, \qquad g(0) = 0.$$

Solve the differential equation by a power series,

$$g(x) = \sum_{n=0}^{\infty} b_n x^{2n+1},$$

and find the recurrence relation for the coefficients,

$$b_0 = \frac{2}{\sqrt{\pi}}, \qquad b_n = \frac{1}{n+\frac{1}{2}} b_{n-1}, \qquad n = 1, 2, \ldots .$$

Conclude that

$$F(x) = \sum_{n=0}^{\infty} a_n \tag{10}$$

where

$$a_0 = \frac{2}{\sqrt{\pi}} x e^{-x^2}, \qquad a_n = \frac{x^2}{n+\frac{1}{2}} a_{n-1}, \qquad n = 1, 2, \ldots . \tag{11}$$

All terms in the series (9) thus are positive. Implement the series, using the machine-independent convergence test $s_n = s_{n-1}$, and show that the series may safely be used for all values of $x$ so that the computation of $e^{-x^2}$ does not cause exponent underflow.

(c)   Use a bisection algorithm (see §2.2) to find the smallest machine number so that, on your calculator, $F(x) = 1$.

---

## §1.5   NUMERICAL INSTABILITY

If an intermediate result of a numerical calculation is contaminated by a rounding error, this error will influence all subsequent results that depend on the intermediate result. The rounding error **propagates**. The effect of smearing discussed in §1.4 is just a very special case of this. The rounding error would propagate even if all subsequent calculations were performed with infinite precision. In reality, however, each new intermediate result will introduce a new rounding error. All these errors will influence the final result. In simple situations (such as the evaluation of a sum), the final error will simply equal the sum of all intermediate errors. Due to statistical effects, the intermediate errors will perhaps even cancel each other, at least partially. In other cases (such as iteration), the errors in intermediate results will have a negligible effect on the final result. Algorithms with this desirable property are called **stable**.

Numerical instability is present if the intermediate errors have a very strong influence on the final result.

**DEMONSTRATION 1.5-1:** We consider once more the problem of evaluating the integrals,

$$y_n := \int_0^1 \frac{x^n}{a+x} \, dx,$$

for a fixed value of $a > 1$ and for $n = 0, 1, \ldots, 10$. Because a whole sequence of values $y_n$ is to be found, it seems reasonable to look for a recurrence relation for the $y_n$. Such a relation is easily constructed. Evidently,

$$
\begin{aligned}
y_n &= \int_0^1 \frac{x^n}{x+a} \, dx \\
&= \int_0^1 \frac{x^{n-1}(x+a-a)}{x+a} \, dx \\
&= \int_0^1 x^{n-1} \, dx - a \int_0^1 \frac{x^{n-1}}{x+a} \, dx \\
&= \frac{1}{n} - ay_{n-1}.
\end{aligned}
$$

Thus,

$$y_n = \frac{1}{n} - ay_{n-1}, \tag{II}$$

and because

$$y_0 = \log \frac{1+a}{a}$$

is known, all further $y_n$ may, theoretically, be computed from (II). Numerical results for $a = 10$ follow:

| $n$ | $y_n$ from (II) |
|-----|-----------------|
| 0   | 0.095310180     |
| 1   | 0.046898202     |
| 2   | 0.031017980     |
| 3   | 0.023153533     |
| 4   | 0.018464667     |
| 5   | 0.015353330     |
| 6   | 0.013133367     |
| 7   | 0.011523476     |
| 8   | 0.009765241     |
| 9   | 0.013458701     |
| 10  | -0.034587011    |
| 11  | 0.436779201     |
| 12  | -4.484458676    |

Again, the values are meaningless, at least from $n = 8$ onward because the algorithm (II) is extremely *unstable* for values $a \gg 1$.

To see that instability exists, assume that the starting value $y_0$ is in error by the quantity $\varepsilon_0$. We assume that all subsequent arithmetical operations are performed exactly. Denoting by $\tilde{y}_n$ the values computed with the wrong starting value, we have

$$\tilde{y}_0 = y_0 + \varepsilon_0,$$

$$\tilde{y}_n = -a\tilde{y}_{n-1} + \frac{1}{n},$$

$n = 1, 2, \ldots$ . Subtracting the corresponding relations for the exact values,

$$y_n = -ay_{n-1} + \frac{1}{n},$$

we obtain for the discrepancies caused by the wrong starting value,

$$r_n := \tilde{y}_n - y_n,$$

the recurrence relation

$$r_0 = \varepsilon_0, \qquad r_n = -ar_{n-1}, \qquad n = 1, 2, \ldots,$$

which is easily solved to yield

$$r_n = (-a)^n \varepsilon_0, \qquad n = 0, 1, 2, \ldots .$$

Thus, at each step of the computation, the discrepancy is magnified by the factor $a$; for example, for $a = 10$ by a factor 10. Thus, precisely one digit will be lost when passing from one value to the next, which is confirmed by the values given above.

We should conclude from the foregoing demonstration that in numerical computations even formulas and algorithms that have the appearance of mathematical elegance should be used with caution. But how are we to find accurate values of the integrals $y_n$? We now present two methods that work.

**DEMONSTRATION 1.5-2:** A recurrence relation being unstable in the direction of increasing $n$ does not preclude the possibility of its being stable in the direction of decreasing $n$. Solving (II) for $y_{n-1}$ yields

$$y_{n-1} = \frac{1}{a}\left(\frac{1}{n} - y_n\right). \tag{III}$$

If used in this form, the relation also requires a starting value. Such a value seems hard to find, because all $y_n$ where $n > 0$ are unknown; they are precisely the values which we are trying to find. However, we know that $y_n \to 0$ for $n \to \infty$. Being generous, let us put, say,

$$y_{20} := 0,$$

and let us use (III) for $n = 20, 19, 18, \ldots$ with this patently wrong starting value. The results are

| $n$ | $y_n$ from (III) |
|---|---|
| 10 | 0.008327966 |
| 9 | 0.009167203 |
| 8 | 0.010194391 |
| 7 | 0.011480561 |
| 6 | 0.013137658 |
| 5 | 0.015352901 |
| 4 | 0.018464710 |
| 3 | 0.023153529 |
| 2 | 0.031017980 |
| 1 | 0.046898202 |
| 0 | 0.095310180 |

The last value agrees with the correct starting value in Demonstration 1.5-1, and, miraculously, all other values are correct to all digits given in spite of the fact that they were calculated with the grossly wrong starting value $y_{20} = 0$.

The mathematical explanation of this phenomenon is not too difficult. Let us, as always, denote by $y_n$ the (mathematically exact) solution of the recurrence relation (II), calculated with the exact starting value $y_0$, and let $y_n^{(m)}$ ($n = 0, 1, \ldots, m$) denote the solution of the recurrence (III) with starting value $y_m = 0$. Subtracting the relations,

$$y_{n-1}^{(m)} = \frac{1}{a}\left(\frac{1}{n} - y_n^{(m)}\right),$$

$$y_{n-1} = \frac{1}{a}\left(\frac{1}{n} - y_n\right),$$

we find for the discrepancy

$$d_n^{(m)} := y_n^{(m)} - y_n$$

the relation

$$d_{n-1}^{(m)} = -\frac{1}{a} d_n^{(m)},$$

which immediately implies

$$d_n^{(m)} = \left(-\frac{1}{a}\right)^{m-n} d_m^{(m)},$$

or since $d_m^{(m)} = y_m^{(m)} - y_m = -y_m$,

$$d_n^{(m)} = -\left(-\frac{1}{a}\right)^{m-n} y_m.$$

Two conclusions may be drawn:

(i)  For every fixed value of $n$, $\lim_{m\to\infty} d_n^{(m)} = 0$, that is,

$$\lim_{m\to\infty} y_n^{(m)} = y_n.$$

Thus, the algorithm described above will always yield the correct value of $y_n$, if the backward recursion (III) is started with a sufficiently large $m$.

(ii)  In view of equation (4) of §1.4 there holds for every value of $m \geq n$ the error estimate

$$|y_n^{(m)} - y_n| \leq \frac{a^{n-m-1}}{m+1}.$$

Thus, for instance, if $m = 20$, we have for $n = 10$ the estimate

$$|y_n^{(m)} - y_n| \leq a^{-11},$$

and for $a = 10$, the error will be less than one unit in the eleventh digit.

The foregoing analysis does not take into account rounding errors, but it may be shown that the influence of these errors likewise is damped very strongly by using the recurrence relation in the backwards direction. In fact, we may conceive the starting value $y_m = 0$ as being generated by a colossal rounding error, and we have just seen that its effect is harmless.

Algorithm (III) requires us to calculate all values $y_n$ within a certain range, even if only one or a few such values are actually wanted. The following demonstration shows how to compute an individual $y_n$ in a numerically stable manner without recursion.

**DEMONSTRATION 1.5-3:**   If $a > 1$, then for all $x$ such that $-1 \leq x \leq 1$

$$\frac{x^n}{a+x} = \frac{1}{a}\frac{x^n}{1+\dfrac{x}{a}} = \frac{1}{a}\sum_{k=0}^{\infty}\frac{(-1)^k x^{n+k}}{a^k}.$$

Integrating term by term, we obtain

$$y_n(a) = \int_0^1 \frac{x^n}{a+x}\, dx = \frac{1}{a} \sum_{k=0}^\infty \frac{(-1)^k}{a^k} \int_0^1 x^{n+k}\, dx$$

or

$$y_n = \sum_{k=0}^\infty \frac{(-1)^k}{(n+k+1)a^k}. \tag{IV}$$

The series is alternating, and the absolute values of its terms form a monotonically decreasing sequence. It follows that the partial sums of the series alternatingly yield upper and lower bounds for the sum of the series, and that the truncation error is always less than the last term taken into consideration. For $a = 10$, stopping the summation as soon as the modulus of a term becomes $< 10^{-12}$, the values obtained agree with those obtained by Algorithm (III).

Contrary to Algorithm (I), which provides a closed formula for the desired result, Algorithm (IV) expresses the result in the form of an infinite series, and thus is infinite in character. It becomes finite—and hence executable—only by the use of an appropriate termination criterion. Infinite algorithms *must* always be used if a given problem has no solution in finite terms. Our demonstration shows that, for numerical reasons, an infinite algorithm may be preferable even in a situation where closed solutions or solutions in finite terms exist.

# PROBLEMS

1  We continue to study various ways to compute the functions

$$f_n(x) := \int_0^x t^n e^{-t}\, dt \tag{1}$$

considered in Problem 2 of §1.4.

(a)  Using integration by parts, show that there holds the recurrence relation

$$f_n(x) = n f_{n-1}(x) - x^n e^{-x}, \tag{2}$$

$n = 1, 2, \ldots$.

(b)  Since $f_0(x) = 1 - e^{-x}$ is known, (2) may theoretically be used to compute $f_1(x)$, $f_2(x)$, ... for any given $x$. Try this, and convince yourself that the method does not work.

(c)  Explain the failure by assuming that $f_0(x)$ is affected by an error $\varepsilon$, and by carrying through an analysis similar to that in Demonstration 1.5-1. Assuming no rounding errors, how does the error $\varepsilon$ affect $f_n(x)$?

(d)   Show that by using (2) in the form

$$f_{n-1}(x) = \frac{1}{n}\{f_n(x) + x^n e^{-x}\},$$

$n = m,\ m-1,\ m-2,\ \ldots,\ 0$, and by starting by arbitrarily setting $f_m(x) = 0$, good values of $f_n$ are obtained provided $m$ is chosen sufficiently large.

(e)   Perform an analysis similar to that in Demonstration 1.5-2 to determine how $m$ has to be selected as a function of $n$ and $x$ in order to obtain $f_n(x)$ to full machine accuracy. Test your result by the experiment.

# §1.6   A MODEL FOR THE PROPAGATION OF ROUNDING ERROR*

In our presentation of rounding errors thus far, we have stressed the intuitive and qualitative understanding of rounding. This is the kind of understanding that should be acquired even by the casual user of numerical methods. For a serious numerical analyst, such as the designer of mathematical software, an understanding that is merely intuitive does not suffice. One should be able to *predict qualitatively* how a given algorithm will perform numerically, and to *explain mathematically* why different algorithms for achieving the same result will perform differently. In this section, we present a mathematical model for the propagation of rounding errors. In order to bring out the essential consequences of floating arithmetic, our model makes certain idealizing or simplifying assumptions (labelled *axioms*) which are not always satisfied on real-life computers. As some examples should make clear, our model nevertheless furnishes realistic descriptions of the numerical performance of a large class of algorithms that can be described in terms of operations on real numbers.

We begin by reviewing some notation. We use unadorned lower case latin letters, such as $x$, $y$, $\ldots$, $a$, $b$ to denote real numbers. The rounded machine representation of a real number $x$ will be denoted by $x^*$. When discussing an algorithm that on a machine with an infinite number of decimal places should produce intermediate results $x$, $y$, $\ldots$, and perhaps a final result $z$, we denote by $\tilde{x}, \tilde{y}, \ldots, \tilde{z}$ the corresponding results that are produced on the machine with finite word length that we are discussing. It is clear from many demonstrations in

---

* This section, although important for the serious numerical analyst, may be omitted at first reading.

preceding sections that $\tilde{x} \neq x^*$ in general; usually, $|\tilde{x} - x|$ is much larger than $|x^* - x|$.

We next describe the relation between $x^*$ and $x$. In floating arithmetic it is reasonable to stress relative errors, and thus to express $x^*$ as $x$ times a correction factor, hopefully close to 1. If $b$ is the base of the floating system, if the mantissa is represented in the form

$$y = \sum_{i=0}^{m} y_i b^{-i}, \tag{1}$$

if there is no exponent underflow or overflow, and if the machine rounds correctly, then we have

$$x^* = x(1 + \varepsilon), \tag{2}$$

where

$$-\tfrac{1}{2}b^{-m} \leq \varepsilon \leq \tfrac{1}{2}b^{-m}. \tag{3}$$

More generally, for any mode of rounding, there exists a smallest set $E$, called the **error set** of the machine, so that (2) holds with some $\varepsilon$ in $E$.

It is convenient at this point to recall some notations of set theory. We assume that the reader is familiar with the symbols $\in$ for "element of" and $\subset$ for "subset of." If $X$ is any set of real numbers, we denote by $aX$ the set of all numbers $ax$ where $x \in X$, and by $b + X$ the set of all numbers $b + x$ where $x \in X$. If both $X$ and $Y$ are sets, then $X + Y$ and $XY$ are the sets of all numbers $x + y$ and $xy$, respectively, where $x \in X$ and $y \in Y$. An ambiguity arises if $X = Y$. By $X^2$, we shall always understand the set of all numbers $xy$ where $x \in X$ and $y \in X$. (The set of all numbers $x^2$ where $x \in X$ will not be used.) Similarly, by $X + X$, we mean the set of all numbers $x + y$ where $x \in X$ and $y \in X$. (The set of all numbers $2x$ where $x \in X$ according to the foregoing is denoted by $2X$.)

With these notations (2) is expressed as

$$x^* \in x(1 + E). \tag{4}$$

In order to ensure that (4) is applicable to *all* real numbers $x$, we postulate

**AXIOM I:** *The exponent range is unlimited.*

Thus, we are not concerned with exponent overflow or underflow. Our model of error propagation is applicable only to computations that stay within the exponent range of the computer.

Concerning the error set $E$, we assume

**AXIOM II:** *The error set is symmetric about O.*

This can always be achieved by replacing $E$ by the larger set $E \cup (-E)$. We shall not be concerned with the fine point of what happens to real numbers that lie exactly halfway between two machine numbers. As a consequence of Axiom II there holds, for arbitrary real $a$ and $b$, the simple rule

$$aE + bE = (|a| + |b|)E. \tag{5}$$

Two axioms for the functions provided by the computer are now in order. For a number of mathematical functions $f$ of one or several variables, a computer system usually provides, by means of hardware or of software, functions $\tilde{f}$ intended to simulate or approximate $f$. These functions $\tilde{f}$ will be called **computer functions**. The simplest computer functions are the basic arithmetical operations, $x + y$, $x - y$, $xy$, $x \div y$, but simulations for the more common elementary functions such as $\sqrt{x}$, $\log x$, $e^x$, as well as the basic trigonometric functions and their inverses, are usually also available. A computer function is defined only for values of the argument that are machine numbers, and it can reasonably be defined only for machine numbers that also belong to the domain of definition of the function $f$ to be simulated. (For instance, division by zero is impossible also on a computer.) We assume that computer functions are defined for *all* such machine numbers.

**AXIOM III:** *Computer functions have their maximal domain of definition.*

If a mathematical function $f$ with domain of definition $D(f)$ is simulated by the computer function $\tilde{f}$, this simulation will rarely be perfect because, even for machine numbers $x \in D(f)$, $f(x)$ in general is not a machine number. The best we can hope for is that $\tilde{f}(x) = [f(x)]^*$ for all such $x$, but, as shown by Demonstration 1.2-6, not even this is always true. We postulate, however,

**AXIOM IV:** *Computer functions are evaluated with bounded relative error.*

If a computer function $\tilde{f}$ depends on one variable, this means that there exists a number $\mu_f$, called the **error constant** of $f$, so that

$$\tilde{f}(x) \in f(x)(1 + \mu_f E) \tag{6a}$$

for all $x \in D(\tilde{f})$; similarly, if $f$ depends on two variables, there is a $\mu_f$ so that

$$\tilde{f}(x, y) \in f(x, y)(1 + \mu_f E) \tag{6b}$$

for all $(x, y) \in D(\tilde{f})$. A computer function is said to be **ideally realized** if (6) holds with $\mu = 1$.

On most modern computers the basic arithmetic operations are ideally realized. Elementary functions of one variable are often nearly ideally realized as long as $f(x) \neq 0$, or as $f(x) = 0$ only for machine numbers $x$, as is the case, for instance, for $f(x) = \log x$. It is difficult to satisfy Axiom IV for functions $f$ that have zeros that are not machine numbers, such as $f(x) = \sin x$. We nevertheless retain Axiom IV in the interest of simplicity. In addition, in our examples we shall assume that all computer functions are ideally realized.

There follows a general description of the kind of algorithm that we can analyze by our method. We assume that the algorithm starts with a finite set of input quantities $x_1, x_2, \ldots, x_m$, and then produces, according to some mathematical prescription, a finite sequence of quantities $t_1, t_2, \ldots, t_n$. The desired result is either the final $t_n$ or several or all of the $t_k$. If the algorithm is broken down sufficiently, as it will when the computer program is written, it will be found that each $t_k$ depends on the input quantities and on the preceding $t_i$,

$$t_k = f_k(t_1, t_2, \ldots, t_{k-1}; x_1, x_2, \ldots, x_m), \tag{7}$$

$k = 1, 2, \ldots, n$, where $f_k$ is a computer function. Naturally, in most cases only one or two of the arguments indicated will actually appear in $f_k$.

If the algorithm is executed on the computer, at the very outset there arises the difficulty that the input quantities, $x_1, x_2, \ldots, x_m$, have to be replaced by machine numbers $x_1^*, \ldots, x_m^*$. The falsification of the result of an algorithm due to a change in the input data is a purely mathematical effect which has nothing to do with the finite arithmetic of the computer. We postpone a study of this effect to §1.7 and in the meantime assume that all input quantities are machine numbers.

Even under this assumption the computer will not actually execute the formulas (7). Instead, it will generate the machine numbers $\tilde{t}_1, \tilde{t}_2, \ldots, \tilde{t}_n$ defined by

$$\tilde{t}_k = \tilde{f}_k(\tilde{t}_1, \ldots, \tilde{t}_{k-1}; x_1, \ldots, x_m), \tag{8}$$

$k = 1, 2, \ldots, n$. Axiom IV permits us to state that

$$\tilde{t}_k \in f_k(\tilde{t}_1, \ldots, \tilde{t}_{k-1}; x_1, \ldots, x_m)(1 + \mu_k E), \tag{9}$$

where $\mu_k$ is the error constant of $f_k$, but this is of no immediate help for our goal which is to estimate $|\tilde{t}_k - t_k|$. To make progress, we state

**HYPOTHESIS D:** *All functions $f_k$ are locally differentiable.*

The word, differentiable, here is used in the sense that increments can be approximated by differentials. By locally differentiable we mean that $f_k$ should be differentiable on a neighborhood of the point $(t_1, \ldots, t_{k-1}; x_1, \ldots, x_m)$ containing $(\tilde{t}_1, \ldots, \tilde{t}_{k-1}; x_1, \ldots, x_m)$.

With Hypothesis D we have put at our disposal the resources of the differential calculus. Suppose that, for an index $k \geq 1$, we have determined positive constants $\gamma_1, \ldots, \gamma_{k-1}$, so that

$$\tilde{t}_i \in t_i + \gamma_i E, \qquad i = 1, \ldots, k-1.$$

This means that there exist numbers $\varepsilon_i \in E$, so that

$$\tilde{t}_i = t_i + \gamma_i \varepsilon_i, \qquad i = 1, \ldots, k-1.$$

We then may write (9) in the form,

$$t_k = f_k(t_1 + \gamma_1 \varepsilon_1, \ldots, t_{k-1} + \gamma_{k-1}\varepsilon_{k-1}; x_1, \ldots, x_m)(1 + \mu_k \varepsilon_k),$$

where also $\varepsilon_k \in E$. Considering the expression on the right as a function of $\varepsilon_1, \ldots, \varepsilon_k$, using the fact that

$$f_k(t_1, \ldots, t_{k-1}; x_1, \ldots, x_m) = t_k,$$

and approximating the increment by the differential, we get

$$\tilde{t}_k = t_k + \sum_{i=1}^{k-1} p_{ki}\gamma_i \varepsilon_i + t_k \mu_k \varepsilon_k + \mathrm{o}(\varepsilon),$$

where the $p_{ki}$ denote values of certain partial derivatives, and where $\varepsilon^{-1}\mathrm{o}(\varepsilon) \to 0$ as $\varepsilon := \max |\varepsilon_i| \to 0$. We know that all $\varepsilon_i \in E$ and thus, using (5), may assert that

$$\tilde{t}_k \in t_k + \gamma_k E + \mathrm{o}(\varepsilon) \tag{10}$$

where

$$\gamma_k := \sum_{i=1}^{k-1} |p_{ki}| \gamma_i + |t_k| \mu_k. \tag{11}$$

We have not completed the induction step because of the presence of the $\mathrm{o}(\varepsilon)$ term in (10). Indeed, as long as we consider a fixed computer it is difficult to give a precise meaning to the term $\mathrm{o}(\varepsilon)$, because any interpretation of this symbol involves a passage to the limit $\varepsilon \to 0$. We may, however, imagine a family of computers with error sets $E := [-\varepsilon, \varepsilon]$ where $\varepsilon \to 0$. (The error constants $\mu_f$ for the evaluation of the machine functions $f$ are assumed to be the same for all machines of the family.) It then makes sense to postulate the existence of numbers $\gamma_i < \infty$ so that

$$\limsup_{\varepsilon \to 0} \frac{|\tilde{t}_i - t_i|}{\varepsilon} \leq \gamma_i, \qquad i = 1, \ldots, k-1. \tag{12}$$

Nothing is assumed for $k = 1$. If (12) is true for some $k \geq 1$, the foregoing computations with differentials are valid and show that the inequality (12) also holds for $i = k$ with the value of $\gamma_k$ given by (11). In this way, any algorithm satisfying Hypothesis D has associated with it in a canonical manner a sequence of numbers $\gamma_1, \ldots, \gamma_n$, called the **coefficients of error propagation** of the algorithm.

Because higher order terms are neglected, it cannot be asserted mathematically that on any fixed machine with error set $E$

$$\tilde{t}_k \in t_k + \gamma_k E, \qquad k = 1, 2, \ldots, n. \tag{13}$$

In an engineering sense, if the error sets are sufficiently small, the statement (13) may nevertheless be assumed to be true with a high degree of probability. The same could not be said, in general, if the $\gamma_k$ were replaced by smaller positive numbers.

If all $t_k \neq 0$, it may be convenient to work with **coefficients of relative error propagation** $\rho_k$ defined by

$$\gamma_k =: |t_k| \rho_k, \qquad k = 1, 2, \ldots, n.$$

Equation (13) then appears in the form

$$\tilde{t}_k \in t_k(1 + \rho_k E), \tag{14}$$

which is to be interpreted similarly.

In the analysis of a concrete algorithm it is not always necessary to compute explicitly the partial derivatives $p_{ki}$ required to evaluate (11). One simply proceeds formally by expanding

$$f_k(t_1 + \gamma_1 E, \ldots, t_{k-1} + \gamma_{k-1} E; x_1, \ldots, x_m)(1 + \mu_k E)$$

in powers of $E$ as if $E$ were an indeterminate, and by observing the two rules

$$aE + bE = (|a| + |b|)E, \qquad E^2 = 0. \tag{15}$$

The coefficient of $E$ in this expansion then is $\gamma_k$.

Some examples will make the procedure clear.

**EXAMPLE 1.6-1:**   *Evaluation of a sum.* Let the sum

$$s := \sum_{k=1}^{n} a_k$$

to be evaluated. The input quantities here are the $a_k$; we assume that they are machine numbers. The usual summation algorithm is mathematically defined by $t_k := s_k$, where

$$s_1 := a_1,$$

$$s_k := f_k(s_{k-1}; a_k) := s_{k-1} + a_k, \qquad k = 2, \ldots, n; \qquad (16)$$

the value $s = s_n$ is desired. Since $a_1$ is a machine number, we clearly have $\tilde{s}_1 = s_1$, hence $\gamma_1 = 0$. The numerically calculated quantities $\tilde{s}_2, \ldots, \tilde{s}_n$ satisfy, using (9),

$$\tilde{s}_k \in (\tilde{s}_{k-1} + a_k)(1 + E).$$

Assuming

$$\tilde{s}_{k-1} \in s_{k-1} + \gamma_{k-1} E,$$

this yields

$$\tilde{s}_k \in (s_k + \gamma_{k-1} E)(1 + E) \doteq s_k + (|s_k| + \gamma_{k-1})E.$$

(We use the symbol $\doteq$ to indicate that terms $O(E^2)$ have been neglected.) Thus, we have

$$\gamma_k = \gamma_{k-1} + |s_k|, \qquad k = 2, \ldots, n,$$

which clearly implies

$$\gamma_k = |s_2| + |s_3| + \cdots + |s_k|.$$

We thus can assert that, neglecting terms in $E^2$, the numerical value $\tilde{s}$ of the sum $s$ satisfies

$$\tilde{s} \in s + (|s_2| + |s_3| + \cdots + |s_n|)E.$$

Some special cases of this result are of interest.

(a)  Let all $a_k > 0$. Then, all $s_k > 0$, and the final coefficient of error propagation is

$$\gamma_n = s_2 + s_3 + \cdots + s_n$$

$$= na_1 + (n-1)a_2 + (n-2)a_3 + \cdots + a_n - a_1.$$

Because the early terms are weighted more heavily in this expression than the later ones, this clearly shows the advantage of beginning the evaluation of a long sum of positive numbers with the *small* terms. For instance, if

$$a_k = \frac{1}{k^2 + k} = \frac{1}{k} - \frac{1}{k+1}$$

as in Demonstration 1.2-2, then

$$s_k = 1 - \frac{1}{k+1},$$

and

$$\gamma_n = n - 1 - \left(\frac{1}{3} + \frac{1}{4} + \cdots + \frac{1}{n+1}\right)$$

$$\sim n - \log n,$$

indicating that the error grows linearly with $n$ as $n \to \infty$. However, if the same sum is evaluated backwards, then

$$s_k = a_n + a_{n-1} + \cdots + a_{n-k+1}$$

$$= \left(\frac{1}{n} - \frac{1}{n+1}\right) + \cdots + \left(\frac{1}{n-k+1} - \frac{1}{n-k+2}\right)$$

$$= \frac{1}{n+1-k} - \frac{1}{n+1},$$

and

$$\gamma_n = 1 + \frac{1}{2} + \cdots + \frac{1}{n-1} - \frac{n-1}{n+1}$$

$$\sim \log n + \text{const.},$$

indicating that the error grows only logarithmically as $n \to \infty$.

(b) *Sums with large intermediate partial sums.* If $s \neq 0$, then

$$\tilde{s} \in s(1 + \rho E),$$

where $\rho$, the coefficient of relative error propagation, is

$$\rho = \frac{|s_2| + |s_3| + \cdots + |s_n|}{|s|}.$$

This coefficient will be large if some intermediate partial sums $|s_k|$ are large compared to the final sum $s$. Our model thus gives a quantitative explanation of the effect of **smearing** (see §1.4), as it occurs, for instance, if the exponential series

$$e^{-x} = \sum_{n=0}^{\infty} (-1)^n \frac{x^n}{n!}$$

is summed for large values of $x$.

(c)  *All $a_k$ approximately equal.* If all terms of a sum have the same order of magnitude, say if $a_k \sim a > 0$ for all $k$, then the order of summation is clearly irrelevant. In view of $s_k \sim ka$, $s \sim na$, the coefficient of relative error propagation is

$$\rho_n = \frac{1}{n} \sum_{k=2}^{n} k = \frac{n+1}{2} - \frac{1}{n},$$

which grows linearly with $n$.

We briefly mention another mode of summation which, at least in the case under discussion, yields a smaller coefficient of relative error propagation. Assume that $n = 2^l$ (this can always be achieved by filling the sum with zeros); for definiteness, we let $n = 2^3 = 8$. Mathematically, the algorithm is then described by

$$t_1 := a_1 + a_2, \qquad t_2 := a_3 + a_4, \qquad t_3 := a_5 + a_6, \qquad t_4 := a_7 + a_8;$$
$$t_5 := t_1 + t_2, \qquad t_6 := t_3 + t_4;$$
$$s = t_7 := t_5 + t_6.$$

(Note that there are 7 additions, as in the ordinary algorithm.) Our model of error propagation yields

$$\tilde{t}_k \in t_k(1 + E), \qquad k = 1, 2, 3, 4;$$
$$\tilde{t}_k \in t_k(1 + E)(1 + E) \doteq t_k(1 + 2E), \qquad k = 5, 6;$$
$$\tilde{t}_7 \in t_7(1 + 2E)(1 + E) \doteq t_7(1 + 3E).$$

Thus, there results a coefficient of relative error propagation $\rho = 3$; if $n = 2^l$, the coefficient would clearly be

$$\rho = l = \log_2 n.$$

Thus, as $n \to \infty$, the coefficient grows only logarithmically, instead of linearly, as in the ordinary mode of summation.

**EXAMPLE 1.6-2:**  *Length of a polygon.* Here we return to the problem considered in Demonstration 1.3-3, which was to evaluate the expression

$$s := \sqrt{a^2 + b^2 - 2ab \cos \phi} \tag{17}$$

with particular attention paid to situations where $a \sim b$ and $\phi$ is small. The expression (17) is evaluated algorithmically by forming

$$t_1 := a^2$$

$$t_2 := b^2$$

$$t_3 := t_1 + t_2$$

$$t_4 := \cos \phi$$

$$t_5 := ab$$

$$t_6 := t_4 t_5$$

$$t_7 := 2t_6$$

$$t_8 := t_3 - t_7$$

$$s = t_9 \leq \sqrt{t_8}.$$

In computation this becomes

$$\tilde{t}_1 \in a^2(1 + E) = t_1(1 + E)$$

$$\tilde{t}_2 \in b^2(1 + E) = t_2(1 + E)$$

$$\tilde{t}_3 \in (\tilde{t}_1 + \tilde{t}_2)(1 + E) \subset (t_1 + t_2)(1 + E)^2 \doteq t_3(1 + 2E)$$

$$\tilde{t}_4 \in \cos\phi(1 + E) = t_4(1 + E)$$

$$\tilde{t}_5 \in ab(1 + E) = t_5(1 + E)$$

$$\tilde{t}_6 \in \tilde{t}_4 \tilde{t}_5(1 + E) \subset t_4 t_5(1 + E)^3 \doteq t_6(1 + 3E)$$

$$\tilde{t}_7 \in 2\tilde{t}_6(1 + E) \doteq t_7(1 + 4E)$$

$$\tilde{t}_8 \in (\tilde{t}_3 - \tilde{t}_7)(1 + E) \subset [t_3(1 + 2E) - t_7(1 + 4E)](1 + E)$$

$$= [t_3 - t_7 + (2t_3 + 4t_7)E](1 + E)$$

$$\doteq t_8 + (t_8 + 2t_3 + 4t_7)E = t_8 + 3(t_3 + t_7)E$$

$$\tilde{t}_9 \in \sqrt{\tilde{t}_8}(1 + E) \subset \sqrt{t_8 + 3(t_3 + t_7)E}(1 + E)$$

$$\doteq \sqrt{t_8}\left(1 + \frac{3}{2}\frac{t_3 + t_7}{t_8}E\right)(1 + E)$$

$$\doteq t_9\left[1 + \left(1 + \frac{3}{2}\frac{t_3 + t_7}{t_8}\right)E\right]$$

Thus, when (17) is used the coefficient of relative error propagation is

$$\rho = 1 + \frac{3}{2}\frac{t_3 + t_7}{t_8}.$$

If $a \sim b \sim r$, $\phi \sim 0$, then

$$t_3 = a^2 + b^2 \sim 2r^2, \qquad t_7 = 2ab \cos \phi \sim 2r^2,$$

so that

$$\rho = 1 + 6\frac{r^2}{s^2} = O\left(\left(\frac{r}{s}\right)^2\right).$$

A more stable formula for $s$ was alleged to be

$$s = \sqrt{(a-b)^2 + 4ab\left(\sin \frac{\phi}{2}\right)^2}. \tag{18}$$

This is translated into algorithmic terms as follows:

$$t_1 := a - b$$
$$t_2 := t_1^2$$
$$t_3 := \tfrac{1}{2}\phi$$
$$t_4 := \sin t_3$$
$$t_5 := t_4^2$$
$$t_6 := ab$$
$$t_7 := t_5 t_6$$
$$t_8 := 4t_7$$
$$t_9 := t_2 + t_8$$
$$s = t_{10} := \sqrt{t_9}$$

In computation this becomes

$$\tilde{t}_1 \in (a - b)(1 + E) \doteq t_1(1 + E)$$
$$\tilde{t}_2 \in \tilde{t}_1^2(1 + E) \subset [t_1(1 + E)]^2(1 + E) \doteq t_2(1 + 3E)$$
$$\tilde{t}_3 \in \tfrac{1}{2}\phi(1 + E) = t_3(1 + E)$$
$$\tilde{t}_4 \in \sin \tilde{t}_3(1 + E) \subset \sin[t_3(1 + E)](1 + E)$$
$$\qquad \doteq [\sin t_3 + (\cos t_3)t_3 E](1 + E)$$
$$\qquad \doteq t_4 + (t_4 + t_3|\cos t_3|)E.$$

Here we can simplify by using

$$|t_3 \cos t_3| \le \sin t_3 = t_4, \qquad 0 \le t_3 = \tfrac{1}{2}\phi \le \frac{\pi}{2}.$$

This yields

$$\tilde{t}_4 \in t_4(1 + 2E)$$

$$\tilde{t}_5 \in \tilde{t}_4^2(1 + E) \doteq t_5(1 + 5E)$$

$$\tilde{t}_6 \in ab(1 + E) = t_6(1 + E)$$

$$\tilde{t}_7 \in \tilde{t}_5\tilde{t}_6(1 + E) \doteq t_7(1 + 7E)$$

$$\tilde{t}_8 \in 4\tilde{t}_7(1 + E) \doteq t_8(1 + 8E)$$

$$\tilde{t}_9 \in (\tilde{t}_2 + \tilde{t}_8)(1 + E) \subset [t_2(1 + 3E) + t_8(1 + 8E)](1 + E)$$

$$\doteq t_9 + (4t_9 + 5t_8)E$$

$$\tilde{t}_{10} \in \sqrt{\tilde{t}_9}(1 + E) \subset \left\{ t_9 \left[ 1 + \left( 4 + \frac{5t_8}{t_9} \right)E \right] \right\}^{1/2} (1 + E)$$

$$\doteq t_{10} \left[ 1 + \left( 2 + \frac{5}{2}\frac{t_8}{t_9} \right)E \right] (1 + E)$$

$$\doteq t_{10} \left[ 1 + \left( 3 + \frac{5}{2}\frac{t_8}{t_9} \right)E \right]$$

Using (18), we thus get a coefficient of relative error propagation of

$$\rho = 3 + \frac{5}{2}\frac{t_8}{t_9}.$$

In view of $t_8 \le t_9$,

$$\rho \le 5.5.$$

Thus for (18) the coefficient of relative error propagation is bounded, independently of $a$, $b$, $\phi$, or $s$.


# PROBLEMS

---

1  Compute the coefficients of relative error propagation for the arithmetic and the geometric mean of two numbers $x$ and $y$, and show that the results are compatible with Demonstration 1.2-4.

2  Compute the coefficients of relative error propagation for the two methods of computing $\sqrt{x} - \sqrt{y}$ discussed in Demonstration 1.3-1, assuming that $x$ and $y$ are integers.

3  Let $p > 0$, $q > 0$. The smaller root of the quadratic equation

$$x^2 - 2px - q = 0$$

may be represented either as

$$x_1 = p - \sqrt{p^2 + q}$$

or as

$$x_1 = -\frac{q}{p + \sqrt{p^2 + q}}$$

(compare Demonstration 1.3-2). Show that for the first formula, if $p^2/q$ is large, the coefficient of relative error propagation equals

$$\rho \sim \frac{4p^2}{q},$$

whereas if the second formula is used, it satisfies

$$\rho < 3.$$

4  Compute the coefficients of relative error propagation $\rho_n$ for the two methods of computing the integrals

$$y_n := \int_0^1 \frac{x^n}{a + x} \, dx$$

described in the Demonstrations 1.5-1 and 1.5-2.

(a)  If forward recursion is used, show that for $a > 1$

$$\rho_k \geq (2a + 1)\frac{a^k - 1}{a - 1}, \qquad k = 1, 2, \ldots .$$

(b)  If backward recursion is used with the initial term $y_m = 0$, show that for $a > 1$

$$\rho_{m-k} \leq \frac{2a + 1}{a - 1}, \qquad k = 0, 1, \ldots, m,$$

confirming that the backward use of the recurrence relation is stable.

---

# §1.7  THE CONDITION OF A PROBLEM

Most numerical processes proceed according to the following general outline:
Some data are given (input).
The data are treated according to some predetermined plan (algorithm).
The result appears (output).

**EXAMPLE 1.7-1:** *Evaluation of a function.* For the problem of evaluating a function $f$ of a single real variable, the input is any real number $x$ in the domain of definition of $f$. The plan to treat the data here is some method to compute $f$. (This usually can be done in many different ways, if only approximately.) The output is the value $f(x)$ of $f$ at $x$.

**EXAMPLE 1.7-2:** *Determination of the zeros of a polynomial.* The data are the coefficients of the polynomial or some other quantities that define the polynomial uniquely; the output is a set of complex numbers, alleged to be the zeros of the polynomial.

**EXAMPLE 1.7-3:** *Solution of a set of linear equations.* The data are the coefficients of the system of equations and the non-homogeneous terms. These data are being worked on, for instance, by the Gaussian algorithm described in §4.1. The output is either a message indicating that the given system cannot be solved or an alleged solution of the system.

**EXAMPLE 1.7-4:** *Solution of a differential equation.* The input (in the simplest case) here is a program for the evaluation of the function $f$ in the differential equation $y' = f(x, y)$, plus the given initial condition. The output consists of numerical approximations to the solution on a set of preselected points.

In this book our concern is only with problems where the result depends *continuously* on the data. In the mathematical literature such problems are called **well-posed**. Certain applications, such as mathematical models for geophysical prospecting, or for the exploration of the human brain by means of computer tomography, require the solution of an **ill-posed** problem, where the result does not depend continuously on the data. The treatment of ill-posed problems requires advanced techniques that are beyond the scope of this work.

If a problem is well-posed, and we are dealing with continuous functions, all seems well from the point of view of the pure mathematician. Researchers not concerned with numbers frequently do not realize the *low degree of continuity* which may be inherent, even in well-posed problems.

**DEMONSTRATION 1.7-5:** Let

$$f(x) := \begin{cases} 0, & x = 0, \\ \left(\log \dfrac{1}{x}\right)^{-1/8}, & 0 < x < 1. \end{cases}$$

Mathematically, this function is continuous at all points $x \geq 0$. Thus, the values $f(x)$ approach 0 as $x \to 0$. If we look at numerical values, however, we observe the following:

| $x$ | $f(x)$ |
|---|---|
| 0.1 | 0.90099637 |
| 0.01 | 0.82621731 |
| 0.001 | 0.78538551 |
| 0.0001 | 0.75764461 |
| 0.00001 | 0.73680368 |
| $10^{-10}$ | 0.67565196 |
| $10^{-20}$ | 0.61957558 |
| $10^{-40}$ | 0.56815331 |
| $10^{-80}$ | 0.52099888 |

For $x = 10^{-99}$, the smallest positive number that is distinguishable from 0 on a HP-33E pocket calculator, we find

$$f(10^{-99}) = 0.50730441$$

Thus, the fact that $\lim_{x \to 0} f(x) = 0$ is in no way borne out numerically, not even for numbers that are far beyond the precision of any physical measurement.

**DEMONSTRATION 1.7-6:** *Zeros of a polynomial.* Let

$$p(z) := z^{10} - 10z^9 + 45z^8 - 120z^7 + 210z^6 - 252z^5 + 210z^4$$
$$- 120z^3 + 45z^2 - 10z + 1.$$

We have $p(z) = (z - 1)^{10}$, the polynomial $p$ thus has the (tenfold) zero $z = 1$. Let us now change the constant coefficient 1 into $1 - 10^{-10}$. Then, $p(z) = 0$ means the same as

$$(z - 1)^{10} = 10^{-10};$$

thus, the zeros now are

$$z = 1 + 10^{-1}e^{2\pi ki/10}, \qquad k = 0, 1, \ldots, 9.$$

A change of $10^{-10}$ in one of the coefficients thus causes all zeros of the polynomial to change by an amount that is 1,000,000,000 times as large as the change in the coefficient.

**DEMONSTRATION 1.7-7:** *System of linear equations.* We consider the system

$$x + y = 1$$
$$x + (1 - a)y = 0$$

where $a$ is a parameter. The system has no solution for $a = 0$; for $a \neq 0$ the solution is

$$x = 1 - \frac{1}{a}, \qquad y = \frac{1}{a}.$$

Let us consider the parameter $a$ as the *datum* for this problem. As long as $a \neq 0$, both components of the solution depend continuously on $a$. However, the degree of this continuity can be very bad. If $a$ is replaced by $\tilde{a}$, both components of the solution change by the amount

$$\left| \frac{1}{\tilde{a}} - \frac{1}{a} \right| = \frac{|\tilde{a} - a|}{|a\tilde{a}|},$$

which can be arbitrarily large, even if $|\tilde{a} - a|$ is small, if $a$ and $\tilde{a}$ are close to zero.

**DEMONSTRATION 1.7-8:** *Solution of a differential equation.* Let the solution $y$ of the differential equation

$$y'' = y$$

be subject to the initial conditions

$$y(0) = a, \qquad y'(0) = b.$$

We consider the problem of evaluating the solution $y(x)$ at a given point $x > 0$ as a function of the data $a$ and $b$. If $a = 1$, $b = -1$, then the solution of the problem clearly is

$$y(x) = e^{-x}.$$

Considered as a function of $x$, the solution thus decreases rapidly to zero. However, if the data are changed to

$$a = 1, \qquad b = -1 + \delta$$

where $|\delta|$ may be arbitrarily small, the solution becomes

$$y(x) = e^{-x} + \frac{\delta}{2}(e^x - e^{-x}).$$

It differs by

$$\frac{\delta}{2}(e^x - e^{-x}) = \delta \sinh x$$

from the solution found previously. Not only can the change of the solution be large, even for arbitrarily small $|\delta|$, if $x$ is large, but the mathematical *character* of the solution has changed completely: While previously the solution tended to zero, it now tends to infinity for $x \to \infty$; all this inspite of the fact that the dependence of $y(x)$ on the data $a$ and $b$ is clearly continuous.

In the light of these examples and demonstrations it obviously becomes necessary to introduce a measure for the degree of continuity of a problem. Such a measure is inherent in the very definition of continuity. Let $X$ be a space of data; the elements $x$ of $X$ may be numbers, points in Euclidean space, matrices, functions, or even more complicated mathematical entities. Talking about continuity makes sense only if we are able to measure distances between the elements of $X$. We thus assume that the space $X$ is endowed with a **distance function** $\rho(x, y)$ measuring the distance between the elements $x$ and $y$ of $X$. The function $\rho$ is assumed to have the usual properties of a *metric* (see Rudin [1964], p. 27). If, for instance, $X$ is the space of real numbers, the simplest distance function is $\rho(x, y) := |x - y|$, and for convenience we use the absolute value bar notation for distance also in more general cases.

Let the process by which the data $x$ are mapped on the result $y$ be denoted by $P$, so that $y = P(x)$. The space of results, like the space of data, is assumed to be metric. If the process $P$ is continuous at the point $x$, then the mathematical definition of continuity requires that for every $\varepsilon > 0$ there exists $\delta(\varepsilon) > 0$ so that

$$|P(\tilde{x}) - P(x)| < \varepsilon \quad \text{whenever} \quad |\tilde{x} - x| < \delta(\varepsilon).$$

The larger the function $\delta(\varepsilon)$ can be chosen, the more continuous is the process $P$. In such a case even large changes in the data will produce only small changes in the results. In this context, the term **condition of a problem** is used. If the function $\delta(\varepsilon)$ can be chosen large, the condition of the problem $P$ is good; equivalently, the problem is said to be **well-conditioned** or **good-natured**. (See Figure 1.7a.) On the other hand, the condition is bad, or the problem is said to be **ill-conditioned**, if $\delta(\varepsilon)$ must be chosen small. (See Figure 1.7b.)

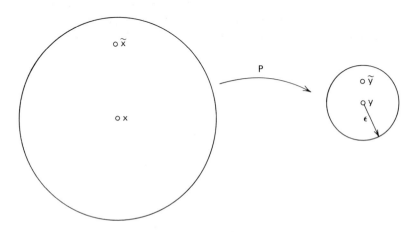

**Fig. 1.7a.**  Well-conditioned problem.

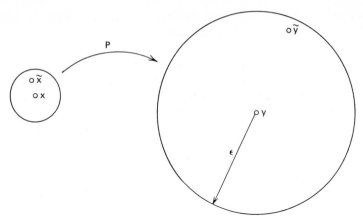

**Fig. 1.7b.**  Ill-conditioned problem.

We examine the condition of some of the problems stated in the preceding demonstrations.

**EXAMPLE 1.7-9:**  If the zeros of the polynomial $p$ considered in Demonstration 1.7-6 are to change by no more than $\varepsilon$, the constant coefficient of the polynomial must be changed by no more than $\delta(\varepsilon) = \varepsilon^{10}$. For small $\varepsilon$, this function is exceedingly small; thus, the condition of the zeros with regard to the constant coefficient is extremely bad. It might be thought that the bad condition of this problem has to do with the fact that the zero considered has a very high multiplicity; however, it has been shown (Wilkinson [1963]) that there exist polynomials with all simple zeros where the problem of zero determination is equally badly conditioned.

**EXAMPLE 1.7-10:**  In the differential equation problem of Demonstration 1.7-8, it easily follows from the formulas given that if the solution $y(x)$ at the point $x$ is to change by no more than $\varepsilon$, the initial condition $y'(0) = -1$ must be changed by no more than

$$\delta(\varepsilon) = \frac{\varepsilon}{\sinh x},$$

which can be made arbitrarily small by choosing $x$ large. For instance, for $x = 10$ we have

$$\delta(\varepsilon) = 0.9 * 10^{-4}\varepsilon.$$

Thus, we are dealing again with an ill-conditioned problem.

**EXAMPLE 1.7-11:**  Considering the number $a$ as input for the problem of solving the linear system encountered in Demonstration 1.7-7, we find that if the components of the solution are to change by not more than $\varepsilon$,

$$|\tilde{x} - x| \le \varepsilon, \qquad |\tilde{y} - y| \le \varepsilon,$$

then the input $a$ must satisfy

$$\left|\frac{1}{\tilde{a}} - \frac{1}{a}\right| \le \varepsilon,$$

which can be shown to be equivalent to

$$|\tilde{a} - a| \le \frac{a^2}{1 + \varepsilon a}\, \varepsilon.$$

Thus, the function $\delta(\varepsilon)$ here is given by

$$\delta(\varepsilon) = \frac{a^2}{1 + \varepsilon a}\, \varepsilon.$$

This function can be made arbitrarily small by choosing $a$ small; our little problem thus is arbitrarily ill-conditioned.

In the examples considered so far, the condition of a problem has been described by means of the size of the function $\delta(\varepsilon)$. While, in the examples given, the meaning of *size* was intuitively clear, the size of a function (with emphasis on the values for $\varepsilon \to 0$) is not a concept that is very well defined mathematically. For instance, for the purpose of comparing the conditions of different problems, or of the same problem at different data points, it would be better to describe the condition of a problem by means of a single number.

Such a description is possible if the function $P$ underlying the process is assumed to be differentiable. If $P$ is differentiable at the point $x$, and if $y = P(x)$, then the change experienced by $y$ if $x$ is changed by $dx$ may be approximated, in the sense of the differential calculus, by the differential of $y$,

$$dy = P'(x)\, dx,$$

where $P'(x)$, in general, is a linear operator acting on $dx$. Obviously, the size $|P'(x)|$ of the linear operator $P(x)$ may be said to describe the condition of the problem at the point $x$, and thus may be called the **condition number** of the problem at the point $x$. The problem is well-conditioned if the condition number is small, and ill-conditioned if the condition number is large. In some cases, it may be desirable to compare $dy$ with the size of the solution $y = P(x)$ at the point $x$. Then, it is appropriate to consider the **relative condition number**

$$c_R := \frac{|P'(x)|}{|P(x)|}.$$

A problem will be said to be **relatively well-conditioned** if it has a relative condition number $\leq 1$.

**EXAMPLE 1.7-12:**    If the problem $P$ consists in evaluating a real function $f$ of a real variable, then the condition number of $P$ at $x$, provided $f$ is differentiable at $x$, is simply $|f'(x)|$. For instance, the problem of evaluating the function

$$f(x) = \left(\log \frac{1}{x}\right)^{-1/8}$$

considered in Demonstration 1.7-5 at the point $x$ has the condition number

$$|f'(x)| = \frac{1}{8x}\left(\log \frac{1}{x}\right)^{-9/8},$$

and the relative condition number

$$c_R = \left|\frac{f'(x)}{f(x)}\right| = \frac{1}{8x \log \frac{1}{x}}.$$

At $x = 0$, both condition numbers are $\infty$, and thus at this point the condition of the problem is extremely bad, which is in agreement with the numerical results presented in Demonstration 1.7-5. At $x = 1$, the condition numbers also are infinite. However, it should also be noted that for $0.1537 \leq x \leq 0.6360$, approximately, we have $c_R \leq 1$. Thus, in this interval, the problem of evaluating $f$ is well-conditioned.

**EXAMPLE 1.7-13:**    If the polynomial

$$p(x) = \sum_{k=0}^{n} a_k x^k$$

has the simple zero $z$, so that $p(z) = 0$, $p'(z) \neq 0$, the condition number of $z$ with regard to a change in a coefficient $a_m$ may be calculated as follows: Denoting the changed polynomial by $\tilde{p}(x)$, we have

$$\tilde{p}(x) = p(x) + da_m x^m,$$

where $da_m$ is the change in $a_m$. Thus, from the fact that

$$0 = \tilde{p}(\tilde{z}) = \tilde{p}(z + dz) = p(z + dz) + da_m(z + dz)^m$$
$$= p(z) + p'(z)\, dz + z^m\, da_m + O(dz^2) + O(dz\, da_m),$$

it follows in view of $p(z) = 0$ that

$$dz = -\frac{z^m}{p'(z)}\, da_m.$$

Thus, the condition number with regard to a change in $a_m$ is

$$c = \frac{|z|^m}{|p'(z)|}.$$

If, in particular, the coefficient of $z^0$ is changed, the condition number is

$$c = \frac{1}{|p'(z)|}.$$

These condition numbers clearly become infinite if $p'(z) = 0$; that is, if the zero $z$ has a multiplicity $> 1$. This is the situation that obtains for the zero $z = 1$ of the polynomial $p$ considered in Demonstration 1.7-6.

**EXAMPLE 1.7-14:** Here, we let the result of the process $P$ be a vector,

$$\mathbf{y} = \begin{pmatrix} y_1 \\ \vdots \\ y_m \end{pmatrix}$$

where each component is a function of the input variables $x_1, \ldots, x_n$,

$$y_i = f_i(x_1, \ldots, x_n), \qquad i = 1, \ldots, m. \tag{1}$$

The differential of $\mathbf{y}$ with respect to a change in the input vector

$$\mathbf{x} = \begin{pmatrix} x_1 \\ \vdots \\ x_n \end{pmatrix}$$

then is given by

$$d\mathbf{y} = \mathbf{J}\, d\mathbf{x},$$

where $\mathbf{J}$ is the Jacobian matrix of the system of functions (1),

$$\mathbf{J} = \left( \frac{\partial f_i}{\partial x_j} \right),$$

and the condition of the problem is just the size of the matrix $\mathbf{J}$, measured in some way.

An especially simple situation of this kind arises in the problem of solving a system of $n$ linear equations with $n$ unknowns (see §4.1), which in matrix form may be written

$$\mathbf{A}\mathbf{y} = \mathbf{x}.$$

The solution is given mathematically by

$$\mathbf{y} = \mathbf{A}^{-1}\mathbf{x}. \tag{2}$$

(See however §4.1 with regard to the numerical usefulness of this formula.) Because the relation between $\mathbf{x}$ and $\mathbf{y}$ is already linear, the Jacobian here simply is

$$\mathbf{J} = \mathbf{A}^{-1},$$

and in order to define a condition number we again have to measure the size of a matrix. Such measures are known as **matrix norms**. We refer to more advanced texts on numerical linear algebra (Forsythe & Moler [1967]; Stewart [1973]) for a discussion of this concept.

In none of the foregoing examples and demonstrations was there any mention of the *algorithm* used to solve the problem posed. Indeed, the condition of a problem is a purely mathematical property of the problem; it exists in the problem before we attempt to solve the problem numerically.

Ill-conditioned problems are notorious, however, for the difficulties they present when an accurate numerical solution is attempted. One way to see this is by adopting a mode of thought known as **backward error analysis** (Wilkinson [1963]). In this type of error analysis, any rounding error that is committed at some stage of an algorithmic process is replaced by a change in the data which has the same effect as the rounding error. Thus, the total effect of rounding is replaced by an equivalent change in the data. If the problem is ill-conditioned, then small errors, and thus small equivalent changes in the data, will cause the solution to change substantially. To obtain accurate solutions in such cases, one will have to change the mathematical model so that its condition is improved.

**DEMONSTRATION 1.7-15:** Let us solve the initial value problem

$$y'' = y, \qquad y(0) = 1, \qquad y'(0) = -1$$

by discretization, which means that the solution is sought only on a discrete set of points

$$x_n := nh, \qquad n = 0, 1, \ldots,$$

where $h > 0$ is the discretization step. The idea is that if $h \to 0$, the accurate solution is obtained. Let $y_n$ be an approximation to the exact solution at the point $x_n$. We simulate the differential equation by replacing $y''(x_n)$ by

$$\frac{1}{h^2}(y_{n+1} - 2y_n + y_{n-1}),$$

thereby committing an error that tends to zero like $h^2$ as $h \to 0$. This yields the difference equation

$$\frac{1}{h^2}(y_{n+1} - 2y_n + y_{n-1}) = y_n,$$

which may be written in the form

$$y_{n+1} = (2 + h^2)y_n - y_{n-1}$$

and thus may be solved recursively. The process is started with the values

$$y_0 = 1 \text{ (exact starting value)}$$

$$y_{-1} = e^h \text{ (exact value of the solution at the point } x_{-1} = -h).$$

The following values result if the step $h = 0.1$ is chosen:

| $x_n$ | $y_n$ | $e^{-x_n}$ |
|---|---|---|
| 0.0 | 1.000000 | 1.000000 |
| 0.1 | 0.904829 | 0.904837 |
| 0.2 | 0.818706 | 0.818731 |
| 0.3 | 0.740771 | 0.740818 |
| . | . . . | . |
| 2.0 | 0.133783 | 0.135335 |
| 2.1 | 0.120718 | 0.122456 |
| 2.2 | 0.108859 | 0.110803 |
| 2.3 | 0.098089 | 0.100259 |
| . | . . . | . |
| 4.0 | 0.005830 | 0.018316 |
| 4.1 | 0.002769 | 0.016573 |
| 4.2 | -0.000265 | 0.014996 |
| 4.3 | -0.003302 | 0.013569 |
| . | . . . | . |
| 10.0 | -5.038236 | 0.000045 |
| 10.1 | -5.567889 | 0.000041 |
| 10.2 | -6.153221 | 0.000037 |
| 10.3 | -6.800086 | 0.000034 |

At the beginning, the numerical values $y_n$ faithfully approximate the values of the exact solution $e^{-x}$. At $x = 4$, however, the relative error of the $y_n$ is already in excess of 200%, and from $x = 4.2$ onward even the sign of the $y_n$ is wrong. From then on, the $y_n$ grow exponentially. Obviously, a component $\delta \sinh x$ has been picked up, accidentally with $\delta < 0$. From the fact that $\sinh(10) = 1.101 * 10^4$, we may conclude that $\delta = 4.574 * 10^{-4}$. After studying linear difference equations in §3.1, we shall be able to explain this value of $\delta$.

We should recognize that the numerical method used in this demonstration is not unstable as such. The method works perfectly well on less badly conditioned problems. The method merely reflects the bad condition of the problem. In order to get better solutions, one would have to change the mathematical model, for instance, by replacing the differential equation by

$$y' = -y, \qquad y(0) = 1,$$

which has the same solution.

The notions of numerical stability and of ill-conditioning are basic for the whole of numerical analysis. A word of caution concerning the usage of these terms is in order, however. Frequently the two terms are used interchangeably, for instance by calling certain *algorithms* ill-conditioned. As used here, the term *ill-conditioning* applies only to mathematical models or problems, while the term *numerical instability* applies only to algorithms.

# PROBLEMS

1  Determine the absolute and the relative condition numbers for the arithmetic mean $m$ and the geometric mean $g$ of $n$ positive numbers $x_1$, ..., $x_n$,

$$m = \frac{1}{n}(x_1 + x_2 + \cdots + x_n),$$

$$g = (x_1 x_2 \cdots x_n)^{1/n},$$

with respect to any $x_k$.

2  Compute the condition of the variance $\sigma^2$ of a set of data $x_1, \ldots, x_n$ with respect to any $x_k$. Show that the condition is excellent if $x_k$ equals the mean $\mu$ and gets worse for "outliers" $x_k$, where $|x_k - \mu|$ is large.

3  Consider the solution of the initial value problem

$$y' = -y, \qquad y(0) = 1 \tag{3}$$

for $x \geq 0$.

(a)  If the initial condition is changed to $y(0) = 1 + \delta$, how does this affect the solution? What do you conclude about the condition of this problem?

**(b)** Determine both the relative and the absolute condition numbers of the solution at a fixed $x > 0$ with respect to the initial condition.

**(c)** Solve the problem (3) numerically by approximating $y'(x_n)$ by

$$\frac{1}{2h}(y_{n+1} - y_{n-1})$$

and proceeding as in Demonstration 1.7-15. What do you observe? How do you judge the stability properties of this numerical method?

**4** Here we consider the initial value problem,

$$y'' = -y, \qquad y(0) = 1, \qquad y'(0) = 0.$$

**(a)** How do you judge the condition of this problem with regard to the initial conditions?

**(b)** Show that the method used in Demonstration 1.7-15 produces reasonable values for this problem.

---

### Bibliographical Notes and Recommended Reading

Comprehensive treatments of error propagation, particularly in algorithms of numerical linear algebra, will be found in Wilkinson [1963, 1965], Forsythe and Moler [1967], Stewart [1973]. The approach used here is similar in spirit to that of Stummel [1980]. For a statistical approach to error propagation, see Henrici [1962].

Some of the demonstrations given in this chapter are elaborations of examples to be found in Dahlquist and Björck [1974], Rutishauser [1976], Forsythe, Malcolm, and Moler [1977].

# CHAPTER 2

## Iteration

In Chapters 2 and 3 we discuss various ways to construct sequences, and how to use them for purposes of numerical mathematics. Our first concern is with sequences of *numbers*. We know from analysis that a sequence of numbers is a mapping

$$\mathbb{Z}_+ \to \mathbb{R}$$

from the nonnegative integers into (for instance) the real numbers. The real number associated with the integer $n$ is usually denoted by $a_n$, $x_n$ or the like; the whole sequence is denoted by $\{a_n\}$, $\{x_n\}$, or by indicating the first few elements of the sequence, as in

$$1, \quad \tfrac{1}{2}, \quad \tfrac{1}{3}, \dots,$$

if the law of formation thus can be recognized.

A sequence is said to be defined **recursively** if the $n$-th element of the sequence is not defined by an explicit formula, such as

$$a_n := \frac{1}{n+1},$$

but as a function of one or several preceding elements,

$$a_n := f(n, a_{n-1}, \dots), \tag{1}$$

such as

$$a_n = \frac{n}{n+1} a_{n-1}. \tag{2}$$

Of course, even if a sequence is defined recursively, its first (or first few) elements have to be exhibited explicitly. For instance, (2) will specify a sequence uniquely if it is stated that $a_0 = 1$.

**64**

A sequence is said to be defined by **iteration** if the function $f$ in (1) is independent of $n$.

## §2.1  SCALAR ITERATION

Here we consider sequences $\{x_n\}$ that are generated by iteration, as defined above, where $f$ is a real function of a *single* variable. Let $x_0$ be chosen arbitrarily. We then form

$$x_1 := f(x_0),$$
$$x_2 := f(x_1),$$
$$x_3 := f(x_2),$$

and generally

$$x_n := f(x_{n-1}), \qquad n = 1, 2, \ldots . \tag{1}$$

If the function $f$ is thought of as a black box that on being excited with the input $x$ generates the output $f(x)$, then the sequence $\{x_n\}$ is generated by short-circuiting the box. (See Figure 2.1a.)

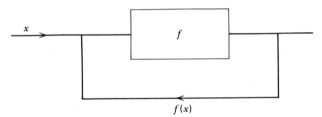

**Fig. 2.1a.**  Iteration.

Mathematically, the sequence defined by (1) is called the **iteration sequence** generated by the function $f$ with starting element $x_0$. (Iteration stems from the latin root, iterare, meaning "to plow once again.") What can be said about such iteration sequences?

For orientation, consider the case where $f$ is *linear*,

$$f(x) := ax + b, \tag{2}$$

with arbitrary real constants $a$ and $b$. Using (1), let us try to find a non-recursive formula for the $n$-th element of the iteration sequence generated by $f$. We have

$$x_1 = ax_0 + b,$$
$$x_2 = ax_1 + b = a(ax_0 + b) + b = a^2 x_0 + (1 + a)b,$$
$$x_3 = ax_2 + b = a[a^2 x_0 + (1 + a)b] + b = a^3 x_0 + (1 + a + a^2)b,$$

which leads us to suspect that

$$x_n = a^n x_0 + (1 + a + a^2 + \cdots + a^{n-1})b, \tag{3}$$

as is easily verified by an induction argument. In place of (3) we may write more compactly

$$x_n = \begin{cases} a^n x_0 + \dfrac{1 - a^n}{1 - a}\, b, & \text{if } a \neq 1, \\[2mm] x_0 + nb, & \text{if } a = 1. \end{cases} \tag{4}$$

The convergence of the sequence $\{x_n\}$ is now discussed very easily. Remembering that

$$\lim_{n \to \infty} a^n = 0 \quad \text{if } |a| < 1,$$

$$\lim_{n \to \infty} a^n = 1 \quad \text{if } a = 1,$$

$$\lim_{n \to \infty} a^n \text{ does not exist if } \quad a = -1 \quad \text{or} \quad |a| > 1,$$

we find that

$$\lim_{n \to \infty} x_n = \frac{1}{1 - a}\, b, \quad \text{if } |a| < 1,$$

$$\lim_{n \to \infty} x_n = x_0, \quad \text{if } a = 1 \quad \text{and} \quad b = 0,$$

$$\lim_{n \to \infty} x_n \text{ does not exist if } a = 1 \quad \text{and} \quad b \neq 0,$$

$$\text{or if } a = -1, \quad \text{or if } |a| > 1.$$

We note that if $|a| < 1$ the limit

$$s := \frac{1}{1 - a}\, b \tag{5}$$

is independent of the starting value $x_0$. It thus depends only on the function $f$. But what is its relation to $f$? We find the answer by considering the *graph* of $f$ (Figure 2.1b).

If, in addition to the line $y = f(x) = ax + b$, we also draw the line $y = x$, the points of the iteration sequence $\{x_n\}$ may be found by an easy graphical construction. It is then seen that the sequence converges to the abscissa of the point of intersection of the two lines $y = f(x)$ and $y = x$; that is, to the unique solution $s$ of the equation

$$x = f(x). \tag{6}$$

Evidently, for the linear function $f$ considered here, this solution is given by (5).

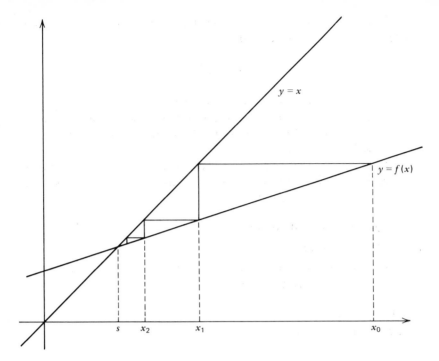

**Fig. 2.1b.**   Iteration of linear function.

That the limit $s$ is a solution of (6) may also be seen in a purely analytical fashion, as follows: by the basic recurrence relation (1),

$$s = \lim_{n \to \infty} x_n = \lim_{n \to \infty} f(x_{n-1}) = \lim_{n \to \infty} f(x_n) = f\left(\lim_{n \to \infty} x_n\right) = f(s).$$

Here, the next to last equality follows from the fact that $f$ is continuous. Note that, in addition to continuity and the fact that $f$ is defined on a closed set, no special properties of the function $f$ have been used to show that if $s := \lim_{n \to \infty} x_n$ exists, $s$ is a solution of equation (6).

For any function $f$, any solution of equation (6) is called a **fixed point** of $f$. (If the function $f$ is interpreted as a map of its domain of definition into the reals, the point $s$ remains fixed under this map.) We thus may summarize the above by stating that the iteration sequence generated by a linear function, whose slope $a$ satisfies $|a| < 1$, converges for any choice of the starting value $x_0$, and the limit is always the (unique) fixed point of the function. How does this generalize to arbitrary functions $f$?

An arbitrary function need not be defined everywhere. Let $D(f)$ denote the domain of definition and $R(f)$ the range (domain of values) of the function $f$. So that the formula $x_n = f(x_{n-1})$ has a meaning for all $n$ and all choices of $x_0 \in D(f)$, it is evidently necessary that

$$R(f) \subset D(f). \tag{I}$$

Thus, for instance, if $D(f)$ is a closed finite interval, $D(f) = [a, b]$, then the values of $f$ must satisfy $a \le f(x) \le b$ for $a \le x \le b$. Expressed geometrically, the graph of $f$ must lie in the "window" of the $(x, y)$-plane described by $a \le x \le b$, $a \le y \le b$. (See Figure 2.1c.)

We assume henceforth that the domain of definition of $f$ is a closed (but not necessarily bounded) interval. Assuming also that (I) holds, we address ourselves to the following questions:

  **(i)**   Does $f$ have a fixed point?
  **(ii)**  Is the fixed point unique?
  **(iii)** Does the iteration sequence generated by $f$ converge for arbitrary choices of the starting value $x_0$?

As Figure 2.1c shows, no fixed point need exist if $f$ is not continuous. We therefore postulate

$$f \text{ is continuous on } D(f). \tag{II}$$

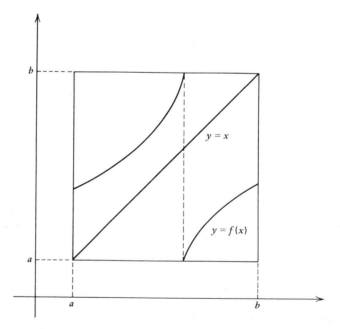

**Fig. 2.1c.**   Condition (I).

If $D(f)$ is a closed and bounded interval, $D(f) = [a, b]$, this assumption already guarantees the existence of a fixed point. This is seen most easily by considering the auxiliary function $g(x) := x - f(x)$. Because $f(a) \geq a$, we have $g(a) \leq 0$. Because $f(b) \leq b$, we have $g(b) \geq 0$. Thus, at least one of the following three conditions is satisfied:

(a) $g(a) = 0$;
(b) $g(b) = 0$;
(c) $g(a) < 0$ and $g(b) > 0$.

It is clear in the cases (a) and (b), and it follows from the intermediate value theorem* and the continuity of $g$ in case (c), that the equation $g(x) = 0$ has a solution in $[a, b]$; that is, that there exists $s \in [a, b]$ so that $s - f(s) = 0$, tantamount to the existence of a fixed point.

The above argument does not work if $D(f)$ is an unbounded interval, and, in fact, no fixed point needs to exist in that case, as is shown by the example $D(f) := [0, \infty)$, $f(x) := x + 1$. Here we need a condition which insures that the curve $y = f(x)$ intersects the straight line $y = x$. We thus want the function $|f|$ to grow less strongly than the function $x$. A simple way to accomplish this would be to assume that $f$ is differentiable, and that the derivative $f'$ satisfies $|f'(x)| \leq L$ for all $x \in D(f)$, where $L$ is a constant satisfying $L < 1$. However, one does not like to assume differentiability if it is not really needed. It is therefore preferable to formulate the required condition in terms of the difference quotient of the function $f$. Avoiding denominators which may become zero, this condition is usually given as follows:

There exists a constant $L$, $0 \leq L < 1$, so that for any two points $x_1$, $x_2$ in the domain of definition of $f$,

$$|f(x_1) - f(x_2)| \leq L|x_1 - x_2|. \tag{III}$$

A condition of the form (III), even if $L$ does not satisfy $L < 1$, is called a **Lipschitz condition**. If $L < 1$, condition (III) means that the map $f$ decreases the distance between any two points in its domain of definition. A function (or map) $f$ satisfying (III) (where $L < 1$) is therefore called a **contracting map**.

If the function $f$ is a contracting map, then it has a fixed point, even if its interval of definition (always assumed to be closed) is unbounded. Assume, for definiteness, that $D(f) = [a, \infty)$. If $f(a) = a$, then $a$ is a fixed point. If $f(a) \neq a$, then $f(a) > a$ by (I). For arbitrary $x > a$,

$$f(x) \leq |f(x)|$$
$$= |f(a) + f(x) - f(a)|$$
$$\leq |f(a)| + |f(x) - f(a)|,$$

---

* See Salas and Hille [1978], p. 80; Anton [1980], p. 185.

and using (III),

$$f(x) \le |f(a)| + L|x - a|$$
$$\le c + L|x|,$$

where $c := |f(a)| + L|a| > 0$. Let now

$$x > \frac{c}{1 - L}.$$

Then, $x > 0$, and $x > c + Lx = c + L|x|$, hence

$$f(x) < x.$$

Thus, the graph of $f$ intersects the line $y = x$ somewhere between the points $a$ and $x$; that is, $f$ has a fixed point. The cases where $D(f) = (-\infty, a)$ and $D(f) = (-\infty, \infty)$ can be dealt with similarly.

Condition (III), which enables us to answer question (i) affirmatively, is sufficient also to give positive answers to the questions (ii) and (iii).

As to (ii), suppose that $f$ has two fixed points $s_1$ and $s_2$, say, $s_1 \ne s_2$. By definition, this means that

$$s_1 = f(s_1),$$
$$s_2 = f(s_2).$$

Subtraction yields

$$s_1 - s_2 = f(s_1) - f(s_2),$$

which by (III) implies

$$|s_1 - s_2| = |f(s_1) - f(s_2)| \le L|s_1 - s_2|.$$

Dividing by $|s_1 - s_2| \ne 0$, we get $1 \le L$, which contradicts $L < 1$. The assumption that $f$ has two distinct fixed points thus cannot be maintained, proving the existence of a unique fixed point.

As to (iii), let $s$ be the unique fixed point of $f$, and let $\{x_n\}$ be the iteration sequence generated by $f$ with starting value $x_0$. We then have for every index $n > 0$, using (III),

$$|x_n - s| = |f(x_{n-1}) - f(s)| \le L|x_{n-1} - s|, \tag{7}$$

and thus by induction,

$$|x_n - s| \le L^n |x_0 - s|.$$

In view of $L < 1$ it thus follows that $\{x_n\}$ converges to $s$. We thus have proved the assertions (a) and (b) of the following theorem:

**THEOREM 2.1:**  *Let $f$ be a real function whose domain of definition $D(f)$ is a closed* (but not necessarily bounded) *interval, and let $f$ satisfy the conditions* (I) *and* (III). *Then,*
    (a)  *$f$ has a unique fixed point $s$;*

(b)  *the iteration sequence* $\{x_n\}$ *generated by* $f$ *converges for any choice of the starting value* $x_0 \in D(f)$*, and its limit is* $s$*;*

(c)  *for* $n = 0, 1, 2, \ldots$ *there holds*

$$|x_n - s| \le \frac{L^n}{1 - L}|x_1 - x_0|. \tag{8}$$

To prove (c), let $m > n$. Then, in view of $|x_{k+1} - x_k| \le L|x_k - x_{k-1}|$,

$$|x_n - x_m| \le |x_n - x_{n+1}| + |x_{n+1} - x_{n+2}| + \cdots + |x_{m-1} - x_m|$$
$$\le \{L^n + L^{n+1} + \cdots + L^{m-1}\}|x_1 - x_0|$$
$$\le \frac{L^n}{1 - L}|x_1 - x_0|,$$

and (8) follows on letting $m \to \infty$.

Some examples on Theorem 2.1 follow.

**DEMONSTRATION 2.1-1:**  Let

$$f(x) := \cos x.$$

This function satisfies the hypotheses of Theorem 2.1, for instance, in the interval $[0, 1]$ where $L = \sin 1 = 0.8415$. The following iteration sequence is obtained for $x_0 = 0$:

| $n$ | | $n$ | |
|---|---|---|---|
| 1 | 1.00000000 | 20 | 0.73893776 |
| 2 | 0.54030231 | | · · · · |
| 3 | 0.85755322 | 30 | 0.73908230 |
| 4 | 0.65428979 | | · · · · |
| 5 | 0.79348036 | 40 | 0.73908508 |
| 6 | 0.70136877 | 41 | 0.73908517 |
| 7 | 0.76395968 | 42 | 0.73908511 |
| | · · · · | 43 | 0.73908515 |
| 10 | 0.73140404 | 44 | 0.73908512 |
| | · · · · | 45 | 0.73908514 |
| 15 | 0.74014743 | 46 | 0.73908513 |
| | · · · · | 47 | 0.73908514 |
| | | 48 | 0.73908513 |
| | | 49 | 0.73908513 |

Convergence (within the accuracy provided by the number of digits reproduced here) has taken place after 49 iterations. We shall see later how to improve the speed of convergence.

**DEMONSTRATION 2.1-2:** What is meant by the symbol

$$\sqrt{2 + \sqrt{2 + \sqrt{2 + \sqrt{2 + \sqrt{2 + \cdots}}}}} \quad ?$$

We interpret this as the limit of the sequence

$$x_1 = \sqrt{2}$$
$$x_2 = \sqrt{2 + \sqrt{2}}$$
$$x_3 = \sqrt{2 + \sqrt{2 + \sqrt{2}}}$$

$$. \quad . \quad . \quad .$$

which is generated by iterating the function

$$f(x) := \sqrt{2 + x}$$

starting with $x_0 = 0$. This function clearly satisfies hypothesis (I) of Theorem 2.1 for the interval $[0, \infty)$. In view of

$$f(x_1) - f(x_2) = \sqrt{2 + x_1} - \sqrt{2 + x_2}$$
$$= \frac{(2 + x_1) - (2 + x_2)}{\sqrt{2 + x_1} + \sqrt{2 + x_2}}$$
$$= \frac{x_1 - x_2}{\sqrt{2 + x_1} + \sqrt{2 + x_2}},$$

it also satisfies (III); the best value of the Lipschitz constant is

$$L = \frac{1}{2\sqrt{2}}.$$

It follows that the sequence $\{x_n\}$ converges to the fixed point of $f$ (we know that there is only one); that is, to the sole nonnegative solution of the equation

$$x = \sqrt{2 + x} \quad (x \geq 0);$$

that is, of

$$x^2 = 2 + x \quad (x \geq 0);$$

that is, to $x = 2$. The numerical test yields

| n |  | n |  |
|---|---|---|---|
| 1 | 1.414213562 | 10 | 1.999997647 |
| 2 | 1.847759065 | 11 | 1.999999412 |
| 3 | 1.961570561 | 12 | 1.999999853 |
| 4 | 1.990369453 | 13 | 1.999999963 |
| 5 | 1.997590912 | 14 | 1.999999991 |
| 6 | 1.999397637 | 15 | 1.999999998 |
| 7 | 1.999849404 | 16 | 1.999999999 |
| 8 | 1.999962351 | 17 | 2.000000000 |
| 9 | 1.999990588 |  |  |

**DEMONSTRATION 2.1-3:**  An expression of the form

$$
\cfrac{1}{a_1 + \cfrac{1}{a_2 + \cfrac{1}{a_3 + \cfrac{1}{a_4 + \ddots}}}} \tag{9}
$$

is called a **continued fraction**. Continued fractions are used in number theory as well as in analysis, where they may be used to approximate functions. The *value* of the continued fraction is defined to be the limit of the sequence

$$
x_1 := \frac{1}{a_1}, \qquad x_2 := \cfrac{1}{a_1 + \cfrac{1}{a_2}}, \qquad x_3 := \cfrac{1}{a_1 + \cfrac{1}{a_2 + 1/a_3}}, \dots,
$$

provided that this limit exists. Here we wish to determine the value of the continued fraction

$$
c(a) = \cfrac{1}{a + \cfrac{1}{a + \cfrac{1}{a + \ddots}}},
$$

where all $a_i = a > 0$. The sequence $\{x_n\}$ here arises by iterating the function

$$f(x) := \frac{1}{a + x}, \qquad x \geq 0,$$

beginning with $x_0 = 0$. This function satisfies condition (I) of Theorem 2.1 for all $a > 0$. Condition (III), on the other hand, is satisfied only if $a > 1$, the best value of the Lipschitz constant then being $L = a^{-2}$. The iteration sequence here nevertheless converges for all $a > 0$, and its limit is the unique fixed point of $f$. For instance, if $a = 1$, the iteration sequence started with $x_0 = 0$ converges to $s = 0.618033989$, which equals $\frac{1}{2}(\sqrt{5} - 1)$, the unique positive solution of $x = 1/(1 + x)$. We thus see that the conditions of Theorem 2.1, although sufficient, are not necessary for the convergence of the iteration sequence.

**DEMONSTRATION 2.1-4:**   *Kepler's equation.* This equation arises in the two-body problem of celestial mechanics, such as in the motion of a planet around the sun, when the influences of all other planets are neglected. As is well known, the orbit of the planet then is an ellipse, one focus of which coincides with the position of the sun. (See Figure 2.1d.)

The following abbreviations are traditional in celestial mechanics:

$W$ = true anomaly = $\measuredangle\ \overrightarrow{OS}, \overrightarrow{SP}$;

$E$ = eccentric anomaly = $\measuredangle\ \overrightarrow{OS}, \overrightarrow{OP'}$;

$M$ = mean anomaly = time, measured in the unit in which
   going around the ellipse once requires the time $2\pi$;

$\varepsilon$ = eccentricity of the ellipse = $\frac{1}{a}\sqrt{a^2 - b^2}$.

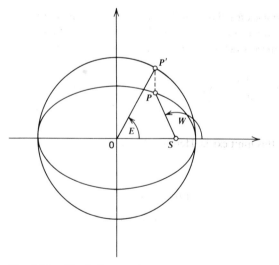

**Fig. 2.1d.**   Kepler's equation.

The problem naturally arises to determine the position of the planet, as given by $W$ or $E$, as a function of time. It follows from the area theorem of mechanics that the connection between $E$ and $M$ is given by the relation

$$M = E - \varepsilon \sin E,$$

called **Kepler's equation**. To determine the position at time $M$, Kepler's equation must be solved for $E$ when $M$ and $\varepsilon$ are given. A closed form solution evidently is not possible. Analytical solutions in the form of series expansions were given by Lagrange in 1770 and by Bessel (in terms of Bessel functions) in 1822, but the form of the terms in these series is rather complicated. However, the numerical solution of Kepler's equation is a simple matter. Writing the equation as

$$E = M + \varepsilon \sin E,$$

it assumes the form

$$E = f(E),$$

where

$$f(E) := M + \varepsilon \sin E.$$

This function clearly satisfies the hypotheses of Theorem 2.1 with $D(f) = (-\infty, \infty)$ and $L = \varepsilon$. Condition (III) is satisfied because $\varepsilon < 1$ for any ellipse.

Some practical remarks concerning iteration follow. In practice it is not always possible to verify all the hypotheses of Theorem 2.1. In such cases one often will have no choice but to iterate blindly. The worst that can happen is that the iteration sequence fails to converge. If the sequence converges, as was seen earlier, its limit necessarily is a fixed point of $f$ for continuous functions $f$.

More serious problems arise from the fact that, mathematically speaking, the fixed point is obtained only by letting $n$ tend to infinity. Since we cannot let the machine run indefinitely, the iteration process must be terminated artificially. When should one terminate? A theoretical answer is given by Theorem 2.1. Assertion (c) states for $n = 1$ that

$$|x_1 - s| \leq \frac{L}{1 - L} |x_1 - x_0|.$$

Because the iteration might as well have started with $x_n$, there holds for every $n$

$$|x_{n+1} - s| \leq \frac{L}{1 - L} |x_{n+1} - x_n|. \tag{10}$$

Thus, theoretically speaking, the error of the $(n + 1)$st approximation can be estimated in terms of the difference of the $n$-th and the $(n + 1)$st approximations. In practice, an estimate of the form (10) ignores both the fact that $L$ frequently is unknown (the estimate becomes arbitrarily bad as $L \to 1$), and the fact that the number system of the computer is discrete. What, then, should one do?

Let $\{x_n\}$ be a mathematically convergent sequence with limit $s$. Denoting the machine representation of $x_n$ by $x_n^*$, one might suspect that, because the number system of the computer is finite, the sequence $\{x_n^*\}$ ultimately becomes stationary; that is, that for some index $n_0$,

$$x_n^* = x_{n_0}^* \text{ for all } n > n_0. \tag{11}$$

One thus would stop the iteration if

$$x_{n+1}^* = x_n^*, \tag{12}$$

because then all later iterates also would equal $x_n^*$.

However, even on a machine that rounds perfectly, the condition (11) need not be satisfied. It is not satisfied, for instance, if the mathematical limit $s$ lies exactly halfway between two machine numbers, and the sequence $\{x_n\}$ has a structure, such as $x_n = s + (-1)^n/n$. Consequently, the numerical values of the iterates may cycle, and condition (12) will never hold. Thus, in general, one can do no better than to use a criterion of the form

$$|x_{n+1} - x_n| < \varepsilon,$$

$\varepsilon$ being a sufficiently small number, depending on the machine.

Condition (11) *is* satisfied if the sequence $\{x_n\}$ is monotone, and if the machine representation of numbers is such that it preserves monotonicity. For sequences generated by iteration, this will be the case if the function $f$ increases monotonically, and if the computer implementation of $f$ is such that it preserves this property. In this special situation, (12) may be used as a termination criterion. This criterion has the desirable property that it is machine-independent, and that it furnishes the fixed point as accurately as is possible on the computer on hand.

**DEMONSTRATION 2.1-5:**  The criterion works in the example of Demonstration 2.1-2. It does not work in the example of Demonstration 2.1-3, where $a := 1.000000001$.

A final point that should be taken into consideration is the possibility that for one reason or another the numerical sequence $\{x_n\}$ will never satisfy the convergence criterion, machine-independent or not. A conscientious iteration program will count the number of iterations performed and will stop the process

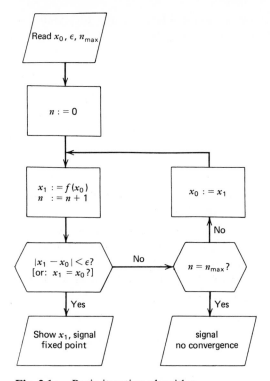

**Fig. 2.1e.** Basic iteration algorithm.

if this number exceeds a predetermined tolerance $n_{max}$. All told, the flow diagram of an iteration program thus might look as shown in Figure 2.1e.

# PROBLEMS

1  Why was it possible to omit hypothesis (II) from the statement of Theorem 2.1?

2  Give an example of a function $f$ that is continuous on the interval $[a, b]$, but does not satisfy a Lipschitz condition on $[a, b]$.

3  What should be the meaning of the symbol

$$2^{-2^{-2^{-2^{-2^{-2^{-2^{-\cdot\,\cdot\,\cdot}}}}}}}\;?$$

Explain why it is reasonable to define this symbol as a real number, and determine that number. What is

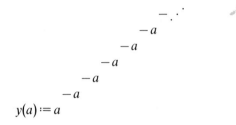

$$-a$$
$$-a$$
$$-a$$
$$-a$$
$$-a$$
$$y(a) := a$$

for $1 < a < e$? Compute $y'(a)$ analytically, and compare with values obtained by forming difference quotients.

4   Find and prove an iteration theorem similar to Theorem 2.1, but where condition (III) is replaced by some condition regarding the monotonicity of $f$. What does your theorem mean for the sequence $x_0 := 1$, $x_{n+1} := \sin x_n$? How does this iteration turn out numerically?

5   Let the real polynomial of degree $n$,

$$p(x) := \sum_{k=0}^{n} a_k x^k,$$

have real zeros $z_i$ satisfying

$$0 < z_1 < z_2 < \cdots < z_n.$$

(a)   Show that $p(x) = 0$ is equivalent to

$$x = f(x) := -\frac{1}{a_n} \sum_{k=0}^{n-1} a_k x^{k+1-n}. \tag{13}$$

(b)   Show that the sequence $\{x_m\}$ obtained by applying iteration to (13) converges to $z_n$ provided that $x_0$ is sufficiently close to $z_n$.

(c)   Show that $p(x) = 0$ may also be written in the form

$$x = -\frac{a_0}{a_1 + a_2 x + \cdots + a_n x^{n-1}}. \tag{14}$$

(d)   Show that iteration applied to (14) yields the smallest zero, provided the initial approximation is close enough. Apply the foregoing results to the specific polynomial

$$p(x) = x^4 - 10x^3 + 35x^2 - 50x + 24$$

(zeros $z_i = 1, 2, 3, 4$) and find practical limits for the starting values $x_0$ for convergence to take place.

6  The cubic equation (see Problem 1, §1.3)

$$x^3 + 3px + 2q = 0$$

may, for $p \gg |q|$, be solved by iteration, by writing it in the form

$$x = -\frac{x^3 + 2q}{3p}.$$

Carry out the iteration in the cases considered previously, and observe the speed of convergence.

## §2.2  SPEED OF CONVERGENCE, CONVERGENCE ACCELERATION

Certain of our demonstrations and exercises show that the convergence of an iteration sequence can be rather slow. In such situations, many iteration steps must be performed in order to determine a fixed point with sufficient accuracy. We now shall examine the speed of the convergence of an iteration sequence more closely. This will enable us to identify conditions under which the convergence is particularly favorable.

In addition to the hypotheses of Theorem 2.1, we now shall assume that $f$ has as many continuous derivatives as are required by the context. Let

$$e_n := x_n - s$$

be the **error** of the $n$-th iteration point. We shall assume that the iteration does not terminate in a finite number of steps; that is, that the error is never accidentally zero. We then have

$$
\begin{aligned}
e_{n+1} = x_{n+1} - s \\
= f(x_n) - s \quad \text{(by the definition of } x_{n+1}) \\
= f(x_n) - f(s) \quad \text{(because } f(s) = s) \\
= f(s + e_n) - f(s) \quad \text{(by definition of } e_n)
\end{aligned}
$$

and because $e_n \neq 0$,

$$\frac{e_{n+1}}{e_n} = \frac{f(s + e_n) - f(s)}{e_n}. \tag{1}$$

Letting $n \to \infty$, we have in view of $e_n \to 0$ and by virtue of the definition of the derivative

$$\lim_{n \to \infty} \frac{e_{n+1}}{e_n} = f'(s). \tag{2}$$

The ratio of two consecutive errors thus approaches a limit. Because $|f'(s)| \leq L < 1$, the modulus of this limit is $< 1$. We now distinguish two cases.

*Case* (i) (*normal case*): $f'(s) \neq 0$. Let

$$q := f'(s)$$

for brevity. Because $q \neq 0$, we may write

$$\frac{e_{n+1}}{e_n} = q(1 + \varepsilon_n), \tag{3}$$

where

$$\varepsilon_n := \frac{1}{q} \left\{ \frac{f(s + e_n) - f(s)}{e_n} - f'(s) \right\} \to 0 \text{ as } n \to \infty. \tag{4}$$

From (3), we have

$$\frac{e_{n+1}}{q e_n} = 1 + \varepsilon_n.$$

Multiplying these relations for $n = 0, 1, \ldots, m - 1$, we get

$$\frac{e_m}{q^m e_0} = (1 + \varepsilon_0)(1 + \varepsilon_1) \cdots (1 + \varepsilon_{m-1}).$$

If $f''$ is continuous in a neighborhood of $s$, it can be shown that the limit of the product on the right exists as $m \to \infty$. [The convergence of the series $\Sigma |\varepsilon_k|$ is a sufficient condition for the convergence of the product. From (4) we have, using Taylor's theorem,

$$\varepsilon_k = \frac{1}{2q} f''(s + \theta_k e_k) e_k,$$

where $0 < \theta_k < 1$, and hence $|\varepsilon_k| < M |e_k|$ where $M$ is a suitable constant. From (2), $|e_k| < Kb^k$ where $|q| < b < 1$. Thus, $\Sigma |\varepsilon_k|$ converges like a geometric series.] Therefore,

$$c := \lim_{n \to \infty} \frac{e_n}{q^n} \tag{5}$$

also exists, which is to say that the asymptotic relation

$$e_n \sim cq^n \tag{6}$$

holds. The error thus tends to zero like the $n$-th term of a geometric series. The type of convergence behavior that is described by (5) or (6) thus is called **geometric convergence**.

Geometric convergence may also be described differently. Taking absolute values and logarithms in (6) we have

$$\log|e_n| \sim \log|c| + n \log|q|.$$

The logarithm of the error thus is asymptotic to a linear function of $n$. Because the logarithm of a small number is roughly proportional to the number of leading zeros in its decimal representation, this also means that, asymptotically speaking, the number of zeros in the decimal representation of $e_n$ increases linearly with $n$. The term **linear convergence** is therefore used as a synonym of geometric convergence.

*Case (ii) (exceptional case):* $f'(s) = 0$. It will suffice to deal with the situation where $f''(s) \neq 0$ (normal case of exceptional case). Proceeding much as above and using Taylor's theorem, we have

$$
\begin{aligned}
e_{n+1} &= x_{n+1} - s \\
&= f(x_n) - s \\
&= f(s + e_n) - f(s) \\
&= f(s) + f'(s)e_n + \tfrac{1}{2}f''(s + \theta_n e_n)e_n^2 - f(s) \\
&= \tfrac{1}{2}f''(s + \theta_n e_n)e_n^2, \qquad 0 < \theta_n < 1.
\end{aligned}
$$

Thus, there exists

$$\lim_{n \to \infty} \frac{e_{n+1}}{e_n^2} = \frac{1}{2}f''(s).$$

Letting $a := \tfrac{1}{2}f''(s)$, we also have

$$e_{n+1} = (1 + \varepsilon_n)ae_n^2 \tag{7}$$

where $\varepsilon_n \to 0$. Thus, the error at the $(n + 1)$st step is approximately proportional to the *square* of the error at the *n*-th step. This type of error behavior is therefore called **quadratic convergence**.

Taking absolute values and logarithms in (7) yields

$$\log|e_{n+1}| = \log(1 + \varepsilon_n) + \log|a| + 2\log|e_n|.$$

If $n$ is large, then $\log|e_n|$ in view of $e_n \to 0$ is a strongly negative number in comparison with which the constant $\log|a|$ becomes negligible. Thus, we see that the logarithm of the absolute value of the error is approximately doubled at each step, and so is the number of leading zeros in the decimal representation of $e_n$. Evidently, this is an especially favorable convergence behavior from the numerical point of view.

It is easy to see that if a continuously differentiable function has a fixed point $s$, and if $f'(s) = 0$, then both hypotheses (I) and (III) of Theorem 2.1 are automatically satisfied for a neighborhood of the point $s$. We thus have

**THEOREM 2.2:** *Let I be a finite or infinite interval, and let the function f be defined on I and satisfy the following conditions:*

(i) *$f, f'$, and $f''$ are continuous on I.*

(ii) *the equation $x = f(x)$ has a solution $s$, located in the interior of I, such that $f'(s) = 0$.*

*Then there exists a number $d > 0$ such that the iteration sequence generated by f converges to s, with quadratic convergence, for any choice of $x_0$ satisfying $|x_0 - s| \le d$.*

The main conclusion of the theorem can be expressed by saying that the iteration sequence always converges when the starting point is "sufficiently close" to the fixed point.

**Convergence Acceleration**

How can the knowledge of the convergence behavior that we have acquired be put to use? Assuming the normal case, $f'(s) \ne 0$, we have seen that the convergence is only linear, and that asymptotically

$$e_n \sim cq^n,$$

that is,

$$x_n \sim s + cq^n \qquad (n \to \infty). \tag{8}$$

If $c$ and $q$ were known, this relation could be used to determine $s$. In reality, $c$ and $q$ are unknown. Consider, however, three consecutive equations (8) and assume, in the sense of a working hypothesis, that equality holds in place of mere asymptotic equality. This yields

$$x_n = s + cq^n$$
$$x_{n+1} = s + cq^{n+1} \tag{9}$$
$$x_{n+2} = s + cq^{n+2}$$

which may be regarded as three equations for the three unknowns $s$, $c$, $q$. Although the equations are not linear, it is to be expected that we can eliminate $c$ and $q$ and solve for $s$. This, in fact, is possible. Evidently,

$$x_{n+1} - s = q(x_n - s),$$
$$x_{n+2} - s = q(x_{n+1} - s);$$

thus,

$$(x_{n+1} - s)^2 = (x_n - s)(x_{n+2} - s).$$

On multiplying out the terms $s^2$ cancel, and there remains

$$(x_{n+2} - 2x_{n+1} + x_n)s = x_n x_{n+2} - x_{n+1}^2, \tag{10}$$

implying

$$s = \frac{x_n x_{n+2} - x_{n+1}^2}{x_{n+2} - 2x_{n+1} + x_n}.$$

This formula (although given in this form in some older textbooks) is numerically useless, because the denominator is the difference of two terms that tend to the same limit, and thus is subject to extreme cancellation. However, on writing (10) as

$$(x_{n+2} - 2x_{n+1} + x_n)s = (x_{n+2} - 2x_{n+1} + x_n)x_n - (x_{n+1} - x_n)^2,$$

we get

$$s = x_n - \frac{(x_{n+1} - x_n)^2}{x_{n+2} - 2x_{n+1} + x_n},$$

which is a numerically more stable formula. From a purely notational point of view, the formula may yet be simplified by means of the **difference operator** $\Delta$ defined by

$$\Delta x_n := x_{n+1} - x_n$$

$$\Delta^2 x_n := \Delta(\Delta x_n)$$

$$= x_{n+2} - 2x_{n+1} + x_n .$$

This yields

$$s = x_n - \frac{(\Delta x_n)^2}{\Delta^2 x_n} . \tag{11}$$

Equation (11) would yield the exact value of the fixed point $s$ if the relations (9) held exactly. In reality, these relations hold only approximately (in a sense that can be made mathematically precise). However, nothing keeps us from evaluating the expression on the right of (11); that is, to compute a new sequence $\{x'_n\}$ from the given iteration sequence $\{x_n\}$ according to the formula,

$$x'_n := x_n - \frac{(\Delta x_n)^2}{\Delta^2 x_n} , \qquad n = 0, 1, 2, \ldots, \tag{12}$$

represented schematically by

Although $x'_n \neq s$ in general, it may be shown under suitable hypotheses (see Henrici [1964], Theorem 4.5) that the sequence $\{x'_n\}$ converges to the fixed point $s$ faster than the original iteration sequence $\{x_n\}$, faster in the sense that

$$\lim_{n \to \infty} \frac{x'_n - s}{x_n - s} = 0.$$

The sequence-to-sequence transformation defined by (12) is called the **Aitken $\Delta^2$ method**\*. The method illustrates a basic principle of numerical compu-

---

\* A. C. Aitken (1895–1967), outstanding numerical analyst and statistician, active in Scotland.

tation which generally if somewhat vaguely may be formulated thus: *Use an approximate knowledge of the error to approximately eliminate the error.* In the situation at hand, approximate knowledge of the error may be defined as knowledge of the asymptotic behavior of the error.

To illustrate Aitken's $\Delta^2$ method, we reconsider some of the demonstrations of section 2.1.

**DEMONSTRATION 2.2-1:**  $f(x) = \cos x$. After 10 iterations, started with $x_0 = 0$, we have

| $n$ | $x_{n+2}$ | $x'_n$ |
|-----|-----------|--------|
| 10 | 0.73560474 | 0.73907638 |
| 11 | 0.74142509 | 0.73908118 |
| 12 | 0.73750689 | 0.73908333 |
| 13 | 0.74014734 | 0.73908432 |
| 14 | 0.73836920 | 0.73908476 |
| . . . | | |
| $\infty$ | 0.73908513 | 0.73908513 |

It is fair to compare $x'_n$ with $x_{n+2}$, because $x_{n+2}$ is required in the computation of $x'_n$. We note that after 14 iterations $x'_n$ has six correct digits, while $x_{n+2}$ has only two.

**DEMONSTRATION 2.2-2:**  $f(x) = \sqrt{2+x}$. Starting with $x_0 = 0$, we get

| $n$ | $x_{n+2}$ | $x'_n$ |
|-----|-----------|--------|
| 0 | 1.84775907 | 2.03942606 |
| 1 | 1.96157056 | 2.00208254 |
| 2 | 1.99036945 | 2.00012537 |
| 3 | 1.99759091 | 2.00000776 |
| 4 | 1.99939764 | 2.00000048 |
| 5 | 1.99984940 | 2.00000003 |
| 6 | 1.99996235 | 2.00000000 |

Similar results are obtained for the function of Demonstration 2.1-3.

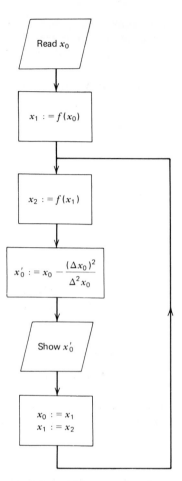

**Fig. 2.2a.**   Aitken acceleration.

The simple version of the $\Delta^2$ method given above transforms a sequence with geometric convergence into a new sequence whose convergence, although faster, will in general still be geometric. The basic flow diagram of the iteration method with Aitken's $\Delta^2$ acceleration (excluding termination and output arrangements) is shown in Figure 2.2a.

However, the basic formula (12) may also be used in a different manner, called **Aitken-Steffensen iteration,** which has the even simpler flow diagram shown in Figure 2.2b. As in the original Aitken process, an accelerated value $x_0'$ is computed as soon as $x_2$ is available, but instead of continuing the iteration sequence started with $x_0$, the value $x_0'$ is used to start a new iteration sequence $\{x_k^{(1)}\}$. A new accelerated value $x_0^{(1)\prime}$ is computed as soon as possible, and is used to start yet another iteration sequence $\{x_k^{(2)}\}$, which again is continued only until a new accelerated value can be computed. In this way, a sequence of accelerated

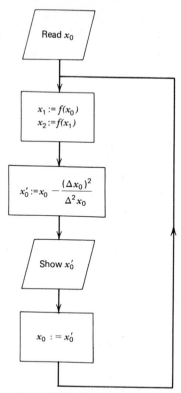

**Fig. 2.2b.** Aitken-Steffensen iteration.

values $x_0^{(k)}$ is obtained which is quite different from the sequence $\{x_k'\}$ constructed earlier. Mathematically, the values $\{x_0^{(k)}\}$ are given by

$$x_0^{(k+1)} := x_0^{(k)} - \frac{\{x_1^{(k)} - x_0^{(k)}\}^2}{x_2^{(k)} - 2x_1^{(k)} + x_0^{(k)}},$$

where $x_1^{(k)} := f(x_0^{(k)})$, $x_2^{(k)} := f(x_1^{(k)})$, $k = 0, 1, \ldots$ . The sequence $\{x_0^{(k)}\}$ thus is identical with the sequence obtained by iterating the function

$$g(x) := x - \frac{\{f(x) - x\}^2}{f(f(x)) - 2f(x) + x},$$

starting with $x_0^{(0)} := x_0$. The function $g$ is undefined for $x = s$, but if we set $g(s) := s$, and if $f$ is sufficiently differentiable, $f'(s) \neq 1$, then $g$ is again differentiable and satisfies

$$g'(s) = 0.$$

(See Henrici [1964], §4.12.) Thus, under the sole condition that $f'(s) \neq 1$, the sequence $\{x_0^{(k)}\}$ according to Theorem 2.2 will always converge if started sufficiently closely to the fixed point $s$, and it will converge to the fixed point with quadratic convergence.

**DEMONSTRATION 2.2-3:** $f(x) = \cos x$.

| $n$ | $x_0^{(k)}$ | |
|---|---|---|
| 0 | 0.00000000 | |
| 1 | 0.68507336 | |
| 2 | 0.73866016 | |
| 3 | 0.73908511 | |
| 4 | 0.73908513 | convergence |

Correct digits are underlined. It is easy to see that their number approximately doubles at each step.

**DEMONSTRATION 2.2-4:** $f(x) = \sqrt{2 + x}$.

| $n$ | $x_0^{(k)}$ | |
|---|---|---|
| 0 | 0.00000000 | |
| 1 | 2.03942606 | |
| 2 | 2.00000802 | |
| 3 | 2.00000000 | convergence |

**DEMONSTRATION 2.2-5:** $f(x) := 10 \cos x$. This function has several fixed points as may be seen from a graph. One of the fixed points is located near $x = 1.4$. Ordinary iteration does not produce this fixed point because the condition $|f'(s)| < 1$ is clearly violated. An Aitken-Steffensen iteration beginning with $x_0 = 1.4$, however, yields the convergent iteration sequence

| $n$ | $x_0^{(k)}$ | |
|---|---|---|
| 1 | 1.427341197 | |
| 2 | 1.427551750 | |
| 3 | 1.427551779 | |
| 4 | 1.427551779 | convergence |

The same fixed point is obtained by starting with $x_0 = 1.3$ or $x_0 = 1.2$. If started with $x_0 = 1.0$, however, the sequence converges to the fixed point $s = 7.068891237$.

In all the foregoing demonstrations, if the computation is carried on too far, an error halt will eventually result, due to the fact that the denominator in (12) numerically becomes zero because of cancellation. A conscientiously written program thus will test for zero before attempting the division in (12).

# PROBLEMS

**1** Construct an example of a bounded divergent iteration sequence for which the Aitken $\Delta^2$ method produces an (apparent) limit.

**2** Is Aitken acceleration applicable to the sequence obtained in Demonstration 1.3-4 (or, if you have done Problem 4 of §1.3, to the sequences considered there)? Verify your answer!

**3** The equation $x^2 = c$ satisfied by the square roots of $c > 0$ may be written in the form,

$$x = \frac{c}{x}. \tag{13}$$

    **(a)** Show that ordinary iteration applied to (13) does never (unless $x_0 = \sqrt{c}$) produce a sequence that converges to $\sqrt{c}$.
    **(b)** Show that Aitken-Steffensen iteration applied to (13) converges for any starting value $x^{(0)} > 0$. Compare the resulting iteration function with that of Demonstration 2.3-1.

**4** Using Aitken-Steffensen iteration, find both fixed points of the function,

$$f(x) := (x - 1)^2.$$

**5** Let

$$f(x) := e^{-10x}.$$

    **(a)** Show that ordinary iteration applied to $f$ furnishes a divergent iteration sequence.
    **(b)** Find the fixed point, using Aitken-Steffensen iteration.

**6\*** Two formulas for computing the Aitken acceleration have been given, namely,

$$x'_n = \frac{x_n x_{n+2} - x_{n+1}^2}{x_{n+2} - 2x_{n+1} + x_n} \tag{14}$$

and

$$x'_n = x_n - \frac{(x_{n+1} - x_n)^2}{x_{n+2} - 2x_{n+1} + x_n}. \tag{15}$$

Assuming that the relation $x_n = s + cq^n$ holds exactly with $s \neq 0$ and $0 < |q| < 1$, compute the *coefficients of relative error propagation* (see §1.6) for both formulas. Show that

$$\rho \sim \text{const} \cdot q^{-n} \qquad (n \to \infty)$$

for (14), while

$$\rho = 1 + \text{const} \cdot q^n \qquad (n \to \infty)$$

for (15), and thus confirm that (15) is the more stable formula.

7   It's going to be a hot day, and Max plans to go swimming in the local swimming pool. The water temperature is 66° at 6 A.M., 70° at 8 A.M., and 72° at 10 A.M. How warm will the water be in the afternoon? [An exponential law of warming is assumed. Do you see the connection with Aitken's $\Delta^2$-method?]

---

## §2.3   SOLUTION OF NONLINEAR EQUATIONS

Let $g$ be a function from $\mathbb{R}$ into $\mathbb{R}$. Any number $x$ so that

$$g(x) = 0 \tag{1}$$

is called a **zero** of $g$. (The older term *root* is also used.) The determination of zeros is a problem that arises frequently in applications. It may suffice to recall the fundamental problem of the differential calculus, the determination of maxima and minima, which requires the calculation of the zeros of the derivative.

In addition to assuming that equation (1) has a solution, we shall always assume that $g$ is continuous. In this case, a zero of $g$ can always be found by the method of **bisection**. Assume $g$ is defined on the closed and bounded interval $[a, b]$, and that, say, $g(a) < 0$ and $g(b) > 0$. Being continuous, $g$ then must have a zero in the interior of $[a, b]$. Now let $x_1 := \frac{1}{2}(a + b)$. Either $g(x_1) = 0$, and we have been lucky, or, if not, then either $g(x_1) > 0$ or $g(x_1) < 0$. In either case, we may replace the original interval $[a, b]$ by an interval of half its length which assuredly contains a zero of $g$. Continuing in this way, the zero can be determined to arbitrary precision. Provided that the function $g$ is programmed so that the computer returns at least its sign correctly, the method of bisection is foolproof, which must be considered its major advantage. In addition, no derivative evaluations are required. The disadvantages of the method are its relatively slow convergence—only one binary digit is gained at each step—and, more seriously, the fact that it possesses no ready generalization to systems of equations. The methods to be discussed below, on the other hand, do generalize to systems. (See §2.5.)

We first show that the problem of solving an equation (1) is equivalent to the problem of determining a fixed point of a suitable function $f$.

(a)   Let $s$ be a solution of (1), and let $k$ be any real number. Then $kg(s) = 0$, and hence,

$$s + kg(s) = s;$$

that is, $s$ is a fixed point of the function

$$f(x) := x + kg(x). \tag{2}$$

(b)  Let $s$ be a fixed point of the function (2), and let $k \neq 0$. Then it follows from

$$f(s) = s + kg(s) = s$$

that $g(s) = 0$; that is, $s$ is a zero of $g$.

More generally, if $h$ is any function that does not take the value 0, then $s$ is a zero of $g$ if, and only if, it is a fixed point of the function

$$f(x) := x + h(x)g(x). \tag{3}$$

We conclude that a zero of $g$ may be found as the limit of an iteration sequence formed with an appropriate function $f$. Convergence to the zero will take place, at least if the initial approximation $x_0$ is close enough, if $f$ is chosen so that

$$|f'(s)| < 1.$$

The convergence will be quadratic if

$$f'(s) = 0. \tag{4}$$

If the form (2) is chosen for $f$, then (4) means that $k$ should have the value $-1/g'(s)$. It is not possible in general to satisfy this condition, unless $s$ is already known. If the form (3) is chosen for $f$, then from

$$f'(x) = 1 + h'(x)g(x) + h(x)g'(x)$$

and the fact that $g(s) = 0$, it follows that (4) is satisfied if the auxiliary function $h$ (which is at our disposal) satisfies

$$h(s) := -\frac{1}{g'(s)}.$$

This condition is satisfied, for instance, if we choose

$$h(x) := -\frac{1}{g'(x)},$$

which yields the iteration function

$$f(x) = x - \frac{g(x)}{g'(x)}.$$

Thus, the iteration relation $x_{n+1} = f(x_n)$ in this case reads

$$x_{n+1} = x_n - \frac{g(x_n)}{g'(x_n)}. \tag{5}$$

The algorithm thus defined for determining the zeros of a given function $g$ is called **Newton's method**. Assuming that $g'(s) \neq 0$, it follows from (4) by Theorem 2.2 that the sequence $\{x_n\}$ generated by Newton's method converges to a zero $s$ of $g$ with quadratic convergence, whenever the initial approximation $x_0$ is chosen sufficiently close to $s$. If $g'(s) = 0$, then the condition (4) for quadratic convergence cannot be satisfied with an iteration function of the form (3).

It is well known that Newton's method has a simple graphical interpretation. Since (5) may be written

$$g'(x_n) = \frac{g(x_n)}{x_n - x_{n+1}},$$

the location of the points $x_n$ and $x_{n+1}$ is as shown in Figure 2.3a.

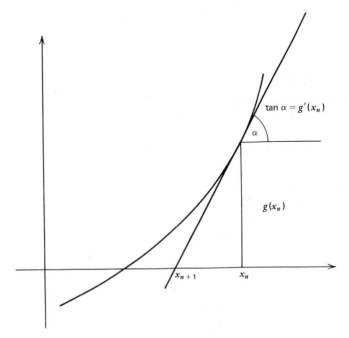

**Fig. 2.3a.**  Newton's method.

The convergence of Newton's method is intuitively evident from Figure 2.3a. The figure tells us nothing about the quadratic character of the convergence, however. Unlike the geometric derivation of the method based on the figure, the analytical derivation which we have given may be extended to more general situations where Newton's method also applies, such as the determination of complex zeros and of solutions of systems of nonlinear equations.

We proceed to give some simple examples for Newton's method.

**DEMONSTRATION 2.3-1:** *Square root iteration.* Let $c > 0$ be given. We seek to determine $s := \sqrt{c}$. The number $s$ evidently is the positive solution of the equation

$$g(x) := x^2 - c = 0.$$

The Newton iteration function here is

$$f(x) := x - \frac{g(x)}{g'(x)} = x - \frac{x^2 - c}{2x} = \frac{1}{2}\left(x + \frac{c}{x}\right).$$

It may be shown (see Problem 2) that the iteration sequence $\{x_n\}$ here converges to $\sqrt{c}$ for every choice of $x_0 > 0$.

Numerical example: For $c = 2$, $x_0 = 1$, we get

$$1.00000000$$
$$1.50000000$$
$$1.41666667$$
$$1.41421569$$
$$1.41421356,$$

thus, to the number of digits given, $x_4$ has already converged.

**DEMONSTRATION 2.3-2:** *Division by iteration.* Let $a > 0$ be given. We seek $s := 1/a$. (This is an unnecessary exercise at a time when all machines have built-in divisions. However, division had to be programmed on some early computers.) The number $s$ solves the equation,

$$g(x) := 1 - ax = 0.$$

The application of Newton's method yields the iteration function

$$f(x) := x - \frac{1 - ax}{-a} = \frac{1}{a}.$$

Although the iteration converges in one step (as indeed it must for a linear function

$g$), it is of no use, because executing the iteration step requires precisely the division which we are trying to avoid. Let us therefore try the equation,

$$g(x) := a - \frac{1}{x} = 0,$$

which likewise has the solution $s = 1/a$. Now the iteration function becomes

$$f(x) = x - \frac{a - \frac{1}{x}}{\frac{1}{x^2}} = x - x^2\left(a - \frac{1}{x}\right)$$

or

$$f(x) = x(2 - ax).$$

No divisions are required for the evaluation of $f$. The resulting iteration algorithm

$$x_{n+1} = x_n(2 - ax_n) \tag{6}$$

may be shown to converge for every $x_0$ satisfying

$$0 < x_0 < \frac{2}{a}.$$

(See Problem 3.) Thus, the correct value may be under- or overestimated by almost 100%.

Numerical example: for $a := e = 2.7182818$, $x_0 = 0.5$, we obtain the convergent sequence

0.32042954
0.36175925
0.36777762
0.36787941
0.36787944

(correct digits are underlined). Starting with $x_0 = 1$ on the other hand yields

−0.71828183
−2.8390035
−27.587197
−2123.9322,

there evidently is no convergence.

### Newton's Method for Polynomials

Let

$$p(x) := a_0 x^n + a_1 x^{n-1} + \cdots + a_{n-1} x + a_n \qquad (7)$$

be a polynomial of degree $n \geq 1$. The coefficients may be thought of as real, although everything which follows also holds for complex coefficients (and complex zeros). We wish to determine the real zeros of $p$ by Newton's method. Once a good starting value $x_0$ has been found, there only remains the problem of evaluating Newton's formula,

$$x_{n+1} = x_n - \frac{p(x_n)}{p'(x_n)}$$

as efficiently as possible. In order to measure the efficiency of various evaluation algorithms which we shall discuss, the following abbreviations will be used:

$\mu$ = one multiplication

$\alpha$ = one addition

$\delta$ = one division.

If $p(x)$ is evaluated directly by the formula (7), then we must build up the powers of $x$ according to $x^k = x * x^{k-1}$, which requires $(n-1)\mu$; form the terms $a_k x^{n-k}$, which requires $n\mu$; and sum the terms, which requires $n\alpha$. Thus, in this method of evaluation, the total amount of computation is $(2n-1)\mu + n\alpha$. Even more operations are required if $p'(x)$ is calculated by this method.

In view of the simplicity of the problem, it is perhaps surprising that there exists an algorithm which calculates $p(x)$, $p'(x)$, and as many higher derivatives as may be required, with a substantially smaller number of operations. This algorithm, called the **Horner algorithm**, is suggested by writing the formula for $p(x)$ as follows: (We consider the case $n = 4$ for concreteness.)

$$p(x) = a_0 x^4 + a_1 x^3 + a_2 x^2 + a_3 x + a_4$$
$$= (((a_0 x + a_1)x + a_2)x + a_3)x + a_4 .$$

Only $4\mu + 4\alpha$ are now required. For general $n$ the process may be algorithmically formulated as follows: Given $a_0, a_1, a_2, \ldots, a_n$, compute $b_0, b_1, b_2, \ldots b_n$ according to

$$b_0 := a_0, \qquad b_k := xb_{k-1} + a_k, \qquad k = 1, 2, \ldots, n. \qquad (8)$$

Schematically, these formulas may be represented thus:

According to the above, $b_n = p(x) =$ value of $p$ at $x$. Thus, $x$ is a zero of $p$ if, and only if, in the Horner algorithm performed with the number $x$, there results $b_n = 0$. The amount of computation required for evaluating $p(x)$ now is only $n\mu + n\alpha$.

Let us now, in the sense of an experiment, apply to the $b_k$ the same algorithm which before we applied to the $a_k$. That is, let us form numbers $c_k$ according to

$$c_0 := b_0, \qquad c_k := xc_{k-1} + b_k, \qquad k = 1, 2, \ldots, n-1. \tag{9}$$

Rather surprisingly, it turns out that

$$c_{n-1} = p'(x);$$

thus by performing a mere additional $(n-1)(\mu + \alpha)$, the value of the derivative at $x$ is obtained. The analytical proof is by differentiating the recurrence relation (8). Observing that the $b_k$ are functions of $x$ while the $a_k$ are not, we have

$$b_0' = 0, \qquad b_k' = xb_{k-1}' + b_{k-1}, \qquad k = 1, 2, \ldots, n.$$

Thus, we see that $b_1' = b_0$, and that the quantities $c_k := b_{k+1}'$ are identical with the $c_k$ as defined by (9). Because $b_n = p(x)$, $c_{n-1} = b_n' = p'(x)$ follows.

Continuing in the same manner, all Taylor coefficients of $p$ at the point $x$ may be obtained (see Henrici [1977], §6.1).

**DEMONSTRATION 2.3-3:**   We seek the Taylor coefficients of the polynomial,

$$p(x) := x^4 - 4x^3 + 3x^2 - 2x + 5$$

at $x = 2$. The Horner scheme turns out as follows:

| 1 | $-4$ | 3 | $-2$ | 5 |
|---|------|-----|------|------|
| 1 | $-2$ | $-1$ | $-4$ | $-\underline{\underline{3}}$ |
| 1 | 0 | $-1$ | $-\underline{\underline{6}}$ | |
| 1 | 2 | $\underline{\underline{3}}$ | | |
| 1 | $\underline{\underline{4}}$ | | | |
| $\underline{\underline{1}}$ | | | | |

The underlined quantities are the Taylor coefficients

$$\frac{1}{k!}p^{(k)}(x)$$

evaluated at $x = 2$ (thus not simply the derivatives!). These are precisely the coefficients that are required when $p(x + h)$ is rearranged in powers of $h$; that is, when the Taylor expansion at the point $x$ is formed. Thus, in the present example,

$$p(2 + h) = -3 - 6h + 3h^2 + 4h^3 + h^4.$$

Yet a further dividend is paid by the Horner algorithm. Assume that the scheme is formed with a zero $x_0$ of $p$. If $p(x_0) = 0$, then $b_n = 0$. We now assert: *The numbers $b_0, b_1, \ldots, b_{n-1}$ are the coefficients of the polynomial obtained from $p(x)$ by dividing by the linear factor $x - x_0$; that is,*

$$b_0 x^{n-1} + b_1 x^{n-2} + \cdots + b_{n-1} = \frac{p(x)}{x - x_0}.$$

Proof:  By the recurrence relations (8) (where $x$ is now to be replaced by $x_0$),

$$\left(b_0 x^{n-1} + b_1 x^{n-2} + \cdots + b_{n-1}\right)(x - x_0)$$
$$= b_0 x^n + (b_1 - b_0 x_0)x^{n-1} + (b_2 - b_1 x_0)x^{n-2} + \cdots$$
$$+ (b_{n-1} - b_{n-2}x_0)x + (b_n - b_{n-1}x_0)$$
$$= a_0 x^n + a_1 x^{n-1} + a_2 x^{n-2} + \cdots + a_{n-1}x + a_n$$
$$= p(x).$$

**DEMONSTRATION 2.3-4:**  The polynomial

$$p(x) := x^4 - 2x^3 + 8x - 16$$

has the zero $x = -2$. We calculate the first row of the Horner scheme with $x = -2$:

| 1 | $-2$ | 0 | 8 | 16 |
|---|------|---|---|----|
| 1 | $-4$ | 8 | $-8$ | 0 ← confirms that |

$$p(-2) = 0.$$

There results

$$\frac{p(x)}{x - (-2)} = x^3 - 4x^2 + 8x - 8.$$

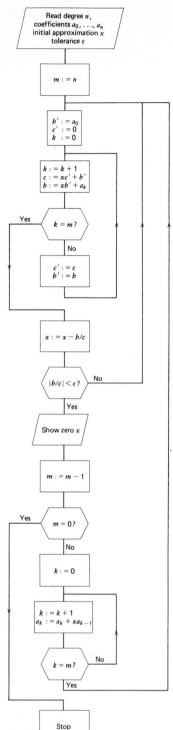

**Fig. 2.3b.** Newton's method for polynomials with deflation.

The property of the Horner scheme last mentioned is important when *all* zeros of a given polynomial are to be determined. After a first zero has been found, it is advantageous not to continue the computation with the full polynomial, but to divide out the linear factor corresponding to the zero that has been found. This process is called **deflation**. By using deflation, the possibility is avoided that the same zero might be determined several times. Moreover, because each deflation lowers the degree of the polynomial by one unit, a substantial amount of work, may be saved.

A flow diagram of an algorithm for finding all (real) zeros of a polynomial based on Newton's method and deflation is shown in Figure 2.3b. In this arrangement of the algorithm, the array of coefficients is the only array that needs to be stored. The quantities, $c := c_{k-1}$, are computed simultaneously with the $b := b_k$; thus, the $b_k$ and $c_k$ need not be stored. In the process of deflation, the new coefficients are written over the old ones. The iteration on the deflated polynomial is started with the zero just found. Thus, if the whole process is started with $x = 0$ and all zeros are real and positive, the zeros usually are determined in increasing order.

**DEMONSTRATION 2.3-5:** Let

$$p(x) := p_4(x) := x^4 - 16x^3 + 72x^2 - 96x + 24,$$

the Laguerre polynomial of degree four. Letting $x_0 = 0$, Newton's method yields the corrections

$$\Delta x_k = 0.25000000$$
$$0.06758937$$
$$0.00493270$$
$$0.00002562$$
$$7.351227 * 10^{-10},$$

and thus the first zero

$$z_1 = 0.32254769.$$

The deflated polynomial is

$$p_3(x) = x^3 - 15.67745231x^2 + 66.94327398x - 74.40760164.$$

Beginning with $x_0 = z_1$, the Newton corrections now are

$$\Delta x_k = 0.95223733$$
$$0.39169998$$
$$0.07638283$$
$$0.00288919$$
$$0.00000408$$
$$4.684247 * 10^{-10};$$

thus, the second zero is

$$z_2 = 1.74576110,$$

and the second deflated polynomial

$$p_2(x) = x^2 - 13.93169121x - 42.62186939.$$

Continuing with $x_0 = z_2$, we now get the corrections

$$\Delta x_k = 2.04480853$$
$$0.65840607$$
$$0.08611858$$
$$0.00152554$$
$$0.00000048$$
$$2.058269* \, 10^{-9},$$

and the zero

$$z_3 = 4.53662030.$$

The last deflation yields

$$p_1(x) = x - 9.39507091.$$

No further iterations are necessary at this point. Letting the computer run anyway, using the starting value $x_0 := z_3$, indeed yields the corrections

$$\Delta x_k = 4.85845062$$
$$0.00000000$$

and the last zero

$$z_4 = 9.39507091.$$

Justifiably, the question may be asked whether the process of successive deflations is numerically stable. Because the zeros that are determined numerically are seldom absolutely accurate, each numerical deflation slightly falsifies the true deflated polynomial, and, hence, all zeros that remain to be determined. J. H. Wilkinson in one of his fundamental contributions (Wilkinson [1959]) has shown that the extent of the numerical instability that is introduced by deflation very much depends on the order in which the zeros are determined. The deflations are numerically harmless if the zeros are determined in the order of increasing absolute values, as was done above.

# PROBLEMS

1 Write a bisection algorithm that discovers the largest number $x$ so that $x = \sin x$ on your calculator, and determine that number. What conclusions do you draw with regard to the accuracy of your calculator and/or its sine routine? [You may assume that if $x_0 \neq \sin x_0$ for some $x_0 > 0$, then $x \neq \sin x$ for all $x > x_0$. This assumption is not correct for all calculators.]

2 Find a closed expression for the elements $x_n$ generated by Newton's square root iteration (Demonstration 2.3-1). Then show that the sequence $\{x_n\}$ converges to $\sqrt{c}$ for arbitrary choices of $x_0 > 0$. [Consider the quantities

$$y_n := \frac{x_n - \sqrt{c}}{x_n + \sqrt{c}}$$

and express $y_{n+1}$ in terms of $y_n$.]

3 If the division iteration in Demonstration 2.3-2 is started with $x_0 = 1$, show that

$$x_n = \frac{1 - (1 - c)^{2^n}}{c}.$$

Deduce the range of values of $c$ for which the sequence $\{x_n\}$ converges to $1/c$, and show that the convergence is always quadratic.

4 Consider a mechanical driving mechanism consisting of two wheels with radii $r$ and $R \geq r$ and a driving belt of length $L$. (See Figure 2.3c.) Given $r$, $R$, and $L$, and assuming that the belt is tight, what is the distance $x$ of the axes of the two wheels?

    (a) Express $L$ as a function of $x$, $L = g(x)$, assuming $x + r = R$ (small wheel not fully contained in large wheel).

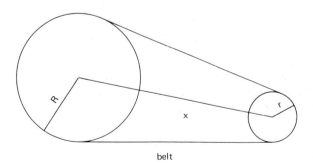

belt

**Fig. 2.3c.** Driving mechanism.

(b)   Set up Newton's method for solving the equation

$$g(x) = L$$

for $x$, when $L$ is given. How would you choose the starting value $x_0$ in order to minimize the number of iterations? Can you guarantee that with your choice of the starting value, Newton's method always converges? [This may require some calculus-type discussion of the graph of the function $g$.]

(c)   Solve the equation for a few sample cases, such as

| R | r | L |
|---|---|---|
| 100 | 20 | 700 |
| 100 | 20 | 1000 |
| 100 | 100 | 1000 |
| 100 | 20 | 629 |

(d*)   What happens if $L = 2\pi R$? [This may require a short theoretical study of an exceptional case of Newton's method.]

5   We have discussed *quadratic* convergence.
   (a)   Define *cubic* convergence.
   (b)   Newton's iteration function,

$$f(x) = x - \frac{x^2 - c}{2x} = \frac{1}{2}\left(x + \frac{c}{x}\right),$$

produces $\sqrt{c}$ with quadratic convergence. We consider a modification of it that has the form

$$f(x) = x - \frac{x^2 - c}{2x} h(x).$$

What are the conditions to be satisfied by $h$ at the point $x = \sqrt{c}$, if cubic convergence is to be achieved?
   (c)   Find an $h$ of the form

$$h(x) = a + bx^{-2}$$

satisfying these requirements.

(d) Test your formulas experimentally and determine the average number of square roots that can be calculated per minute both by the classical Newton method and by your modification of it.

**6** Horner's rule for evaluating the polynomial,

$$p(x) = a_0 x^n + a_1 x^{n-1} + \cdots + a_n$$

at $x = x_1$, proceeds by forming $b_0, b_1, \ldots$ according to

$$b_0 := a_0, \quad b_k := x_1 b_{k-1} + a_k, \quad k = 1, 2, \ldots, n.$$

Now let $c_0, c_1, \ldots$ be formed from the $b_k$ by

$$c_0 := b_0, \quad c_k := x_2 c_{k-1} + b_k, \quad k = 1, 2, \ldots,$$

where $x_2 \neq x_1$.
   (a) What is the meaning of $c_{n-1}$?
   (b) What happens if $x_2 \to x_1$?

**7** For a polynomial of degree $n$, let the $c_k$ be defined by (9). We know that $c_{n-1} = p'(x)$. What is the meaning of $c_n$?

**8** The polynomial,

$$p(x) := x^4 - 1{,}111\pi x^3 + 112{,}110\pi^2 x^2 - 1{,}111{,}000\pi^3 x$$
$$+ 1{,}000{,}000\pi^4,$$

has the zeros

$$z_k = 10^k \pi, \quad k = 0, 1, 2, 3.$$

Determine these zeros by Newton's method, using deflation,
   (a) In increasing order.
   (b) In decreasing order.
Which values are more accurate?

**9** Find all zeros of the Chebyshev polynomial of degree 8,

$$T_8(x) := 128x^8 - 256x^6 + 160x^4 - 32x^2 + 1.$$

[See §5.5 on Chebyshev polynomials. The zeros are given analytically by

$$z_k = \cos \frac{2k+1}{16} \pi, \quad k = 1, 2, \ldots, 8.]$$

**10**   The polynomial

$$p(x) := x^4 - 4\sqrt{2}x^3 + 12x^2 - 8\sqrt{2}x + 4$$

has the fourfold zero $z_1 = \sqrt{2}$. Try to determine this zero by Newton's method. What do you observe with regard to
   **(a)**   Speed of convergence?
   **(b)**   Accuracy?
Try to explain theoretically. Reconstitute $p$ from the zeros which you have found. How does the reconstituted polynomial compare to $p$?

**11**   Let $p$ be a cubic polynomial with three real zeros $z_1, z_2, z_3$. Show that if Newton's method is started with

$$x_0 := \frac{z_1 + z_2}{2},$$

it yields the zero $z_3$ in one step. (H. Walser)

---

## §2.4   ITERATION FOR SYSTEMS OF EQUATIONS

In §2.1 we have seen how the basic iteration algorithm can be used under appropriate hypotheses to solve equations of the form

$$x = f(x).$$

Here, we show how to apply iteration to obtain solutions of *systems* of equations. The algorithms discussed in this section are, in principle, applicable to problems involving any number of equations and unknowns. For greater concreteness, and also to avoid cumbersome notation, however, we shall consider explicitly only two equations with two unknowns. These equations will usually be written in the form

$$\begin{aligned} x &= f(x, y) \\ y &= g(x, y), \end{aligned} \tag{1}$$

where $f$ and $g$ are certain functions of the point $(x, y)$ that are defined in suitable regions of the plane. Each of the two equations $x - f(x, y) = 0$ and $y - g(x, y) = 0$ defines, in general, a curve in the $(x, y)$-plane. The problem of solving the system of equations (1) is equivalent to the problem of finding the point or points of intersection of these curves. It will usually be assumed, and in some cases also proved, that such a point of intersection exists. Its coordinates will be denoted by $s$ and $t$. The quantities $s$ and $t$ then satisfy the relations

$$s = f(s, t)$$
$$t = g(s, t).$$

**EXAMPLE 2.4-1:** Let $f(x, y) := x^2 + y^2$, $g(x, y) := x^2 - y^2$. The system (1) reads here

$$x = x^2 + y^2$$

$$y = x^2 - y^2.$$

The equation, $x^2 + y^2 - x = 0$, defines a circle centered at $(\frac{1}{2}, 0)$; the equation $x^2 - y^2 - y = 0$ a hyperbola centered at $(0, -\frac{1}{2})$. Both the circle and the hyperbola pass through the origin, thus our system has the obvious solution $s = t = 0$, but an inspection of the graph shows that there must be another solution near $x = 0.8$, $y = 0.4$.

It will be convenient to employ *vector notation*. Thus, we not only shall be able to simplify the writing of our equations, but also to aid the understanding of the theoretical analysis and even the programming of our algorithms. We represent the coordinates of the point $(x, y)$ by the column vector

$$\mathbf{x} = \begin{pmatrix} x \\ y \end{pmatrix}.$$

The functions $f$ and $g$ then become functions of the vector $\mathbf{x}$, whose value for a particular $\mathbf{x}$ we denote by $f(\mathbf{x})$ and $g(\mathbf{x})$. If we denote by $\mathbf{f}$ the column vector with components $f$ and $g$, the system (1) can be written more simply as follows:

$$\mathbf{x} = \mathbf{f}(\mathbf{x}). \tag{2}$$

The fact that the vector

$$\mathbf{s} := \begin{pmatrix} s \\ t \end{pmatrix}$$

is a solution of the system (1) is expressed in the form

$$\mathbf{s} = \mathbf{f}(\mathbf{s}).$$

A vector analog of the absolute value of a number (or *scalar*, as numbers are called in this context) will be required. Clearly, the *length* of a vector is such an analog. If $\mathbf{x} = (x, y)^T$, we write

$$\|\mathbf{x}\| := \sqrt{x^2 + y^2}.$$

The number $\|\mathbf{x}\|$ is called the **Euclidean norm** of the vector $\mathbf{x}$. It is nonnegative, and zero only if $\mathbf{x}$ is the zero vector $\mathbf{0} = (0, 0)^T$. Furthermore, if the sum of two vectors and the product of a vector and a scalar are defined in the natural way, we have

$$\|c\mathbf{x}\| = |c|\,\|\mathbf{x}\|,$$

$$\|\mathbf{x}_1 + \mathbf{x}_2\| \le \|\mathbf{x}_1\| + \|\mathbf{x}_2\|.$$

The last relation is called the **triangle inequality**. If $\mathbf{x}_1$ and $\mathbf{x}_2$ are two vectors, then $\|\mathbf{x}_1 - \mathbf{x}_2\|$ is the distance of the points whose coordinates are the components of $\mathbf{x}_1$ and $\mathbf{x}_2$.

We shall have occasion to consider sequences of vectors $\{\mathbf{x}_n\}$. Such a sequence is said to converge to a vector $\mathbf{v}$ if

$$\|\mathbf{x}_n - \mathbf{v}\| \to 0 \qquad \text{for} \quad n \to \infty.$$

The following condition, called the **Cauchy criterion**, is necessary and sufficient for the convergence of the sequence $\{\mathbf{x}_n\}$ to some vector $\mathbf{v}$: Given any number $\varepsilon > 0$, there exists an integer $N$ so that for all $n > N$ and all $m > N$

$$\|\mathbf{x}_m - \mathbf{x}_n\| < \varepsilon.$$

As in the scalar case, the **iteration sequence** generated by the function $\mathbf{f}$ is defined by choosing a starting vector $\mathbf{x}_0$ and calculating the sequence of vectors $\{\mathbf{x}_n\}$ by

$$\mathbf{x}_n := \mathbf{f}(\mathbf{x}_{n-1}), \qquad n = 1, 2, \ldots.$$

As in the scalar situation, there arise the questions of whether the sequence $\{\mathbf{x}_n\}$ is well defined, whether it converges, and whether its limit necessarily is a solution of the equation

$$\mathbf{x} = \mathbf{f}(\mathbf{x}). \tag{3}$$

All these questions are answered by the following result:

**THEOREM 2.4:** *Let $R$ denote the rectangular region $a \le x \le b$, $c \le y \le d$, and let the functions $f$ and $g$ satisfy the following conditions:*

    (i)   *$f$ and $g$ are defined and continuous in $R$;*
    (ii)   *for each $\mathbf{x} \in R$, the point $(f(\mathbf{x}), g(\mathbf{x}))$ also lies in $R$;*
    (iii)   *there exists a constant $L < 1$ so that for any two points $\mathbf{x}_1$ and $\mathbf{x}_2$ in $R$ the following inequality holds:*

$$\|\mathbf{f}(\mathbf{x}_1) - \mathbf{f}(\mathbf{x}_2)\| \le L\|\mathbf{x}_1 - \mathbf{x}_2\|. \tag{4}$$

*Then the following statements are true:*

(a)  *Equation (3) has precisely one solution* $\mathbf{s}$ *in* $R$;

(b)  *for any choice of* $\mathbf{x}_0$ *in* $R$, *the iteration sequence* $\{\mathbf{x}_n\}$ *is defined and converges to* $\mathbf{s}$;

(c)  *for any* $n = 1, 2, \ldots$, *the following inequality holds:*

$$\|\mathbf{x}_n - \mathbf{s}\| \leq \frac{L^n}{1 - L} \|\mathbf{x}_1 - \mathbf{x}_0\|. \tag{5}$$

Condition (4) is again called a **Lipschitz condition**. It here expresses the fact that the mapping $\mathbf{x} \to \mathbf{f}(\mathbf{x})$ diminishes the distance between any two points in $R$ at least by the factor $L$. For this reason, a mapping with the property (4) is again called a **contracting map**.

It should be noted that statement (a) is not as trivial now as it was for one variable, where the existence of a solution could be inferred from an inspection of the graph of the function $f$. Statement (b) guarantees convergence of the algorithm, while (c) gives an upper bound for the error after $n$ steps.

*Proof of Theorem* 2.4:  The proof will be accomplished in several stages. From condition (ii) it is evident that the sequence $\{\mathbf{x}_n\}$ is defined, and that its elements lie in $R$. Proceeding exactly as in the proof of Theorem 2.1, we now can show that (iii) implies

$$\|\mathbf{x}_{n+1} - \mathbf{x}_n\| \leq L^n \|\mathbf{x}_1 - \mathbf{x}_0\|, \qquad n = 0, 1, 2, \ldots. \tag{6}$$

Now let $n$ be a fixed positive integer, and let $m > n$. With a view towards applying the Cauchy criterion, we shall find a bound for $\|\mathbf{x}_m - \mathbf{x}_n\|$. Writing

$$\mathbf{x}_m - \mathbf{x}_n = (\mathbf{x}_{n+1} - \mathbf{x}_n) + (\mathbf{x}_{n+2} - \mathbf{x}_{n+1}) + \cdots + (\mathbf{x}_m - \mathbf{x}_{m-1})$$

and applying the triangle inequality, we obtain

$$\|\mathbf{x}_m - \mathbf{x}_n\| \leq \|\mathbf{x}_{n+1} - \mathbf{x}_n\| + \|\mathbf{x}_{n+2} - \mathbf{x}_{n+1}\| + \cdots + \|\mathbf{x}_m - \mathbf{x}_{m-1}\|$$

and, using (6) to estimate each term on the right,

$$\|\mathbf{x}_m - \mathbf{x}_n\| \leq (L^n + L^{n+1} + \cdots + L^{m-1}) \|\mathbf{x}_1 - \mathbf{x}_0\|$$

$$= \frac{L^n}{1 - L} \|\mathbf{x}_1 - \mathbf{x}_0\|, \tag{7}$$

by virtue of $0 \le L < 1$. Since the expression on the right does not depend on $m$ and tends to zero as $n \to \infty$, we have established that the sequence $\{x_n\}$ satisfies the Cauchy criterion. It thus has a limit $s$. Since $R$ is closed, $s \in R$.

We next show that $s$ is a solution of (3). By virtue of the continuity of $f$ and $g$,

$$\lim_{n \to \infty} f(x_n) = f(s),$$

and thus

$$s = \lim_{n \to \infty} x_n = \lim_{n \to \infty} x_{n+1} = \lim_{n \to \infty} f(x_n) = f(s),$$

as desired.

Uniqueness of the solution $s$ follows from the Lipschitz condition (iii), exactly as it does for one variable. Relation (5) finally follows by letting $m \to \infty$ in (7). This completes the proof of Theorem 2.4. No use has been made of the fact that the number of equations and unknowns is two; both the theorem and its proof remain valid in a Euclidean space of arbitrary dimension, and even in some infinite-dimensional spaces.

The application of Theorem 2.4 requires a knowledge of the Lipschitz constant $L$. It may be shown (see Henrici [1964], §5.3) that if the functions $f$ and $g$ have continuous partial derivatives in the region $R$ defined in Theorem 2.4, then the inequality (4) holds for

$$L := \max_{(x, y) \in R} \sqrt{f_x^2 + f_y^2 + g_x^2 + g_y^2}. \tag{8}$$

**DEMONSTRATION 2.4-2:**   Let

$$f(x, y) := A \cos x - B \sin y, \qquad g(x, y) := A \sin x + B \cos y,$$

where $A$ and $B$ are constants. We find

$$f_x^2 + f_y^2 + g_x^2 + g_y^2 = A^2 + B^2,$$

thus (4) holds with $L := \sqrt{A^2 + B^2}$. The conditions of Theorem 2.4 thus are satisfied, for instance for $a = c = -2$, $b = d = 2$, whenever $A^2 + B^2 < 1$. Convergence of the process for $A := 0.7$, $B := 0.2$, $x_0 = y_0 = 0$ is illustrated by the values given below.

| $x_n$ | $y_n$ |
|-------|-------|
| 0.700000 | 0.200000 |
| 0.495656 | 0.646966 |
| 0.495206 | 0.492509 |
| 0.521342 | 0.508879 |
| 0.509566 | 0.523289 |
| 0.511123 | 0.514695 |
| 0.512083 | 0.516498 |
| 0.511440 | 0.516907 |
| 0.511590 | 0.516474 |
| 0.511614 | 0.516608, and so on. |

Naturally, the condition that the value of $L$ given by (8) satisfies $L < 1$ is not necessary for convergence of the iterative process. For instance, in the problem on hand, convergence also takes place for $A = B = 0.71$.

## PROBLEMS

1   Three nonoverlapping circles $C_1$, $C_2$, $C_3$ with radii $r_1, r_2, r_3$ have their centers at the points $(0, 0)$, $(a, 0)$, $(0, b)$, respectively. The problem is to find the center $(x, y)$ of a circle $C$ which is tangent to the three given circles.

    **(a)**  If the distance of $(x, y)$ from the center of $C_k$ is denoted by $d_k$, and if $C$ does not contain the $C_k$, the equations

$$d_1 - r_1 = d_2 - r_2,$$
$$d_1 - r_1 = d_3 - r_3 \tag{9}$$

are to be satisfied. (If $C$ contains the $C_k$, the signs of the $r_k$ are reversed.) Show that the equations (9) are equivalent to the fixed point equations,

$$x = f(x, y) := \tfrac{1}{2}\{a + a^{-1}(r_1 - r_2)(d_1 + d_2)\},$$
$$y = g(x, y) := \tfrac{1}{2}\{b + b^{-1}(r_1 - r_3)(d_1 + d_3)\}. \tag{10}$$

    **(b)**  Show that the Lipschitz constant $L$ of the map $\mathbf{x} \to \mathbf{f}(\mathbf{x})$ satisfies $L \le L_0$, where

$$L_0 := \frac{(r_1 - r_2)^2}{a^2} + \frac{(r_1 - r_2)^2}{b^2}.$$

[Interpret the partial derivatives of the $d_k$ as sines and cosines.]

(c)  Experiment with the special data

| a | b | $r_1 - r_2$ | $r_1 - r_3$ | |
|---|---|---|---|---|
| 10 | 20 | 2 | 3 | $(L_0 < 1)$ |
| 10 | 10 | $-6$ | $-8$ | $(L_0 = 1)$ |
| 20 | 10 | 18 | 5 | $(L_0 > 1)$ |

2  From the point of view of programming it is advantageous to iterate the system (3) in the form,

$$x_{n+1} = f(x_n, y_n),$$
$$y_{n+1} = g(x_{n+1}, y_n);$$

that is, to use the new value $x_{n+1}$ as soon as it is available. (This is called **successive relaxation**.) Do the example of Demonstration 2.4-2 in this fashion. Do you observe an appreciable difference in the rate of convergence?

3  Make a thorough experimental study of iteration as applied to the *linear* system of two equations with two unknowns,

$$x + ay = c,$$
$$bx + y = d,$$

written in the form

$$x = c - ay,$$
$$y = d - bx.$$

  (a)  Under what conditions on $a$ and $b$ do you observe convergence?
  (b)  Does the symmetry of the matrix of the system $(a = b)$ have an influence on the speed of convergence?
  (c)  Does successive relaxation speed up convergence?
  (d)  It has been observed that convergence can sometimes be accelerated by the use of **overrelaxation**, where the new values $x_{n+1}$ and $y_{n+1}$ are obtained from

$$x_{n+1} = x_n + \omega[f(x_n, y_n) - x_n],$$
$$y_{n+1} = y_n + \omega[g(x_{n+1}, y_n) - y_n],$$

where $\omega$ is a fixed number, called the **overrelaxation factor**. Assuming the linear system to be symmetric $(b = a)$, determine experimentally good values of $\omega$ as a function of $a$.

(e*)   Also try to answer the foregoing questions analytically, which may require some linear algebra.

---

## §2.5   QUADRATIC CONVERGENCE FOR VECTOR ITERATION SEQUENCES

Suppose the functions $f$ and $g$ satisfy the conditions of Theorem 2.4 and have continuous partial derivatives up to order 2 in $R$. The sequence of points $(x_n, y_n)$ of the iteration sequence then converges to a solution $(s, t)$ of the system (1) of §2.4. What is the behavior of the errors

$$d_n := x_n - s,$$

$$e_n := y_n - s$$

as $n \to \infty$?

An application of Taylor's theorem for functions of two variables shows that

$$
\begin{aligned}
d_{n+1} &= x_{n+1} - s \\
&= f(x_n, y) - f(s, t) \\
&= f(s + d_n, t + e_n) - f(s, t) \\
&= f_x(s, t)d_n + f_y(s, t)e_n + O(\|\mathbf{d}_n\|^2),
\end{aligned}
$$

and similarly

$$e_{n+1} = g_x(s, t)d_n + g_y(s, t)e_n + O(\|\mathbf{d}_n\|^2).$$

Here $\mathbf{d}_n$ denotes the vector of errors,

$$\mathbf{d}_n = \begin{pmatrix} d_n \\ e_n \end{pmatrix},$$

and $O(\|\mathbf{d}_n\|^2)$ denotes a quantity bounded by $C\|\mathbf{d}_n\|^2$. In terms of the **Jacobian matrix** of the functions $f$ and $g$,

$$\mathbf{J} := \begin{pmatrix} f_x & f_y \\ g_x & g_y \end{pmatrix},$$

the above relations can be written in abbreviated form thus:

$$\mathbf{d}_{n+1} = \mathbf{J}(s, t)\mathbf{d}_n + O(\|\mathbf{d}_n\|^2). \tag{1}$$

Relation (1) is the multidimensional generalization of relation (1) of §2.2. If $\mathbf{J}(s, t) \neq \mathbf{O}$ (that is, if the elements of the matrix $\mathbf{J}(s, t)$ are not all zero), it shows that, at each step of the iteration, the error is approximately multiplied by a constant matrix. In this sense, we again may speak of *linear convergence*. If $\mathbf{J}(s, t) = \mathbf{O}$ (that is, if all four elements of $\mathbf{J}$ are zero), we see that the norm of the error at the $(n + 1)$st step is of the order of the square of the norm of the error at the $n$-th step. This is similar to what earlier has been called quadratic convergence.

The following analog of Theorem 2.2 holds in the case where $\mathbf{J}(s, t) = \mathbf{O}$:

**THEOREM 2.5:** *Let the functions $f$ and $g$ be defined in a region $R$, and let them satisfy the following conditions:*

*(i) The partial derivatives up to order 2 of $f$ and $g$ exist and are continuous in $R$.*

*(ii) The system (1) of §2.4 has a solution $(s, t)$ in the interior of $R$ such that $\mathbf{J}(s, t) = \mathbf{O}$.*

*Then there exists a number $d > 0$ so that the iteration sequence generated by $(f, g)$ exists and converges to $(s, t)$ with quadratic convergence for any choice of the starting point within the distance $d$ of the solution.*

The conclusion can be expressed by saying that the algorithm always converges, and does so quadratically, if the starting vector is sufficiently close to the solution vector. The proof of Theorem 2.5 is based on the fact that by virtue of the continuity of the first partial derivatives

$$\sqrt{f_x^2 + f_y^2 + g_x^2 + g_y^2} = L < 1$$

in a certain neighborhood of $(s, t) \in R$. It then follows from equation (8) of §2.4 that the conditions of Theorem 2.4 are satisfied in that neighborhood. Further details are omitted.

We apply the foregoing to the problem of solving systems of two equations in two unknowns which are of the form,

$$F(x, y) = 0,$$
$$G(x, y) = 0, \tag{2}$$

where both functions $F$ and $G$ are defined and twice continuously differentiable on a certain rectangle $R$ of the $(x, y)$-plane. We suppose that the system (2) has a solution $(s, t)$ in the interior of $R$, and that the Jacobian determinant

$$D(x, y) := \begin{vmatrix} F_x(x, y) & F_y(x, y) \\ G_x(x, y) & G_y(x, y) \end{vmatrix} \tag{3}$$

is different from zero when $(x, y) = (s, t)$. It then follows from a theorem of multivariate calculus (see Buck [1956], p. 216) that the system (2) has no solution other than $(s, t)$ in a certain neighborhood of the point $(s, t)$.

Newton's method for solving a single equation $F(x) = 0$ could be understood as arising from replacing $F(x + \delta)$ by $F(x) + F'(x)\delta$, and solving the equation

$$F(x) + F'(x)\delta = 0$$

for $\delta$, thus obtaining a supposedly better approximation $x + \delta$ to $s$ than $x$. We now apply the same principle to the system (2). Assuming that $(x, y)$ is a point "near" the desired solution $(s, t)$, we replace the function $F(x + \delta, y + \varepsilon)$ by its first degree Taylor polynomial at the point $(x, y)$; that is, by $F(x, y) + F_x(x, y)\delta + F_y(x, y)\varepsilon$. A similar replacement is made for the function $G(x + \delta, y + \varepsilon)$. Setting the Taylor polynomials equal to zero, we obtain a system of two linear equations for $\delta$ and $\varepsilon$,

$$
\begin{aligned}
F_x(x, y)\delta + F_y(x, y)\varepsilon &= -F(x, y), \\
G_x(x, y)\delta + G_y(x, y)\varepsilon &= -G(x, y).
\end{aligned}
\tag{4}
$$

The determinant of this system is just the quantity $D(x, y)$ defined by (3). Since $D$ is continuous, and $D(s, t) \neq 0$ by hypothesis, it follows that $D(x, y) \neq 0$ for all points $(x, y)$ sufficiently close to $(s, t)$. We thus obtain **Newton's method for systems of equations** in two unknowns, which algorithm may be described as follows:

Choose $(x_0, y_0)$, and determine the sequence of points $(x_n, y_n)$ by

$$
\begin{aligned}
x_{n+1} &= f(x_n, y_n), \\
y_{n+1} &= g(x_n, y_n),
\end{aligned}
\tag{5}
$$

$n = 0, 1, 2, \ldots$, where the functions $f$ and $g$ are defined by

$$f(x, y) := x + \delta(x, y), \qquad g(x, y) := y + \varepsilon(x, y),$$

and where $\delta$ and $\varepsilon$ denote the solution of the linear system (4).

We assert that the sequence $(x_n, y_n)$ converges quadratically to $(s, t)$ for all $(x_0, y_0)$ sufficiently close to $(s, t)$. In order to verify this statement, it suffices by Theorem 2.5 to show that the elements of the Jacobian matrix of the iteration functions $f$ and $g$,

$$\mathbf{J}(x, y) = \begin{pmatrix} f_x(x, y) & f_y(x, y) \\ g_x(x, y) & g_y(x, y) \end{pmatrix}$$

are all zero for $(x, y) = (s, t)$. Omitting arguments, we have

$$f_x = 1 + \delta_x, \qquad f_y = \delta_y,$$
$$g_x = \varepsilon_x, \qquad g_y = 1 + \varepsilon_y. \tag{6}$$

The values of the derivatives $\delta_x$, $\delta_y$, $\varepsilon_x$, $\varepsilon_y$ are best determined from (4) by implicit differentiation. Differentiating the first of these equations, we get

$$F_{xx}\delta + F_x\delta_x + F_{xy}\varepsilon + F_y\varepsilon_x = -F_x,$$
$$F_{xy}\delta + F_x\delta_y + F_{yy}\varepsilon + F_y\varepsilon_y = -F_y.$$

Two similar relations involving the function $G$ are obtained by differentiating the second relation. We now set $(x, y) = (s, t)$ and observe that $\delta(s, t) = \varepsilon(s, t) = 0$. We thus obtain for $(x, y) = (s, t)$

$$F_x\delta_x + F_y\varepsilon_x = -F_x,$$
$$G_x\delta_x + G_y\varepsilon_x = -G_x,$$
$$F_x\delta_y + F_y\varepsilon_y = -F_y,$$
$$G_x\delta_y + G_y\varepsilon_y = -G_y.$$

The determinant of each of these two systems of linear equations is again the Jacobian determinant $D(s, t)$, and hence is different from zero. We now easily find

$$\delta_x = -1, \qquad \delta_y = 0$$
$$\varepsilon_x = 0 \qquad \varepsilon_y = -1.$$

According to (6) this implies that all elements of the matrix $\mathbf{J}$ are zero at the point $(s, t)$, as desired.

For the execution of Newton's method for systems of equations (although not for the above theoretical analysis), it is necessary at each step of the iteration to solve the system (4) for $\delta$ and $\varepsilon$. For systems of order 2, the solution is effected by Cramer's rule, which yields

$$\delta = \frac{GF_y - FG_y}{F_xG_y - F_yG_x}, \qquad \varepsilon = \frac{FG_x - GF_y}{F_xG_y - F_yG_x}.$$

For systems of higher order, however, Cramer's rule should never be used, and the linear system resulting from Newton's method is best solved by the process of Gaussian elimination described in §4.1.

**DEMONSTRATION 2.5-1:** We use Newton's method to find the point of intersection $\neq (0, 0)$ of the circle

$$F(x, y) := x - x^2 - y^2 = 0$$

and the hyperbola

$$G(x, y) := y - x^2 + y^2 = 0.$$

(See Example 2.4-1.) We have already seen that this point lies in the first quadrant. Starting with $x_0 = y_0 = 0.5$, we find

| $x_n$ | $y_n$ |
|-----------|-----------|
| 0.50000000 | 0.50000000 |
| 1.00000000 | 0.50000000 |
| 0.81250000 | 0.43750000 |
| 0.77371988 | 0.42055723 |
| 0.77184895 | 0.41964566 |
| 0.77184451 | 0.41964338 |
| 0.77184451 | 0.41964338 |

The quadratic nature of the convergence is clearly evident.

**Bairstow's Method**

In §2.3 we have seen how to apply the scalar version of Newton's method to the problem of finding real zeros of real polynomials. We left unresolved the problem of how to find complex zeros. If complex arithmetic is used, such zeros can be found, in principle, by the complex version of the ordinary Newton's method (see Henrici [1977], §6.12, for a theoretical justification). Here we show how to find complex zeros by using *real arithmetic only*. The problem will be attacked as a special case of the problem of determining a *quadratic factor* of a given polynomial.

Let $p$ be a polynomial of degree $n$,

$$p(x) = a_0 x^n + a_1 x^{n-1} + \cdots + a_n.$$

If $q$ is any polynomial of degree $m \leq n$, we may attempt to divide $p$ by $q$. The result will be a *quotient polynomial*, which has the precise degree $n - m$ and which we denote by $s$, and a *remainder polynomial*, which has a degree $<m$ and which we denote by $r$. The four polynomials in question are linked by the identity,

$$p(x) = q(x)s(x) + r(x). \tag{7}$$

The polynomials $s$ and $r$ are uniquely determined by $p$ and $q$; they may be constructed, for instance, by comparing coefficients in the identity (7). We call $q$ a **factor** of $p$ if, and only if, $r$ is the zero polynomial. If $q$ is a factor of $p$, then clearly every zero of $q$ is a zero of $p$; indeed, every zero of multiplicity $\mu$ of $q$ is a zero of multiplicity $\geq \mu$ of $p$.

In the method about to be discussed, $q$ is assumed to be a quadratic polynomial, which we choose to write in the form

$$q(x) = x^2 - ux - v.$$

Assuming $n \geq 2$, our goal is to determine the coefficients $u$ and $v$ so that, in the representation (7), $r = 0$. What are the equations to be solved? Let

$$s(x) = b_0 x^{n-2} + b_1 x^{n-3} + \cdots + b_{n-2}, \qquad r(x) = r_0 x + r_1.$$

The identity (7) then reads

$$a_0 x^n + a_1 x^{n-1} + \cdots + a_n = (x^2 - ux - v)(b_0 x^{n-2}$$
$$+ b_1 x^{n-3} + \cdots + b_{n-2}) + r_0 x + r_1;$$

it is seen to be satisfied if, and only if, the $b_i$ and the $r_j$ satisfy the conditions

$$b_k - ub_{k-1} - vb_{k-2} = a_k, \qquad k = 0, 1, \ldots, n-2$$

(here we are assuming $b_{-1} = b_{-2} = 0$) and

$$r_0 - ub_{n-2} - vb_{n-1} = a_{n-1},$$
$$r_1 - vb_{n-2} = a_n.$$

We thus see that the $b_i$ are determined by the recurrence relation $b_{-2} = b_{-1} = 0$,

$$b_k = a_k + ub_{k-1} + vb_{k-2}, \tag{8}$$

$k = 0, 1, \ldots, n-2$; moreover, if the relation (8) is used also for $k = n-1$ and for $k = n$, we see that

$$r_0 = b_{n-1},$$
$$r_1 = b_n - ub_{n-1}.$$

Clearly, the polynomial $r$ vanishes identically, and $q$ is a quadratic factor of $p$, if, and only if, the equations

$$b_{n-1} = 0, \qquad b_n = 0 \tag{9}$$

are satisfied, where it should be borne in mind that $b_{n-1}$ and $b_n$, like all coefficients $b_k$, are functions of $u$ and $v$.

**Bairstow's method** consists in solving the system of two equations (9) in the two unknowns $u$ and $v$ by Newton's method. To this end, we require the partial derivatives

$$\frac{\partial b_{n-1}}{\partial u}(u, v), \qquad \frac{\partial b_{n-1}}{\partial v}(u, v),$$

$$\frac{\partial b_n}{\partial u}(u, v), \qquad \frac{\partial b_n}{\partial v}(u, v).$$

We shall obtain recurrence relations for these derivatives by differentiating the recurrence relations (8).

We first differentiate with respect to $u$ and observe that $\partial b_0 / \partial u = 0$. Writing,

$$c_k := \frac{\partial b_{k-1}}{\partial u} \tag{10}$$

for notational convenience, we obtain

$$c_0 = b_0,$$
$$c_1 = b_1 + uc_0,$$
$$c_2 = b_2 + uc_1 + vc_0,$$

and generally

$$c_k = b_k + uc_{k-1} + vc_{k-2},$$

where $k = 0, 1, 2, \ldots, n - 1$; $c_{-2} = c_{-1} := 0$. We see that the $c_k$ are generated from the $b_k$ exactly as the $b_k$ were generated from the $a_k$. From (10) we have

$$\frac{\partial b_{n-1}}{\partial u} = c_{n-2}, \qquad \frac{\partial b_n}{\partial u} = c_{n-1}.$$

We next differentiate with respect to $v$. We observe that now $\partial b_0 / \partial v = \partial b_1 / \partial v = 0$. Thus, writing

$$d_k := \frac{\partial b_{k+2}}{\partial v}, \tag{11}$$

we obtain $d_{-2} = d_{-1} = 0$,

$$d_0 = b_0,$$
$$d_1 = b_1 + u d_0,$$
$$d_2 = b_2 + u d_1 + v d_0,$$

and generally

$$d_k = b_k + u d_{k-1} + v d_{k-2},$$

$k = 0, 1, \ldots, n - 2$. These recurrence relations and initial conditions are exactly the same as those for the $c_k$; hence we have $d_k = c_k$, $k = 0, 1, \ldots, n - 2$, and from (11) we obtain

$$\frac{\partial b_{n-1}}{\partial v} = c_{n-3}, \qquad \frac{\partial b_n}{\partial v} = c_{n-2}.$$

If the increments of $u$ and $v$ are denoted by $\delta$ and $\varepsilon$, respectively, their values as determined by Newton's method must satisfy

$$c_{n-2} \delta + c_{n-3} \varepsilon = -b_{n-1},$$
$$c_{n-1} \delta + c_{n-2} \varepsilon = -b_n,$$

and hence are given by

$$\delta = \frac{b_n c_{n-3} - b_{n-1} c_{n-2}}{c_{n-1}^2 - c_{n-1} c_{n-3}}, \qquad \varepsilon = \frac{b_{n-1} c_{n-1} - b_n c_{n-2}}{c_{n-2}^2 - c_{n-1} c_{n-3}} \tag{12}$$

The whole procedure of Bairstow's method is summarized as follows: Given the polynomial

$$p(x) = a_0 x^n + a_1 x^{n-1} + a_2 x^{n-2} + \cdots + a_n$$

and an arbitrary tentative quadratic factor $x^2 - u_0 x - v_0$, determine a sequence $\{x^2 - u_m x - v_m\}$ of quadratic factors as follows: For $m = 0, 1, 2, \ldots$, construct the sequence $\{b_k\} = \{b_k^{(m)}\}$ from $b_{-2} = b_{-1} = 0$,

$$b_k = a_k + u_m b_{k-1} + v_m b_{k-2}, \qquad k = 0, 1, \ldots, n,$$

and the sequence $\{c_k\} = \{c_k^{(m)}\}$ from $c_{-2} = c_{-1} = 0$,

$$c_k = b_k + u_m c_{k-1} + v_m c_{k-2}, \qquad k = 0, 1, \ldots, n - 1.$$

Then, set $u_{m+1} := u_m + \delta$, $v_{m+1} := v_m + \varepsilon$, where $\delta$ and $\varepsilon$ are given by (12).

The flow diagram in principle looks like the diagram shown in Figure 2.3b. The arrays of the $b_k$ and $c_k$ need not be stored if the $c_k$ are computed simultaneously with the $b_k$. After a quadratic factor has been determined to sufficient accuracy, the coefficients of the deflated polynomial

$$s(x) = \frac{p(x)}{q(x)} = b_0 x^{n-2} + b_1 x^{n-3} + \cdots + b_{n-2}$$

are found by once more using the recurrence relations (8).

To obtain zeros of $p$ from a quadratic factor

$$q(x) = x^2 - ux - v,$$

one uses the usual formulas for solving a quadratic equation. The zeros are complex if

$$\left(\frac{u}{2}\right)^2 + v < 0;$$

and the zeros then are

$$z_1 = \frac{u}{2} + i\sqrt{-v - \left(\frac{u}{2}\right)^2}, \qquad z_2 = \frac{u}{2} - i\sqrt{-v - \left(\frac{u}{2}\right)^2}.$$

If the zeros are real, one should beware of the possibility of cancellation which was illustrated in Demonstration 1.3-2. Cancellation is avoided if the zero of larger absolute value,

$$z_1 = \frac{u}{2} + \text{sign}\left(\frac{u}{2}\right)\sqrt{\left(\frac{u}{2}\right)^2 + v},$$

is determined first, and the second zero is found from the relation

$$z_2 = -\frac{v}{z_1}.$$

**DEMONSTRATION 2.5-2⊕:**   To determine all zeros of the polynomial

$$p(x) = 3x^6 + 9x^5 + 9x^4 + 5x^3 + 3x^2 + 8x + 5.$$

Starting with the quadratic factor $x^2 - 0x - 0$, we obtain

| $u$ | $v$ |
|---|---|
| 0.00000000 | 0.00000000 |
| $-0.03225806$ | $-1.58064516$ |
| $-0.17463231$ | $-0.59651964$ |
| $-0.38932129$ | 0.81459497 |
| 0.23086993 | $-0.91040568$ |
| 0.27592249 | $-0.07962437$ |
| $-0.09207601$ | $-0.69109116$ |
| $-0.20477128$ | 0.53724443 |

The example nicely illustrates the vagaries of a quadratically convergent method if no good starting values are known. After seven iterations, there is no hint of convergence. Let us try instead the factor $x^2 + 2x + 2$ with the zeros $-1 \pm i$. We obtain,

| $u$ | $v$ |
|---|---|
| $-2.00000000$ | $-2.00000000$ |
| $-1.72517321$ | $-1.53117783$ |
| $-1.49118278$ | $-1.37710392$ |
| $-1.55372589$ | $-1.49553798$ |
| $-1.54158725$ | $-1.48710864$ |
| $-1.54156814$ | $-1.48731583$ |
| $-1.54156818$ | same |
| same | |

Thus, we have found the quadratic factor

$$x^2 + 1.54156818x + 1.48731583;$$

its zeros are

$$z_{1, 2} = -0.77078409 \pm i0.94509667.$$

Deflation yields the polynomial

$$3x^4 + 4.37529546x^3 - 2.20676375x^2 + 1.89443057x + 3.36176077.$$

We now start with the quadratic factor $x^2 - x + 1$. The results are

| $u$ | $v$ |
|---|---|
| 1.00000000 | $-1.00000000$ |
| 1.07179417 | $-0.74678870$ |
| 1.13231304 | $-0.83633924$ |
| 1.12725609 | $-0.83164579$ |
| 1.12723862 | $-0.83164428$ |
| same | same |

The zeros of the new factor are

$$z_{3,4} = 0.56361931 \pm i0.71692228.$$

We deflate once more to get the quadratic polynomial

$$3x^2 + 7.75701131x + 4.04230611.$$

There is no need to use Bairstow's method on this polynomial, but in an automatic process one would iterate anyway, starting perhaps with $x^2 - 0x - 0$. This would yield

| $u$ | $v$ |
|---|---|
| 0.00000000 | 0.00000000 |
| $-2.58567044$ | 5.33825624 |
| same | $-1.34743537$ |
| | same |

The final zeros are

$$z_5 = -1.86203425, \qquad z_6 = -0.72363619.$$

# PROBLEMS

1 *Delayed exponential population growth.* A population $y$ whose rate of increase at time $t$ is proportional to the population at time $t$ satisfies (if the unit of time is suitably chosen) the differential equation

$$y'(t) = y(t) \tag{13}$$

with the well-known solution $y(t) = Ae^t$. If the rate of increase at time $t$ is proportional to the population at time $t - 1$, then (13) must be replaced by the differential-difference equation

$$y'(t) = y(t - 1). \tag{14}$$

(a) Show that $y(t) := e^{zt}$ is a solution of (14) if, and only if, $z$ satisfies the transcendental equation

$$z = e^{-z}, \tag{15}$$

and determine the sole real solution of (15).

(b)   By separating real and imaginary parts in

$$x + iy = e^{-(x+iy)} = e^{-x}(\cos y - i \sin y),$$

show that (15) has an infinity of complex solutions $z_k$, $\overline{z_k}$, which are located near the points

$$-\log[(2k - \tfrac{1}{2})\pi] \pm i(2k - \tfrac{1}{2})\pi, \tag{16}$$

$k = 1, 2, \ldots$ . [Let $F(x, y) := x - e^{-x} \cos y$, $G(x, y) := y + e^{-x} \sin y$, and draw rough graphs of the curves $F(x, y) = 0$ and $G(x, y) = 0$.]

(c)   Determine the complex solutions of (15) by Newton's method, using the points (16) as starting values.

(d)   Generalize the above to the differential-difference equation

$$y'(t) = \alpha y(t - 1),$$

where $\alpha$ is a real parameter.

2   The polynomial

$$p(z) = 5z^4 + 4z^3 + 3z^2 + 2z + 1$$

has precisely one zero in each quadrant of the complex plane.

(a)   Determine the zeros by the scalar Newton's method, using complex arithmetic. Use $z_0 = \pm 1 + i$ as a starting value. [Because the zeros occur in conjugate pairs, only the zeros in the upper half-plane need to be determined.]

(b⊕)   Find the zeros by Bairstow's method.

3⊕   For appropriate values of the capacities, inductances, and resistances the electric circuit shown in Figure 2.5 has the transfer function

$$g(s) := \cfrac{1}{R + \cfrac{1}{\cfrac{1}{s + \cfrac{25}{s}} + \cfrac{1}{s + 2 + \cfrac{2}{s}}}}.$$

(a)   Show by a graph how the poles of the transfer function move in the complex plane as a function of $R$ for $0 \le R < \infty$.

[The poles are the zeros of the denominator when $g$ is expressed as the ratio of two polynomials. Determine the poles for a sequence of discrete values $\{R_i\}$, using Bairstow's method. The poles are easily

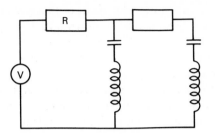

**Fig. 2.5.** Electric circuit.

recognized for $R = 0$. From then on, the factors found for $R_{i-1}$ may be taken as first approximations to the factors for $R_i$.]

(b*) For which value of $R$ has $g(s)$ a double pole?

**4⊕** Find all zeros of the polynomial

$$p(x) = x^4 - 8x^3 + 39x^2 - 62x + 51,$$

using the fact that

$$p(x) = (x^2 - 2x + 2)(x^2 - 6x + 25) + 1.$$

**5⊕** Find all zeros of

$$p(x) = 4x^6 - 5x^5 + 4x^4 - 3x^3 + 7x^2 - 7x + 1.$$

**6** Generalizing the approach outlined in §2.3, one might try to generalize Newton's method to systems by suitably choosing functions $h$ and $k$, so that iteration of the functions

$$f(x, y) := x + h(x, y)F(x, y)$$
$$g(x, y) := y + k(x, y)G(x, y)$$

yields a quadratically convergent process. Show that this is *not* possible in general.

**7*** Determine the matrix $\mathbf{H} = \mathbf{H}(\mathbf{x})$ so that application of iteration to

$$\mathbf{f}(\mathbf{x}) := \mathbf{x} + \mathbf{H}(\mathbf{x})\mathbf{F}(\mathbf{x})$$

yields a quadratically convergent process. Show that the process obtained is identical with Newton's method.

**8*** *Aitken acceleration for systems.* The model assumption analogous to (8) of §2.2 here is

$$\mathbf{x}_n = \mathbf{s} + \mathbf{J}^n \mathbf{e}_0,$$

where **J** is the (unknown) Jacobian of the system $\mathbf{x} = \mathbf{f}(\mathbf{x})$ at the fixed point **s**, and where $\mathbf{e}_0$ is the unknown initial error.

   **(a)**   Assuming that **J** has a single eigenvalue $\lambda$ of maximum modulus, show that it is feasible to apply Aitken's $\Delta^2$ method individually to each component of $\mathbf{x}_n$.

   **(b)**   How can the convergence of the sequence $\{x_n\}$ be accelerated if **J** has two conjugate complex leading eigenvalues?

   **(c)**   Experiment with cases of Problem 1 of §2.4 where convergence is slow.

   [Some linear algebra is required to do this problem.]

---

### Bibliographical Notes and Recommended Reading

For a rigorous treatment of iteration, solution of equations, and convergence acceleration see Ostrowski [1966], Ortega and Rheinboldt [1970], and Householder [1970]. In his earlier text (Henrici [1964]), the author has given a somewhat broader elementary treatment of these same topics. The subject of polynomial equations is dealt with extensively, although from a pre-computer point of view, in many classical texts on the "theory of equations"; see for example, Uspensky [1948]. For the concepts of calculus required in this chapter, see Salas and Hille [1978] or Anton [1980].

# CHAPTER 3

# Recursion

Here we deal with scalar iteration sequences that are defined by a relation of the special form

$$x_n := f(x_{n-1}, x_{n-2}, \ldots, x_{n-k}), \qquad n = k, k+1, \ldots, \tag{1}$$

where $f$ is a *linear and homogeneous* function of its $k$ variables, $k$ being a fixed integer. Thus for symmetry (1) can be written as

$$a_0 x_n + a_1 x_{n-1} + \cdots + a_k x_{n-k} = 0, \tag{2}$$

where $a_0 \neq 0$. In addition to their direct algorithmic applications in Bernoulli's method and in the quotient-difference algorithm, such sequences are essential for the understanding of algorithmic processes in seemingly unrelated fields of numerical analysis, such as the numerical solution of ordinary and partial differential equations.

Technically, sequences defined by a relation of the form (2) could be subsumed under the case of vector iteration by considering an appropriate linear iteration function for the vectors

$$\mathbf{x}_n = \begin{pmatrix} x_n \\ x_{n-1} \\ \vdots \\ x_{n-k+1} \end{pmatrix}.$$

However, the direct treatment offered below seems to bring out the essential features more quickly.

## §3.1 LINEAR DIFFERENCE EQUATIONS

A recurrence relation of the form (1) is also called a **difference equation** for the sequence $\{x_n\}$. The difference equation is said to be of **order** $k$, because $x_n$

**125**

depends on the $k$ preceding elements of the sequence. The difference equation (2) is called **linear**, and it is said to have **constant coefficients** because the coefficients $a_0, a_1, \ldots, a_k$ are independent of $n$. The difference equation (2) is **homogeneous** inasmuch as it is satisfied by the zero sequence. The non-homogeneous linear difference equation of order $k$ with constant coefficients has the form,

$$a_0 x_n + a_1 x_{n-1} + \cdots + a_k x_{n-k} = b_n, \tag{3}$$

where $\{b_n\}$ is a given sequence. The difference equations (2) and (3) have the order $k$, and not any smaller order, if $a_0 a_k \neq 0$. This will always be assumed in the following, and, if convenient, we may also assume that $a_0 = 1$.

We can aid the understanding by two notational simplifications. First, we need not refer to a sequence by explicitly exhibiting its elements, as in the symbol $\{x_n\}$, but may instead denote a sequence by a single symbol such as x. The $n$-th element of the sequence x will be denoted by $x_n$, or occasionally by $(\mathbf{x})_n$. Secondly, if x is any sequence, we shall denote by $L\mathbf{x}$ the sequence whose $n$-th element is

$$(L\mathbf{x})_n = a_0 x_n + a_1 x_{n-1} + \cdots + a_k x_{n-k}.$$

With these notations, the problem of solving the difference equation (3) is the same as the problem of finding a sequence x so that

$$L\mathbf{x} = \mathbf{b}. \tag{4}$$

We note that the operator $L$ defined above is a *linear* operator. Defining the product $a\mathbf{x}$ of a scalar $a$ and of a sequence x by

$$(a\mathbf{x})_n := a(\mathbf{x})_n,$$

and the sum of two sequences x and y by

$$(\mathbf{x} + \mathbf{y})_n := (\mathbf{x})_n + (\mathbf{y})_n,$$

we have for arbitrary scalars $a$ and $b$ and for any two sequences x and y

$$L(a\mathbf{x} + b\mathbf{y}) = aL\mathbf{x} + bL\mathbf{y}. \tag{5}$$

The comprehension of the above notation may be helped by considering analogous simplifications in the theory of functions. Writing x in place of $\{x_n\}$ is much like writing $f$ in place of $f(x)$ to denote a function. The operator $L$ plays a role similar to that, say, of the differentiation operator $D$, which associates with a function $f$ its derivative $Df := f'$. Relation (5) is the analog of the familiar fact that differentiation is a linear operation; that is, that $D(af + bg) = aDf + bDg$.

We now turn to the task of solving the difference equation (2).

To solve a difference equation is, in a way, a trivial matter. Given the starting values $x_0, x_1, \ldots, x_{k-1}$, one uses (1) to compute

$$x_k = f(x_{k-1}, x_{k-2}, \ldots, x_0).$$

One then makes the substitutions,

$$x_0 := x_1$$
$$x_1 := x_2$$
$$\ldots$$
$$x_{k-1} := x_k$$

and repeats the process. In this sense, as many elements of the solution sequence $\{x_n\}$ can be determined as may be required. We shall call this process of solution the **numerical method** of solving the difference equation.

What we have in mind here, however, is an **analytical**, closed form solution of the linear difference equation (2). While for nonlinear difference equations the existence of a closed form solution is a rare exception*, such a solution exists for the linear difference equation with constant coefficients, at least in the homogeneous case. The fact that any numerical solution can also be expressed in closed form will have remarkable algorithmic implications.

We first consider the homogeneous equation

$$Lx = 0 \tag{H}$$

where $0$ denotes the zero sequence. The basic fact is that the totality of all solutions $x$ of (H) forms a linear space, $\mathcal{L}$. Indeed, if $x$ is a solution and $a$ is any scalar, then

$$L(ax) = aLx = a0 = 0,$$

showing that the sequence $ax$ is again a solution. Furthermore, if $x^{(1)}$ and $x^{(2)}$ are two solution sequences, then in view of the linearity of the operator $L$,

$$L(x^{(1)} + x^{(2)}) = Lx^{(1)} + Lx^{(2)} = 0 + 0 = 0,$$

showing that the sequence $x^{(1)} + x^{(2)}$ is a solution.

What is the dimension of the space $\mathcal{L}$? By definition, the dimension of a linear space is the maximum number of linearly independent elements in $\mathcal{L}$.

---

* For one such exception, see Problem 4 of §1.3.

Here the elements $\mathbf{x}^{(1)}, \ldots, \mathbf{x}^{(m)}$ are called **linearly independent**, if any relation of the form

$$c_1 \mathbf{x}^{(1)} + c_2 \mathbf{x}^{(2)} + \cdots + c_m \mathbf{x}^{(m)} = \mathbf{0} \tag{6}$$

implies that all $c_j = 0$.

It is easy to see that the dimension of our linear space $\mathscr{L}$ is at least $k$. Consider the $k$ solutions $\mathbf{x}^{(j)}$ ($j = 0, 1, \ldots, k - 1$) of (H) so that

$$x_j^{(j)} = 1, \; x_0^{(j)} = \cdots = x_{j-1}^{(j)} = x_{j+1}^{(j)} = \cdots = x_{k-1}^{(j)} = 0, \tag{7}$$

and suppose that

$$c_0 \mathbf{x}^{(0)} + c_1 \mathbf{x}^{(1)} + \cdots + c_{k-1} \mathbf{x}^{(k-1)} = \mathbf{0}. \tag{8}$$

Consider, for $j = 0, 1, \ldots, k - 1$, the $j$-th element in (8). There follows

$$c_j \cdot 1 = 0,$$

and, hence, that all $c_j = 0$. We conclude that the $k$ solutions $\mathbf{x}^{(j)}$ described above are linearly independent. Suppose, on the other hand, that $\mathbf{x}^{(0)}, \mathbf{x}^{(1)}, \ldots, \mathbf{x}^{(k)}$ are *any* $k + 1$ solutions of (H). We form the linear combination

$$\mathbf{y} := c_0 \mathbf{x}^{(0)} + c_1 \mathbf{x}^{(1)} + \cdots + c_k \mathbf{x}^{(k)}$$

and try to determine the constants $c_0, c_1, \ldots, c_k$ so that the first $k$ elements of the sequence $\mathbf{y}$ are zero, without all the $c_j$ being zero. This yields the $k$ homogeneous equations

$$\sum_{j=0}^{k} c_j x_i^{(j)} = 0, \qquad i = 0, 1, \ldots, k - 1.$$

By a basic result in the theory of linear equations, a homogeneous linear system of $k$ equations for $k + 1$ unknowns always has a non-trivial solution. Let $(c_0, c_1, \ldots, c_k)$ be such a solution, and let $\mathbf{y}$ be defined with these $c_j$. Then, the first $k$ elements of the sequence $\mathbf{y}$ are zero. As a linear combination of solutions of (H), $\mathbf{y}$ is a solution of (H). Having zero starting values, it is the zero solution, $\mathbf{y} = \mathbf{0}$. It follows that the $k + 1$ solutions $\mathbf{x}^{(0)}, \ldots, \mathbf{x}^{(k)}$ are linearly dependent. We thus have shown that $\mathscr{L}$ has dimension $k$.

Again, by linear algebra, there follows the existence of a *basis* of $k$ fixed solutions so that every solution of (H) can be expressed as a linear combination of them. One such basis evidently is given by the $k$ solutions satisfying (7), but the sequences of this basis cannot easily be expressed in closed form. Another basis may be found by trying to find, analogous to the situation that prevails in the theory of linear *differential* equations with constant coefficients, solutions that are exponential functions in the variable under consideration. This variable here being $n$, we try to find solutions $\mathbf{x} = \{x_n\}$ that are of the form

$$x_n = z^n, \tag{9}$$

where $z$ is to be determined. For the sequence defined by (9), we have

$$(\mathbf{Lx})_n = a_0 x_n + a_1 x_{n-1} + \cdots + a_k x_{n-k}$$
$$= a_0 z^n + a_1 z^{n-1} + \cdots + a_k z^{n-k}$$
$$= z^{n-k} p(z),$$

where $p$ is a polynomial,

$$p(z) := a_0 z^k + a_1 z^{k-1} + \cdots + a_k,$$

called the **characteristic polynomial** of the difference equation (H). Evidently, the sequence (9) is a nonzero solution of (H) if, and only if, $z$ is a zero of the characteristic polynomial of (H).

Normally, the characteristic polynomial $p$ has $k$ zeros $z_1, z_2, \ldots, z_k$ which are all distinct. In this case, our method yields $k$ different solutions of (H),

$$\mathbf{x}^{(j)} = \{x_n^{(j)}\} = \{z_j^n\}, \qquad j = 1, 2, \ldots, n,$$

which obviously are linearly independent. It follows that every solution of (H) can be expressed in the form

$$x_n = c_1 z_1^n + c_2 z_2^n + \cdots + c_k z_k^n;$$

where $c_1, c_2, \ldots, c_k$ are suitable constants. If the characteristic polynomial has repeated zeros, our method does not provide a sufficient number of linearly independent solutions. In those instances, additional solutions may be obtained by multiplying by suitable powers of $n$. More precisely, let $z_1, z_2, \ldots, z_m$ be the *distinct* zeros of $p$, and let the multiplicity of $z_j$ be $\mu_j$. (Clearly, $\mu_1 + \mu_2 + \cdots + \mu_m = k$.) Then, the sequences

$$\mathbf{x}^{(j,\, l)} := \{x_n^{(j,\, l)}\} = \{n^{l-1} z_j^n\}, \tag{10}$$
$$j = 1, 2, \ldots, m; \qquad l = 1, 2, \ldots, \mu_j$$

are all solutions of (H). We omit the easy proof of this fact, as well as the more difficult proof that these solutions are linearly independent. From the linear independence, there follows

**THEOREM 3.1:** *Let* $x^{(j, l)}$ *denote the* $n$ *special solutions of* (H) *defined by* (10). *Then the sequence*

$$x = \sum_{j=1}^{m} \sum_{l=1}^{\mu_j} c_{jl} x^{(j, l)} \tag{11}$$

*is a general solution of* (H) *in the sense that* (i) *for every choice of the constants* $c_{jl}$, $x$ *is a solution, and* (ii) *every solution* $x$ *of* (H) *can be represented in the form* (11) *for a suitable choice of the constants* $c_{jl}$.

**EXAMPLE 3.1-1:** Let the sequence $x = \{x_n\}$ be defined by $x_0 := 0$, $x_1 := 1$,

$$x_n := \tfrac{1}{2}(x_{n-1} + x_{n-2}), \qquad n = 2, 3, \dots. \tag{12}$$

The characteristic polynomial of the difference equation is

$$p(z) = z^2 - \tfrac{1}{2}z - \tfrac{1}{2};$$

its zeros are

$$z_1 = 1, \qquad z_2 = -\tfrac{1}{2}.$$

Thus, the general solution of (12) is

$$x_n = c_1 1^n + c_2(-\tfrac{1}{2})^n = c_1 + (-1)^n c_2 2^{-n}.$$

The initial conditions furnish the relations

$$x_0 = c_1 + c_2 = 0,$$
$$x_1 = c_1 - \tfrac{1}{2}c_2 = 0,$$

taking the form of a system of two linear equations with two unknowns. The solution is

$$c_1 = \tfrac{2}{3}, \qquad c_2 = -\tfrac{2}{3},$$

therefore

$$x_n = \tfrac{2}{3}[1 - (-\tfrac{1}{2})^n], \qquad n = 0, 1, 2, \dots.$$

The result permits to conclude, for instance, that

$$\lim_{n \to \infty} x_n = \frac{2}{3},$$

which result may also be observed experimentally.

**EXAMPLE 3.1-2:** *The Fibonacci sequence.* This sequence is defined by $x_0 := 0$, $x_1 := 1$,

$$x_n := x_{n-1} + x_{n-2}, \qquad n = 2, 3, \dots. \tag{13}$$

What is the $n$-th term of this sequence? The characteristic polynomial of the difference equation (13) is

$$p(z) = z^2 - z - 1;$$
$$z_1 := \tfrac{1}{2}(1 + \sqrt{5}), \qquad z_2 := \tfrac{1}{2}(1 - \sqrt{5}).$$

The general solution of (13) therefore is

$$x_n = c_1 [\tfrac{1}{2}(1 + \sqrt{5})]^n + c_2 [\tfrac{1}{2}(1 - \sqrt{5})]^n.$$

The initial conditions furnish the relations

$$x_0 = c_1 + c_2 = 0,$$
$$x_1 = c_1 [\tfrac{1}{2}(1 + \sqrt{5})] + c_2 [\tfrac{1}{2}(1 - \sqrt{5})] = 1,$$

yielding

$$c_1 = 1/\sqrt{5}, \qquad c_2 = -1/\sqrt{5}.$$

Thus, the $n$-th term of the Fibonacci sequence is given by the explicit formula,

$$x_n = \frac{1}{\sqrt{5}} \{[\tfrac{1}{2}(1 + \sqrt{5})]^n - [\tfrac{1}{2}(1 - \sqrt{5})]^n\},$$

as may be verified for the first few values of $n$.

**EXAMPLE 3.1-3:** To find the general solution of the difference equation

$$x_n - 4x_{n-1} + 6x_{n-2} - 4x_{n-3} + x_{n-4} = 0. \tag{14}$$

The characteristic polynomial is

$$p(z) = z^4 - 4z^3 + 6z^2 - 4z + 1 = (z - 1)^4;$$

it has the sole zero $z_1 = 1$, of multiplicity $\mu_1 = 4$. Hence, the general solution of (14) is

$$x_n = c_0 + c_1 n + c_2 n^2 + c_3 n^3.$$

More generally, the corresponding difference equation of order $k$,

$$\sum_{j=0}^{k} (-1)^j \binom{k}{j} x_{n-j} = 0,$$

has as its general solution the general polynomial of degree $k - 1$ in the variable $n$.

It remains to discuss the non-homogeneous equation

$$Lx = b, \tag{NH}$$

where $b$ is a given sequence. Let $y$ be a fixed special solution of (NH), for instance, a solution determined analytically by some special trick, or a solution determined numerically. If $x$ is any other solution of (NH), then the sequence $x - y$ satisfies

$$L(x - y) = Lx - Ly = b - b = 0$$

hence it is a solution of (H), and thus can be represented in the form (11). We conclude that the general solution of (NH) is obtained by superimposing onto the general solution of (H) a special solution of (NH).

Special solutions of (NH) may sometimes be found by a device that frequently works for homogeneous linear *differential* equations with constant coefficients: Try to find a solution that has the same general appearance as the non-homogeneous term.

**EXAMPLE 3.1-4:**  Let $b = \{b_n\} = \{u^n\}$, where $u$ is a real or complex number, $u \neq 0$. We try to find a special solution $y$ of (NH) of the form

$$y_n = cu^n,$$

where $c$ is to be determined. Substituting into (NH) yields

$$a_0 y_n + a_1 y_{n-1} + \cdots + a_k y_{n-k} = c\left(a_0 u^n + a_1 u^{n-1} + \cdots + a_k u^{n-k}\right) = u^n$$

or, denoting by $p$ the characteristic polynomial of (H),

$$cu^{n-k} p(u) = u^n,$$

and this is satisfied for all $n$ if $cu^{-k} p(u) = 1$. Thus, unless $u$ is a zero of the characteristic polynomial, our method works and yields

$$c = \frac{u^k}{p(u)}.$$

If $u$ is a zero of $p$ of multiplicity $\mu$, it will be found that (NH) has a special solution of the form

$$y_n = cn^\mu u^n.$$

# PROBLEMS

---

**1** Aitken's $\Delta^2$ method, if applied to the sequence in Example 3.1-1, yields the correct limit in one step. Verify and explain.

**2** Explain what happened in Demonstration 1.7-15 by solving the difference equation

$$\frac{1}{h^2}(y_{n+1} - 2y_n + y_{n-1}) = y_n$$

*analytically* for the starting values $y_0 = 1$, $y_{-1} = e^h$.

**3** Explain the failure of the method described in part (c) of Problem 3, §1.7, by solving the difference equation analytically.

**4** All solutions of the *differential* equation

$$y'' + ay' + by = 0$$

tend to zero for $x \to \infty$ if, and only if, $a > 0$, as well as $b > 0$. Find and prove an analogous theorem for the solutions of the *difference* equation

$$y_n + ay_{n-1} + by_{n-2} = 0.$$

[Treat separately the two cases where the zeros of the characteristic polynomial are real or imaginary.] Plot the result in the $(a, b)$-plane.

**5** Let $x_0$, $x_1$ be arbitrary positive numbers, and let

$$x_{n+1} := \tfrac{1}{2}(\sqrt{x_n} + \sqrt{x_{n-1}}), \qquad n = 1, 2, \dots. \tag{15}$$

    **(a)** Convince yourself by *experimentation* that the sequence $\{x_n\}$ has a limit $s$ that does not depend on the choice of the starting values.
    **(b)** Try to prove *mathematically* that $s = \lim_{n\to\infty} x_n$.
    **(c)** *How fast* do the $x_n$ tend to $s$? Let $e_n := x_n - s$, and by expanding (15) in powers of $e_n$ and retaining linear terms only, obtain a

linear difference equation for $e_n$. What is the best (that is, smallest) constant $q$ such that

$$|e_n| \le cq^n$$

for a suitable $c$ and for all $n$? Verify your answer experimentally.

6   We have repeatedly considered the recurrence relation

$$x_{k+1} := x_k \sqrt{\frac{2x_k}{x_k + x_{k-1}}}, \qquad k = 1, 2, \dots. \tag{16}$$

Is this recurrence numerically stable?

[Let $x_k := s(1 + \varepsilon_k)$, where $s := \lim x_k$, substitute in (16), expand in powers of the $\varepsilon$, and retain linear terms only. There results a linear difference equation for the sequence $\varepsilon_n$. Solve it, and draw your conclusions.]

7   Difference equations have remarkable applications in the theory of economics. Let $y_n$ be the national income in the year $n$, $a$ the marginal propensity to consume, and $b$ the ratio of private investment to increase in consumption. Assuming that government expenditure is constant and equal to 1, a certain economic theory states that the following difference equation holds:

$$y_n = ay_{n-1} + ab(y_{n-1} - y_{n-2}) + 1.$$

(a)   Solve this equation with $a := 0.5$, $b := 1$ under the initial condition $y_0 = 2$, $y_1 = 3$. How frequent are depressions in this economy?

(b)   For what values of $a$ and $b$ is the economy thus described stable?

8   The values of the functions $\cos n\phi$ and $\sin n\phi$ furnished by the calculator can be unreliable for large values of $n\phi$.

(a)   Verify this statement, and explain.

(b)   Show that more reliable (although not perfect) values of $\cos n\phi$ and $\sin n\phi$ can be generated by letting $x := \cos \phi$ and solving the recurrence relation

$$y_n = 2xy_{n-1} - y_{n-2} \tag{17}$$

with appropriate starting values $y_0$ and $y_1$. How does one have to choose the starting values to obtain $y_n = \cos n\phi$? How to obtain $y_n = \sin n\phi$?

(c)   Letting $\phi = \pi/3$, compute the numbers $\cos n\phi$ both directly and by means of the recurrence relation and make a statistical analysis of the errors in both cases.

**9** *Chebyshev polynomials.* The solution $y_n$ of the recurrence relation (17) computed with the starting values $y_0 = 1$, $y_1 = x$ is a polynomial of degree $n$, commonly called the $n$-th Chebyshev polynomial and denoted by $T_n(x)$.

    **(a)** Calculate the zeros of $T_n(x)$, and show that they are all real and located in the interval $(-1, 1)$.

    **(b)** Show that the relative extrema of $T_n(x)$ all have the same absolute value.

    **(c)** Sketch the graph of the functions $T_0$, $T_1$, ..., $T_6$ in the interval $[-1, 1]$.

    **(d)** Show that for $n = 1, 2, \ldots$, the leading coefficient of $T_n(x)$ is $2^{n-1}$.

**10** *Chebyshev polynomials of the second kind.* If the recurrence relation (17) is solved under the initial conditions $y_0 = 0$, $y_1 = 1$, then $y_n$ turns out to be a polynomial of degree $n - 1$, which is commonly called the $n$-th Chebyshev polynomial of the second kind and denoted by $U_n(x)$.

    **(a)** Show that for $-1 < x < 1$,

$$U_n(x) = \frac{\sin n\phi}{\sin \phi}, \qquad \text{where} \quad \phi := \arccos x.$$

    **(b)** Calculate the $n - 1$ zeros of $U_n(x)$.

    **(c)** Show that $U_n$ has the parity of $n - 1$.

    **(d)** Show that

$$U_n(1) = n, \qquad n = 0, 1, \ldots .$$

    **(e)** By differentiating the recurrence relation with respect to $x$, and solving the resulting non-homogeneous relation, show that

$$U_n'(1) = \tfrac{1}{3}(n^3 - n), \qquad n = 0, 1, 2, \ldots .$$

**11** In a certain problem on binary trees (see Amer. Math. Monthly *85*, p. 827), numbers $w_k$ are studied that satisfy the difference equation

$$w_{k+1} - w_k - 2w_{k-1} = 2^k + 4.$$

Solve the difference equation under the initial condition $w_1 = -1$, $w_2 = 2$, and verify that the analytical solution represents an integer for $k \geq 1$.

## §3.2 BERNOULLI'S METHOD

Here we return to the problem of finding zeros of polynomials. The methods which we have discussed so far—Newton's method, Bairstow's method—can be very effective if close approximations to the zeros are already known. If no such approximations are known, then, as shown in Demonstration 2.5-2, their behavior can be subject to vagaries. If the zeros are real, crude approximations can be found by the method of bisection. No simple analog of this method is available for complex zeros, however.

In this section we shall exploit the connection between the solution of a linear difference equation (H) and the zeros of its characteristic polynomial to obtain a method to determine zeros of a polynomial without knowing crude first approximations, even if these zeros are complex.

Let

$$p(z) = a_0 z^k + a_1 z^{k-1} + \cdots + a_k$$

be a polynomial of degree $k \geq 1$, $a_0 a_k \neq 0$, and let

$$a_0 x_n + a_1 x_{n-1} + \cdots + a_k x_{n-k} = 0 \tag{H}$$

be the difference equation which has $p$ as its characteristic polynomial. We have seen that given any starting values $x_0, x_1, \ldots, x_{k-1}$, the corresponding solution of (H) can always be found numerically by the recurrence relation,

$$x_n := -\frac{1}{a_0}(a_1 x_{n-1} + a_2 x_{n-2} + \cdots + a_k x_{n-k}),$$

$n = k, k + 1, k + 2, \ldots$. The question now is: Is it possible to extract, from a solution that is constructed numerically, information on the zeros of $p$?

A satisfactory answer can certainly be given if the degree of $p$ equals 1,

$$p(z) = a_0 z + a_1.$$

Here the difference equation is

$$a_0 x_n + a_1 x_{n-1} = 0,$$

and the solution with starting value $x_0$ is given by

$$x_n = z_1^n x_0,$$

where

$$z_1 = -\frac{a_1}{a_0}.$$

is the zero of $p$. To recover $z_1$ from the sequence $x_n$, we therefore merely need to form the quotients

$$q_n := \frac{x_{n+1}}{x_n}, \qquad n = 0, 1, 2, \ldots . \tag{1}$$

We then have

$$z_1 = q_n$$

for every $n$.

Let us, in the sense of an experiment, form the same quotients when the degree $k$ of $p$ is $>1$. To keep the discussion simple, we make

**ASSUMPTION ($A_1$):**   *The zeros $z_1, z_2, \ldots, z_k$ of $p$ all have multiplicity 1.*

We then know that any solution of (H) can be expressed in the form

$$x_n = c_1 z_1^n + c_2 z_2^n + \cdots + c_k z_k^n. \tag{2}$$

The values of the constants $c_j$ can be computed if the zeros $z_j$ are known. Since these zeros are unknown, the $c_j$ are also unknown. Nevertheless, the quotients

$$q_n = \frac{x_{n+1}}{x_n}$$

in view of (2) are analytically represented by

$$q_n = \frac{c_1 z_1^{n+1} + c_2 z_2^{n+1} + \cdots + c_k z_k^{n+1}}{c_1 z_1^n + c_2 z_2^n + \cdots + c_k z_k^n}. \tag{3}$$

In order to discuss the behavior of the $q_n$, we make two further assumptions.

For the first assumption we require the notion of a dominant zero. If a polynomial $p$ of degree $k$ has the zeros $z_1, z_2, \ldots, z_k$, not necessarily distinct, then the zero $z_j$ is called **dominant**, if its modulus is exceeded by the modulus of no other zero; that is, if

$$|z_j| = \max_{1 \leq i \leq k} |z_i|.$$

A polynomial can have several dominant zeros, of course; examples are $p(z) = z^2 + 2z + 1$ or $p(z) = z^{297} - 5$. However, we now postulate

**ASSUMPTION ($A_2$):**   *The polynomial $p$ has a single dominant zero.*

The zeros of a polynomial satisfying Assumption ($A_2$) will always be numbered so that $z_1$ is the dominant zero. If the coefficients of $p$ are real, then $z_1$ will by necessity be real.

**ASSUMPTION ($A_3$):**   *The starting values $x_0$, $x_1$, ..., $x_{k-1}$ of the solution $\{x_n\}$ of (H) are chosen so that $c_1 \neq 0$.*

It seems difficult to verify this assumption without knowing the zeros. However, it may be argued that, in a probabilistic sense, Assumption ($A_3$) will be satisfied for almost all choices of the starting values.

Under Assumption ($A_3$), the expression (3) may be written

$$q_n = z_1 \frac{1 + \dfrac{c_2}{c_1}\left(\dfrac{z_2}{z_1}\right)^{n+1} + \cdots + \dfrac{c_k}{c_1}\left(\dfrac{z_k}{z_1}\right)^{n+1}}{1 + \dfrac{c_2}{c_1}\left(\dfrac{z_2}{z_1}\right)^{n} + \cdots + \dfrac{c_k}{c_1}\left(\dfrac{z_k}{z_1}\right)^{n}}. \tag{4}$$

Under Assumption ($A_2$), as $n \to \infty$,

$$\left(\frac{z_j}{z_1}\right)^n \to 0 \qquad \text{for} \quad j = 2, 3, \ldots, k,$$

and it follows that

$$\lim_{n \to \infty} q_n = z_1. \tag{5}$$

Thus, we have found the following *preliminary version* of what is known as **Bernoulli's method**:

To find the dominant zero of a polynomial

$$p(z) = a_0 z^k + a_1 z^{k-1} + \cdots + a_k$$

satisfying Assumption ($A_2$), construct a numerical solution $\{x_n\}$ of the difference equation (H) which has $p$ as its characteristic polynomial, and simultaneously form the quotients

$$q_n := \frac{x_{n+1}}{x_n}.$$

Our analysis has shown that under Assumptions ($A_1$) and ($A_3$) the $q_n$ are defined, at least if $n$ is sufficiently large, that they tend to the dominant zero of $p$, and that the convergence is geometric.

It is easily seen that Assumption ($A_1$) is not essential. Under Assumption ($A_2$), the dominant zero is simple in any case. If one of the nondominant zeros has a multiplicity $\mu > 1$, then numerator and denominator in the expression (4) will contain terms like

$$\frac{c_j}{c_1} n^{\mu - 1} \left(\frac{z_j}{z_1}\right)^n.$$

In view of $|z_j/z_1| < 1$, these terms still tend to zero, and the convergence statement (5) still holds. If $p$ has $\mu > 1$ dominant zeros which all coincide (such as $p(z) = (z - 2)^2(z + 1)$), then $q_n$ will be of the form

$$q_n = \frac{c_1(n + 1)^{\mu-1} z_1^{n+1} + \cdots}{c_1 n^{\mu-1} z_1^n + \cdots}$$

$$\sim z_1\left(1 + \frac{\mu - 1}{n}\right).$$

The $q_n$ will still converge to $z_1$, but the speed of convergence will be intolerably slow in view of the fact that the error now is $O(1/n)$ in place of $O((z_2/z_1)^n)$.

### Bernoulli Algorithm, Streamlined Version

Before trying to soften the Assumptions $(A_2)$ and $(A_3)$, we reformulate Bernoulli's algorithm to make it computationally more efficient. Evidently, the $x_n$ in themselves are of no interest. The algorithm should be formulated in such a manner that it generates directly the sequence $\{q_n\}$. This is easily possible as follows: Writing the basic recurrence relation as

$$x_{n+k+1} = -\frac{1}{a_0}(a_1 x_{n+k} + a_2 x_{n+k-1} + \cdots + a_k x_{n+1})$$

and dividing it by $x_{n+k}$, there follows

$$q_{n+k} = -\frac{1}{a_0}\left(a_1 + a_2\frac{x_{n+k-1}}{x_{n+k}} + \cdots + a_k\frac{x_{n+1}}{x_{n+k}}\right).$$

In view of

$$\frac{x_{n+k}}{x_{n+j}} = \frac{x_{n+k}}{x_{n+k-1}}\frac{x_{n+k-1}}{x_{n+k-2}} \cdots \frac{x_{n+j+1}}{x_{n+j}} = q_{n+k-1} \cdots q_{n+j}$$

$(j = 1, 2, \ldots, k - 1)$, this is the same as

$$q_{n+k} = -\frac{1}{a_0}\left(a_1 + a_2\frac{1}{q_{n+k-1}} + a_3\frac{1}{q_{n+k-1}q_{n+k-2}} + \cdots \right.$$

$$\left. + a_k\frac{1}{q_{n+k-1} \cdots q_{n+1}}\right).$$

At first sight, this expression seems awkward to evaluate, but by writing it as

$$q_{n+k} = -\frac{1}{a_0}\left(a_1 + \frac{1}{q_{n+k-1}}\left(a_2 + \frac{1}{q_{n+k-2}}\left(a_3 + \cdots \right.\right.\right.$$
$$\left.\left.\left. + \frac{1}{q_{n+2}}\left(a_{k-1} + \frac{a_k}{q_{n+1}}\right)\cdots\right)\right)\right),$$

we see that it may be evaluated in a Horner-like fashion in merely $k\delta + (k-1)\alpha$. Since there is no need to store the whole sequence $\{q_n\}$, the formula to be evaluated in the algorithm is

$$q_k := -\frac{1}{a_0}\left(a_1 + \frac{1}{q_{k-1}}\left(a_2 + \frac{1}{q_{k-2}}\left(a_3 + \cdots + \frac{1}{q_2}\left(a_{k-1} + \frac{1}{q_1}a_k\right)\cdots\right)\right)\right);$$

(6a)

as soon as $q_k$ is computed, we make the shift

$$q_j := q_{j+1}, \qquad j = 1, 2, \ldots, k-1. \tag{6b}$$

**Two Dominant Zeros**

Dropping Assumption $(A_2)$, we now turn our attention to polynomials $p$ with several dominant zeros. The problem of determining all zeros of a polynomial from a solution $\{x_n\}$ of the corresponding difference equation is easily solved if the polynomial is known to have degree 2, even if the coefficients of that polynomial are unknown. Assuming the polynomial in the form

$$p(z) = z^2 - bz + c,$$

the corresponding difference equation is

$$x_{n+2} - bx_{n+1} + cx_n = 0. \tag{$H_2$}$$

Considering that $(H_2)$ holds for any two consecutive indices $n$, we obtain a system of two linear equations for the two unknowns $b$ and $c$, which we may write as

$$bx_{n+1} - cx_n = x_{n+2},$$
$$bx_{n+2} - cx_{n+1} = x_{n+3}. \tag{7}$$

Clearly, this system may be solved for $a$, $b$ if the determinant

$x_n x_{n+2} - x_{n+1}^2 \neq 0$. Assuming this to be the case, we express the solution in terms of the quantities

$$q_j := \frac{x_{n+j+1}}{x_{n+j}}, \qquad j = 0, 1, 2,$$

that we presume to be already available. If the auxiliary quotient

$$q_1' := \frac{q_2 - q_1}{q_1 - q_0} q_1 \tag{8}$$

is introduced, a short calculation shows that the solution of (7) is given by the formulas

$$b = q_1 + q_1', \qquad c = q_0 q_1'. \tag{9}$$

Once the coefficients $b$ and $c$ are known, the zeros of $p$ may be found by the usual formulas for solving a quadratic.

Now let the polynomial $p$ have degree $k > 2$, but let it have precisely two dominant zeros, so that

$$|z_1| = |z_2| > |z_j|, \qquad j = 3, 4, \ldots, k.$$

The solution of the corresponding difference equation then has the form

$$x_n = c_1 z_1^n + c_2 z_2^n + O(\theta^n |z_1|^n),$$

where $0 < \theta < 1$, and the resulting quotients

$$q_j := \frac{x_{n+j+1}}{x_{n+j}}$$

are

$$q_j = \frac{c_1 z_1^{n+j+1} + c_2 z_2^{n+j+1} + O([\theta|z_1|]^{n+j+1})}{c_1 z_1^{n+j} + c_2 z_2^{n+j} + O([\theta|z_1|]^{n+j})}.$$

In view of $0 < \theta < 1$, the $O$-terms become small compared to the leading terms if $n$ is sufficiently large. Sharpening Assumption $(A_3)$ to the assumption that both $c_1$ and $c_2$ are different from zero, we see that for large values of $n$ the $q_n$ behave like they would for a polynomial of degree 2 with zeros $z_1$ and $z_2$. As in the case of a single dominant zero, we assume that the quantities $q_k, q_{k-1}, \ldots, q_1$ are available. If for each $n$ we then form

$$b_n := q_{k-1} + q_{k-1}', \qquad c_n := q_{k-2} q_{k-1}', \tag{10}$$

where

$$q'_{k-1} := \frac{q_k - q_{k-1}}{q_{k-1} - q_{k-2}} q_{k-1},$$

then the $b_n$ and the $c_n$ tend to limits

$$b := \lim_{n \to \infty} b_n, \qquad c := \lim_{n \to \infty} c_n \qquad (11)$$

with geometric convergence, and the zeros $z_1$ and $z_2$ may be found as the solutions of the quadratic equation

$$z^2 - bz + c = 0. \qquad (12)$$

Using Cardano's formulas (see Uspensky [1948]), it would also be possible to deal in a similar manner with polynomials that have three dominant zeros, but the resulting formulas are too unwieldy for practical use.

### Balanced Starting Values

It seems unsatisfactory that the success of Bernoulli's method as described so far seems to depend on Assumption $(A_3)$ which contains an element of chance. Even if this assumption is satisfied, it could happen that $c_1$ is very small in comparison with some of the other $c_j$, in which case the convergence of the sequence $\{q_n\}$ would be delayed considerably. It is therefore a remarkable fact that a simple algorithm exists which furnishes starting values $x_0, \ldots, x_{k-1}$ with the following desirable properties:

(i)   In the representation (2) of $x_n$, all $c_j$ are $\neq 0$; in fact,

$$c_j = 1, \qquad j = 1, 2, \ldots, k.$$

(ii)   Even if $p$ has zeros of multiplicity $\mu > 1$, no powers of $n$ appear in the representation (4). Thus, both Assumptions $(A_1)$ and $(A_3)$ can be dispensed with entirely.

Let

$$p(z) := a_0 z^k + a_1 z^{k-1} + \cdots + a_k$$

be the polynomial whose zeros we wish to determine, $a_0 a_k \neq 0$, and let its zeros (repeated or not) be numbered so that

$$|z_1| \geq |z_2| \geq |z_3| \geq \cdots \geq |z_k| > 0.$$

Along with $p$, we consider the **reciprocal polynomial**

$$\hat{p}(z) := a_k z^k + a_{k-1} z^{k-1} + \cdots + a_0$$

with the same coefficients, but in reverse order. It is easy to see that the zeros in $\hat{p}$ are $z_j^{-1}, j = 1, \ldots, k$; indeed,

$$\hat{p}(z_j^{-1}) = a_k z_j^{-k} + a_{k-1} z_j^{-k+1} + \cdots + a_0$$
$$= z_j^{-k}(a_k + a_{k-1} z_j + \cdots + a_0 z_j^k)$$
$$= z_j^{-k} p(z_j) = 0.$$

We now consider the function

$$f(z) := -\frac{\hat{p}'(z)}{\hat{p}(z)}.$$

Because $\hat{p}(0) = a_0 \neq 0$, $f$ is analytic at $z = 0$. Let its Taylor series at $z = 0$ be

$$f(z) = \sum_{n=0}^{\infty} x_{n+1} z^n;$$

the notation for the coefficients is chosen judiciously. We shall compute the Taylor coefficients $x_n$ in two different ways. On the one hand, $-f$ is the logarithmic derivative of

$$\hat{p}(z) = \text{const} \cdot (z - z_1^{-1}) \cdots (z - z_k^{-1});$$

therefore

$$f(z) = -\sum_{j=1}^{k} \frac{1}{z - z_j^{-1}} = \sum_{j=1}^{k} \frac{z_j}{1 - z z_j}.$$

Using the geometric series,

$$\frac{z_j}{1 - z z_j} = z_j + z_j^2 z + z_j^3 z^2 + \cdots$$
$$= \sum_{n=0}^{\infty} z_j^{n+1} z^n,$$

and there follows

$$x_n = \sum_{j=1}^{k} z_j^n, \qquad n = 1, 2, \ldots. \tag{13}$$

On the other hand, the $x_n$ can be determined by comparing coefficients in the identity

$$\hat{p}(z) f(z) = -\hat{p}'(z);$$

that is, in

$$(a_0 + a_1 z + a_2 z^2 + \cdots + a_k z^k)(x_1 + x_2 z + x_3 z^2 + \cdots)$$
$$= -a_1 - 2a_2 z - 3a_3 z^2 - \cdots - ka_k z^{k-1}.$$

This yields

$$\left. \begin{array}{ll} a_0 x_1 & = -a_1 \\[2mm] a_0 x_2 + a_1 x_1 & = -2a_2 \\[2mm] a_0 x_3 + a_1 x_2 + a_2 x_1 & = -3a_3 \\[2mm] \qquad\qquad \cdot\quad\cdot\quad\cdot & \\[2mm] a_0 x_k + a_1 x_{k-1} + \cdots + a_{k-1} x_1 & = -ka_k \end{array} \right\} \qquad (14)$$

For $n > k$, we get

$$a_0 x_n + a_1 x_{n-1} + \cdots + a_k x_{n-k} = 0.$$

We see that the sequence $\{x_n\}$ is a solution of our difference equation (H), whose starting values are chosen to satisfy (14). In other words, if the starting values are chosen according to (14), then, because of (13), a solution of (H) results for which the representation (2) holds with all $c_j = 1$. This is true, even if some $z_j$ occur repeatedly.

For the streamlined version of the Bernoulli method, the starting formulas should also be arranged to produce directly the $q_j$. Dividing the $j$-th relation (14),

$$a_0 x_{j+1} = -a_1 x_j - a_2 x_{j-1} - \cdots - a_j x_1 - (j+1)a_{j+1},$$

by $x_j$ $(j = 1, 2, \ldots, k-1)$, we get

$$a_0 \frac{x_{j+1}}{x_j} = -a_1 - a_2 \frac{x_{j-1}}{x_j} - \cdots - a_j \frac{x_1}{x_j} - (j+1)a_{j+1} \frac{1}{x_j}.$$

Using $x_1 = -a_1/a_0$ and

$$x_j = x_1 q_1 q_2 \cdots q_{j-1} = -\frac{a_1}{a_0} q_1 q_2 \cdots q_{j-1},$$

this becomes

$$q_j = -\frac{1}{a_0}\left(a_1 + \frac{1}{q_{j-1}} a_2 + \frac{1}{q_{j-1}q_{j-2}} a_3 + \cdots + \frac{1}{q_{j-1} \cdots q_1} a_j\right.$$
$$\left. -\frac{j+1}{q_{j-1} \cdots q_1} \frac{a_0}{a_1} a_{j+1}\right).$$

This may be evaluated as

$$q_j = -\frac{1}{a_0}\left(a_1 + \frac{1}{q_{j-1}}\left(a_2 + \frac{1}{q_{j-2}}\left(a_3 + \cdots \right. \right. \right.$$
$$\left. \left. \left. + \frac{1}{q_1}\left(a_j - \frac{(j+1)a_0}{a_1} a_{j+1}\right) \cdots\right)\right)\right), \tag{15}$$

$j = 1, 2, \ldots, k - 1$, which becomes a special case of (6a) after proper identifications.

## Small Zeros Determined First

Bernoulli's method as described above finds the dominant zero, or the pair of dominant zeros, of the given polynomial $p$. If all zeros of $p$ are desired, one will use the zero or the quadratic factor found to *deflate* $p$, as shown in the §§2.3 and 2.5. Bernoulli's method will then be used again on the deflated polynomial. Thus, the zeros will be found in decreasing order of magnitude. It has already been pointed out that, by proceeding in this manner, the problem of error propagation in the successive deflations can be severe, and that from the point of view of numerical stability it would be better to first determine the zeros of smallest modulus.

It is easy to modify Bernoulli's method to yield the zero (or zeros) of smallest modulus. We already have made use of the fact that if the given polynomial

$$p(z) = a_0 z^k + a_1 z^{k-1} + \cdots + a_k$$

has the zeros $z_1, z_2, \ldots, z_k$, numbered so that

$$|z_1| \geq |z_2| \geq \cdots \geq |z_k| > 0,$$

then the reciprocal polynomial

$$\hat{p}(z) := a_k z^k + a_{k-1} z^{k-1} + \cdots + a_0 \tag{16}$$

has the zeros

$$\hat{z}_j := (z_j)^{-1}, \qquad j = 1, 2, \ldots, k.$$

Thus, all we have to do is to apply Bernoulli's method to $\hat{p}$ and to form the reciprocal quotients

$$\hat{q}_n := \frac{x_n}{x_{n+1}}.$$

If $\hat{p}$ has a single dominant zero, that is, if $p$ has a single zero of smallest modulus, then the sequence of the $\hat{q}_n$ will generally tend to that zero,

$$\lim_{n \to \infty} \hat{q}_n = z_k$$

with geometric convergence. By adapting (6), the $\hat{q}_n$ may be found directly from

$$\hat{q}_k = -\frac{a_k}{a_{k-1} + \hat{q}_{k-1}(a_{k-2} + \hat{q}_{k-2} + \hat{q}_{k-2}(a_{k-3} + \cdots + \hat{q}_1 a_0))},$$
$$\hat{q}_j := \hat{q}_{j+1}, \quad j = 1, 2, \ldots, k-1. \tag{17}$$

The method for finding quadratic factors also may be adapted. If $p$ has precisely two zeros of smallest modulus, these will correspond to a pair of dominant zeros $\hat{z}_k = z_k^{-1}$ and $\hat{z}_{k-1} = z_{k-1}^{-1}$ of $\hat{p}$. These zeros may be found as solutions of a quadratic equation

$$z^2 - bz + c = 0.$$

Consequently, $z_k$ and $z_{k-1}$ are the solutions of a quadratic equation

$$z^2 - \hat{b}z + \hat{c} = 0, \tag{18}$$

where $\hat{b} := b/c$, $\hat{c} := 1/c$. The coefficients $\hat{b}$ and $\hat{c}$ can be found in terms of the reciprocal quotients $\hat{q}$ as follows: We have

$$\hat{b} = \lim_{n \to \infty} \hat{b}_n, \qquad \hat{c} = \lim_{n \to \infty} \hat{c}_n, \tag{19}$$

where, writing $\hat{q}_j$ for $\hat{q}_{n+j}$ everywhere,

$$\hat{b}_n := \hat{q}_{k-1} + \hat{q}'_{k-1}, \qquad \hat{c}_n := \hat{q}_k \hat{q}'_{k-1}, \tag{20}$$

and

$$\hat{q}'_{k-1} := \frac{\hat{q}_{k-1} - \hat{q}_{k-2}}{\hat{q}_k - \hat{q}_{k-1}} \hat{q}_{k-1}. \tag{21}$$

Finally, balanced starting values are obtained from the formula

$$\hat{q}_j :=$$
$$-\frac{a_k}{a_{k-1} + \hat{q}_{j-1}(a_{k-2} + \hat{q}_{j-2}(a_{k-3} + \cdots + \hat{q}_1(a_{k-j} - (j+1)a_k a_{k-j-1}/a_{k-1}) \cdots))} \tag{22}$$

which results by adapting (15).

The method thus described for finding the zeros of smallest modulus will be called the **modified Bernoulli method**.

## Flow Diagram

The flow diagram for Bernoulli's method is, in principle, simple. Figure 3.2 shows the flow diagram for the modified Bernoulli method as described above.

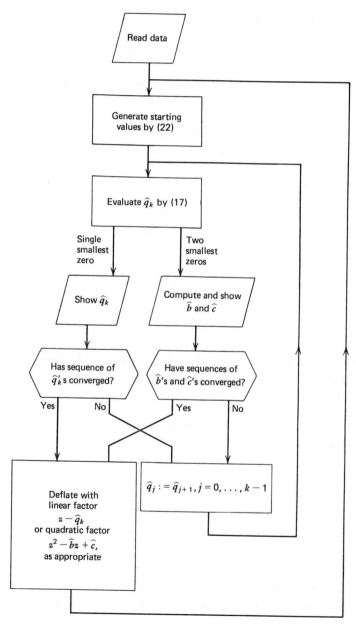

**Fig. 3.2.** Flow diagram for modified Bernoulli method.

**Demonstrations**

The following examples illustrate the modified Bernoulli method. Of course, analogous examples could be made up for the ordinary method.

**DEMONSTRATION 3.2-1:**   Let

$$p(z) := z^4 - 16z^3 + 72z^2 - 96z + 24.$$

(See Demonstration 2.3-5.) The modified algorithm yields the starting values

$$q_0 = 0.40000000$$
$$q_1 = 0.33333333$$
$$q_2 = 0.32432432$$

The further $q$-values are

$$0.32286213$$
$$0.32260486$$
$$0.32255819$$
$$0.32254963$$
$$0.32254805$$
$$0.32254776$$
$$0.32254770$$
$$\to z_1 = 0.32254769$$

Convergence to the smallest zero is relatively fast due to the favorable ratio $z_1/z_2$. Deflation yields a polynomial which is identical with the first deflated polynomial in Demonstration 2.3-5. Starting again yields

$$2.31855532$$
$$1.94146602$$

At $n = 17$, we get

$$1.74576119$$
$$1.74576113$$
$$1.74576111$$
$$\to z_2 = 1.74576110,$$

again in agreement with the zero obtained by Newton. Restarting with the deflated polynomial of degree 2 yields the sole starting value

$$5.45525124,$$

and at $n = 22$

$$4.53662054$$
$$4.53662041$$
$$4.53662034$$
$$4.53662031$$
$$4.53662029$$
$$\to z_3 = 4.53662029$$

A last deflation yields the linear polynomial $z - 9.39507092$, where the starting algorithm calculates the correct zero

$$z_4 = 9.39507092.$$

**DEMONSTRATION 3.2-2:** Here we consider the contrived polynomial

$$p(z) = (z - 1)^3(z + 2)$$
$$= z^4 - z^3 - 3z^2 + 5z - 2$$

with the triple zero $z = 1$. If Bernoulli's method is started with random values $q_0$, $q_1$, $q_2$, we get a sequence which, although theoretically convergent, converges in fact very slowly. After about 90 iterations, we still only get

$$0.98020917$$
$$0.98040307$$
$$0.98059322$$
$$0.98077970$$
$$\cdot \quad \cdot \quad \cdot$$

Starting with balanced starting values on the other hand furnishes the sequence

$$0.76923077$$
$$1.13043478$$
$$0.93877551$$
$$1.03157895$$
$$0.98445596$$

(thus, $q_5$ is better than the $q_{90}$ obtained with unbalanced starting values!),

$$1.00783290$$
$$0.99609883$$
$$1.00195440$$
$$0.99902376$$
$$1.00048837$$
$$0.99975588$$
$$\cdot \quad \cdot \quad \cdot$$

The convergence being geometric, Aitken acceleration is applicable. Applied to the last three values, it yields 1.00000008.

**DEMONSTRATION 3.2-3:**   Let

$$p(z) = z^4 - 8z^3 + 40z^2 - 68z + 74.$$

Using balanced starting values and letting Bernoulli's method run in the ordinary mode yields a $q$-sequence which shows no indication of convergence. We thus try for a quadratic factor and find

| $b_n$ | $c_n$ |
|------------|------------|
| 2.06221561 | 3.06595688 |
| 2.06318883 | 2.99924683 |
| 2.03340618 | 2.99798172 |
| 2.01379686 | 2.96845360 |
| 2.01262371 | 2.96159804 |

After 10 more iterations, the results are

| | |
|------------|------------|
| 2.01120345 | 2.96062979 |
| 2.01120345 | 2.96062980 |

Thus, the quadratic factor formed with the two zeros of smallest modulus is $z^2 - 2.01120345z + 2.96062980$, and the zeros are $z_{1,2} = 1.00560173 \pm i1.39620735$. A deflation yields the quadratic polynomial $z^2 - 5.98879655z + 24.99468192$ with the zeros $z_{3,4} = 2.99439828 \pm i4.00353105$.

**DEMONSTRATION 3.2-4$^\oplus$:**   For the polynomial,

$$p(z) = z^6 - 3z^5 + 9z^4 - 13z^3 + 18z^2 - 12z + 8$$
$$= (z^2 - z + 2)^3$$

Bernoulli's method with random starting values converges extremely slowly due to the multiple quadratic factor. Using balanced starting values and aiming directly at the determination of quadratic factors, we get

$$b_6 = 1.00000000, \qquad c_6 = 2.00000001.$$

Thus, the correct factor is obtained (up to rounding errors) at the first step.

**DEMONSTRATION 3.2-5$^\oplus$:**   The polynomial

$$p(z) = z^8 - 8z^7 + 28z^6 - 56z^5 + 70z^4 - 56z^3 + 28z^2 - 8z + 2$$
$$= (z - 1)^8 + 1$$

has the exact quadratic factors $z^2 - bz + c$, where $b = c = 4(\sin(2m - 1)\pi/16)^2$, $m = 1, 2, 3, 4$. The modified Bernoulli method with balanced starting values yields, always iterating to convergence and deflating,

| $m$ | $b$ | $c$ | $b = c$ (exact) |
|---|---|---|---|
| 1 | 0.15224097 | 0.15224094 | 0.15224094 |
| 2 | 1.23463318 | 1.23463312 | 1.23463314 |
| 3 | 2.76536684 | 2.76536673 | 2.76536686 |
| 4 | 3.84775902 | 3.84775903 | 3.84775907 |

After having acquired numerical experience with both methods, how would we summarize the balance sheets for the Newton-type methods (including Bairstow's) on the one hand and the Bernoulli-type methods on the other? The Newton-type methods have the clear advantage of ultimate quadratic convergence in the case of simple zeros. This is somewhat offset by the fact that the initial behavior of the approximating sequences may be erratic if no good first approximations are known, which may be a problem, especially for complex zeros of real polynomials. The quadratic convergence is disturbed and instabilities occur in the determination of multiple zeros. Bernoulli's method, on the other hand, converges only linearly, and even as linear convergence goes, the convergence may be slow if more than two zeros have nearly equal moduli. On the other hand, multiple zeros present no problem, and approximations to complex as well as to real zeros are obtained.

Another difference between the two methods that should be noted is that Bernoulli's method depends very much on the representation of the polynomial as a Taylor series. The method cannot be used unless the coefficients $a_0, a_1, \ldots, a_k$ of the polynomial are known. Newton's method, on the other hand, merely requires that the values $p(z)$ and $p'(z)$ can be found for any $z$. How the polynomial $p$ is represented does not matter. In this sense, Newton's method is more general than Bernoulli's. It can be used in situations where the definition of $p$ is highly implicit, which may be an advantage, for example, in the determination of eigenvalues of a matrix if the matrix is available in Hessenberg form.

# PROBLEMS

1  The polynomial

$$p(z) = 5z^4 + 4z^3 + 3z^2 + 2z + 1$$

has four zeros of approximately equal modulus.

(a)  Convince yourself by experimentation that the convergence of Bernoulli's method is intolerably slow in this case.

(b)  Show how Bernoulli's method can nevertheless be used to determine the zeros by expanding $p$ in powers of $h$ where $z = -1 + h$. Use Horner's algorithm to obtain the Taylor expansion of $p$ at $z = -1$.

2  Let $p$ have a single dominant zero of multiplicity 1. Show that the sequence of quotients $\{q_n\}$ obtained in Bernoulli's method may be speeded up by the Aitken $\Delta^2$ method. Experiment with the polynomial of Demonstration 3.2-1.

3  Use Bernoulli's method, combined with convergence acceleration, to determine the largest zero of

$$p(x) = 70x^4 - 140x^3 + 90x^2 - 20x + 1.$$

4  The polynomial

$$p(x) = x^4 - 6x^3 + 23x^2 - 50x + 50$$

has two pairs of complex conjugate zeros. Determine them all, using Bernoulli's method.

5  Discuss the convergence of the streamlined form of Bernoulli's method if (6a) is changed to

$$q_k = -\frac{1}{a_0}\left(a_1 + \frac{1}{q_{k-1}}\left(a_2 + \frac{1}{q_{k-1}}\left(a_3 + \cdots \right.\right.\right.$$
$$\left.\left.\left. + \frac{1}{q_{k-1}}\left(a_{k-1} + \frac{1}{q_{k-1}}a_k\right)\cdots\right)\right)\right).$$

Compare Problem 5 of §2.1.

6  If the polynomial $p$ has a dominant zero $z_1$ of multiplicity $>1$, the convergence of the sequence $\{q_n\}$ obeys a law of the form

$$q_n = z_1 + \frac{a}{n} + O(n^{-2}),$$

where $a$ is unknown. Devise an Aitken-like acceleration algorithm appropriate for this situation, and experiment with the polynomial of Demonstration 3.2-2.

7  Let $p$ be a real polynomial with precisely two dominant zeros

$$z_1 = re^{i\phi}, \qquad z_2 = re^{-i\phi},$$

where $0 < \phi < \pi$.

(a) Show that under Assumption $(A_3)$, the $x_n$ for large $n$ behave like

$$x_n = c r^n \cos(n\phi + \alpha) + \varepsilon_n,$$

where $c$ and $\alpha$ are constants, and $|\varepsilon_n| < (\theta r)^n$, where $0 < \theta < 1$.

(b) Assuming no $x_n$ to be zero, let $\sigma_n$ be the number of sign changes in the finite sequence $\{x_0, x_1, x_2, \ldots, x_n\}$. Show that

$$\lim \frac{\sigma_n}{n} = \frac{\phi}{\pi}.$$

(c) State an analogous result for the quotients $q_n$. [For which $n$ is $q_n$ negative?]

(d) By merely inspecting the signs of the quotients $q_n$, indicate the sectors of the complex plane in which the two dominant zeros of the polynomial

$$p(z) = z^4 + 2z^3 + 4z^2 - 2z - 5$$

are located.

**8** As we have seen in Demonstration 2.5-1, the circle

$$x^2 + y^2 - x = 0$$

and the hyperbola

$$x^2 - y^2 - y = 0$$

intersect at $(0, 0)$, and at one other point. Determine this point by eliminating $y$ from the above equations and solving the resulting equation for $x$ by the Bernoulli method.

---

## §3.3 THE QUOTIENT-DIFFERENCE ALGORITHM

Bernoulli's method furnishes only one or two zeros of a given polynomial at a time, and these zeros, by necessity, are the ones of largest or smallest absolute value. If a zero of intermediate modulus is desired, it is necessary first to compute all larger (or all smaller) zeros, and then to remove them from the polynomial by deflation. We now shall discuss a modern extension of the Bernoulli method, due to Rutishauser*, which has the advantage of providing simultaneous approximations to *all* zeros. Since the prerequisites for this volume do not include complex function theory, we are unable to provide proofs for the convergence theorems in this section. Even though its theoretical background cannot be fully exposed, however, we feel that Rutishauser's algorithm is of sufficient interest to warrant presentation at this point.

### The Quotient-Difference Scheme

The quotient-difference $(qd)$ algorithm can be looked at as a generalization of Bernoulli's method. As in §3.2, we are given a polynomial

$$p(z) = a_0 z^m + a_1 z^{m-1} + \cdots + a_m \tag{1}$$

and form a solution of the associated difference equation

$$a_0 x_n + a_1 x_{n-1} + \cdots + a_m x_{n-m} = 0. \tag{2}$$

The sequence $\{x_n\}$ may be started by using the balanced starting values discussed in §3.2, or more simply by

$$x_{-m+1} = x_{-m+2} = \cdots = x_{-1} = 0, \qquad x_0 = 1. \tag{3}$$

In §3.2 we now formed the quotients

$$q_n := \frac{x_{n+1}}{x_n}. \tag{4}$$

If the polynomial $p$ has a single dominant zero, then, as was shown in §3.2, the sequence $\{q_n\}$ converges to it.

The elements of the sequence $\{q_n\}$ will now be denoted by $q_n^{(1)}$; they form the first column of the two-dimensional array known as the **quotient-difference** $(qd)$ **scheme**. The elements of the remaining columns are conventionally denoted by $e_n^{(1)}, q_n^{(2)}, e_n^{(2)}, q_n^{(3)}, \ldots, e_n^{(m-1)}, q_n^{(m)}$ and are generated by alternatingly forming differences and quotients, as follows:

$$e_n^{(k)} := \left(q_{n+1}^{(k)} - q_n^{(k)}\right) + e_n^{(k-1)}, \tag{5a}$$

$$q_n^{(k+1)} := \frac{e_n^{(k)}}{e_{n-1}^{(k)}} q_n^{(k)}, \tag{5b}$$

---

* H. Rutishauser (b. 1918, d. 1970), Swiss mathematician, pioneer of numerical analysis in the computer age.

where $k = 1, 2, \ldots, m - 1$ and $n = k, k + 1, \ldots$ . In (5a) we set $e_n^{(0)} = 0$ when $k = 1$. The number of $q$ columns formed is equal to the degree of the given polynomial.

**EXAMPLE 3.3-1:** For $m = 4$, the general $qd$ scheme looks thus:

$$
\begin{array}{ccccccccc}
 & q_0^{(1)} & & & & & & & \\
0 & & e_0^{(1)} & & & & & & \\
 & q_1^{(1)} & & q_1^{(2)} & & & & & \\
0 & & e_1^{(1)} & & e_1^{(2)} & & & & \\
 & q_2^{(1)} & & q_2^{(2)} & & q_2^{(3)} & & & \\
0 & & e_2^{(1)} & & e_2^{(2)} & & e_2^{(3)} & & \\
 & q_3^{(1)} & & q_3^{(2)} & & q_3^{(3)} & & q_3^{(4)} & \\
0 & & e_3^{(1)} & & e_3^{(2)} & & e_3^{(3)} & & \\
 & q_4^{(1)} & & q_4^{(2)} & & q_4^{(3)} & & q_4^{(4)} & \\
 & \vdots & \vdots & \vdots & \vdots & \vdots & \vdots & \vdots &
\end{array}
$$

In each column of the scheme, the superscripts are constant, and in each row the subscripts are constant. The rules (5) can be memorized by observing that, in each of the rhombus-like configurations shown in the scheme, either the sums or the product of the *SW* and the *NE* pair of elements are equal. If a rhombus is centered in a $q$ columnn, sums are equal; if it is centered in an $e$ column, products are equal. In view of this interpretation, the formulas (5) occasionally are referred to as the **rhombus rules**.

The $qd$ scheme can be described in yet another way if we introduce, in addition to the forward difference operator $\Delta$ already introduced, the **quotient operator** $Q$ defined by

$$
Qx_n := \frac{x_{n+1}}{x_n} .
$$

The relations (5) then can be written more compactly thus:

$$
e_n^{(k)} := e_n^{(k-1)} + \Delta q_n^{(k)}, \qquad q_n^{(k+1)} = q_n^{(k)} Q e_{n-1}^{(k)}.
$$

Here it must be understood that the operators $\Delta$ and $Q$ act on the *sub*script.

**EXAMPLE 3.3-2:** For the polynomial

$$
p(z) := z^2 - z - 1
$$

the sequence $\{x_n\}$ (started according to (3)) is the *Fibonacci sequence*. The following $qd$ scheme results:

| $x_n$ | $e_n^{(0)}$ | $q_n^{(1)}$ | $e_n^{(1)}$ | $q_n^{(2)}$ | $e_n^{(2)}$ |
|---|---|---|---|---|---|
| 1 | 0 | | | | |
| | | 1.000000 | | | |
| 1 | 0 | | 1.000000 | | |
| | | 2.000000 | | −1.000000 | |
| 2 | 0 | | −0.500000 | | −0.000001 |
| | | 1.500000 | | −0.500001 | |
| 3 | 0 | | 0.166667 | | −0.000001 |
| | | 1.666667 | | −0.666669 | |
| 5 | 0 | | −0.066667 | | 0.000002 |
| | | 1.600000 | | −0.600000 | |
| 8 | 0 | | 0.025000 | | 0.000025 |
| | | 1.625000 | | −0.624975 | |
| 13 | 0 | | −0.009615 | | −0.000049 |
| | | 1.615385 | | −0.615409 | |
| 21 | 0 | | 0.003663 | | −0.000171 |
| | | 1.619048 | | −0.619243 | |
| 34 | 0 | | −0.001401 | | |
| | | 1.117647 | | | |
| 55 | 0 | | | | |
| | | $\vdots$ | | $\vdots$ | |
| | | $\dfrac{1+\sqrt{5}}{2}$ | | ? | |

### Existence of the $qd$ Scheme

Evidently the $qd$ scheme fails to exist if a coefficient $e_n^{(k)}$ with $0 < k < m$ becomes zero, and it is easy to construct examples for which this actually occurs. Another trivial case of nonexistence of the scheme arises when an $x_n$ becomes accidentally zero. It appears to be difficult to state explicit necessary and sufficient conditions on the polynomial $p$ for the scheme to exist. In terms of the sequence $\{x_n\}$, a necessary and sufficient condition is that the determinants

$$H_n^{(k)} := \begin{vmatrix} x_n & x_{n+1} & \cdots & x_{n+k-1} \\ x_{n+1} & x_{n+2} & \cdots & x_{n+k} \\ \cdots\cdots\cdots\cdots\cdots\cdots \\ x_{n+k-1} & x_{n+k} & \cdots & x_{n+2k-2} \end{vmatrix} \tag{6}$$

should be different from zero for $k = 1, 2, \ldots, m$ and for $n = 0, 1, 2, \ldots$. It is possible to state simple *sufficient* (but not necessary) conditions for this to be the case. Among them are the following:

(i)   The zeros $z_1, z_2, \ldots, z_m$ of $p$ are positive, and balanced values are used to start the sequence $\{x_n\}$.

(ii)   The zeros $z_1, z_2, \ldots, z_m$ of $p$ are simple (but not necessarily real) and have distinct absolute values:

$$|z_1| > |z_2| > \cdots > |z_m| > 0. \tag{7}$$

In case (ii) we can assert only that $H_n^{(k)} \neq 0$, and consequently that $e_n^{(k)} \neq 0$, for all sufficiently large values of $n$.

There is a good deal of numerical evidence that the $qd$ scheme exists in many cases even if neither of the above sufficient conditions is satisfied; for instance, if $p$ is a polynomial with real coefficients with pairs of complex conjugate zeros.

### Convergence Theorems

If the $qd$ scheme exists, some remarkable statements are possible about the limits of its elements as $n \to \infty$. The simplest situation arises when the zeros of the polynomial satisfy (7). We then have

**THEOREM 3.3a:**   *Under the conditions just stated,*

$$\lim_{n \to \infty} q_n^{(k)} = z_k, \qquad k = 1, 2, \ldots, m, \tag{8}$$

*that is, the k-th q-column of the qd scheme converges to the k-th zero of the polynomial.*

It follows from (8) by virtue of (5a) that

$$\lim_{n \to \infty} e_n^{(1)} = 0,$$

and from this we easily get by induction

$$\lim_{n \to \infty} e_n^{(k)} = 0, \qquad k = 1, 2, \ldots, m - 1. \tag{9}$$

Thus, under the condition (7), all $e$ columns of the $qd$ scheme tend to zero.

**EXAMPLE 3.3-3:**   In Example 3.3-2 the column $q_n^{(2)}$ tends to $(1 - \sqrt{5})/2$, the smaller zero of $p(z) = z^2 - z - 1$. (Concerning the column $e_n^{(2)}$, see below.)

If several zeros of $p$ have the same absolute value (this happens, for instance, every time when a polynomial with real coefficients has a pair of complex conjugate zeros), the convergence properties of the scheme are more complicated. We still assume that the zeros are numbered so that

$$|z_1| \geq |z_2| \geq |z_3| \geq \cdots \geq |z_m| > 0. \tag{10}$$

For conveniénce in formulating some of the conditions below, we shall put

$$|z_0| := \infty, \qquad |z_{m+1}| := 0.$$

Always assuming that the scheme exists, we then have

**THEOREM 3.3b:**   *For every k so that* $|z_{k-1}| > |z_k| > |z_{k-1}|$,

$$\lim_{n \to \infty} q_n^{(k)} = z_k. \tag{11}$$

*For every k so that* $|z_k| > |z_{k+1}|$,

$$\lim_{n \to \infty} e_n^{(k)} = 0. \tag{12}$$

These facts can be used in the following manner: The $e$ columns which tend to zero (a behavior which is numerically conspicuous) divide the $qd$ tabl. into subtables. All zeros with subscripts that agree with the superscripts of the $q$'s in one subtable have the same modulus. Thus, if a $z_k$ is the only zero of its modulus, this will be evident from the fact that the corresponding subtable contains one $q$ column only, and the value of $z_k$ can be obtained as the limit of that $q$ column.

It is not yet clear from the above how to deal with several zeros having the same absolute value. (Most frequently, this situation occurs in connection with complex conjugate zeros of real polynomials.) Such zeros, too, can be obtained from the $qd$ table. We first consider the general case where $j$ zeros $z_{k+1}, z_{k+2}, \ldots,$ $z_{k+j}$ have the same modulus:

$$|z_k| > |z_{k+1}| = |z_{k+2}| = \cdots = |z_{k+j}| > |z_{k+j+1}|. \tag{13}$$

Here, it is necessary to construct polynomials $p_n^{(l)}$, $l = k$, $k+1$, ..., $k+j$, by means of the recurrence relations

$$p_n^{(k)}(z) := 1, \qquad n = k, k+1, \ldots, ; \tag{14a}$$

$$p_n^{(l)}(z) := z p_n^{(l-1)}(z) - q_{n-1}^{(l)} p_{n-1}^{(l-1)}(z), \tag{14b}$$

$$l = k+1, k+2, \ldots, k+j; \qquad n = l, l+1, \ldots.$$

These polynomials can again be thought of as being arranged in a two-dimensional array. The scheme below shows a segment of this array for $j = 2$.

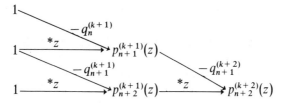

The zeros $z_{k+1}, z_{k+2}, \ldots, z_{k+j}$ can now be obtained from the polynomials $p_n^{(k+j)}$ by virtue of the following theorem.

**THEOREM 3.3c:** *If the zeros satisfy* (13), *then for each fixed z*

$$\lim_{n \to \infty} p_n^{(k+j)}(z) = (z - z_{k+1}) \cdots (z - z_{k+j}), \qquad (15)$$

*that is, the coefficients of the polynomials $p_n^{(k+j)}$ tend for $n \to \infty$ to the coefficients of the polynomial with zeros $z_{k+1}, z_{k+2}, \ldots, z_{k+j}$ and leading coefficient 1.*

For $j = 1$ Theorem 3.3c reduces to relation (8). For $j = 2$ (practically, the most frequent case) the converging polynomials are given by

$$p_{n+2}^{(k+2)}(z) = z[z - q_{n+1}^{(k+1)}] - q_{n+1}^{(k+2)}[z - q_n^{(k+1)}]$$
$$= z^2 - (q_{n+1}^{(k+1)} + q_{n+1}^{(k+2)})z + q_n^{(k+1)}q_{n+1}^{(k+2)}.$$

Relation (15) here means that the limits

$$b_k := \lim_{n \to \infty} (q_{n+1}^{(k+1)} + q_{n+1}^{(k+2)}),$$
$$c_k := \lim_{n \to \infty} q_n^{(k+1)}q_{n+1}^{(k+2)} \qquad (16)$$

exist, and that the polynomial $z^2 - b_k z + c_k$ has the zeros $z_{k+1}$ and $z_{k+2}$. In the special case where $k = 0$, the reader will recognize in (16) the rule for determining a leading quadratic factor by the Bernoulli method.

We finally mention the following fact, which plays the role of a computational check.

**THEOREM 3.3d:** *The quantities $e_n^{(m)}$ calculated from* (5a) *with $k = m$ are identically zero.*

**Numerical Instability**

As described above, the $qd$ scheme is built up proceeding from the left to the right. The sequence $\{x_n\}$ determines the first $q$ column $\{q_n^{(1)}\}$; from it we obtain in succession the columns $\{e_n^{(1)}\}$, $\{q_n^{(2)}\}$, ... by means of the relations (5). The reader will be shocked to learn that this method of generating the $qd$ scheme is not feasible in practice. In fact, the $qd$ algorithm as described represents a classical example of numerical instability due to cancellation. It is easy to see why cancellation happens, at least in the case covered by Theorem 3.3a. If $q_n^{(k)} \to z_k$, then the quantities

$$q_{n+1}^{(k)} - q_n^{(k)}$$

required to evaluate (5a) are differences of nearly equal nonzero numbers, and thus, by §1.3, are afflicted with increasingly large relative errors. Since $e_{n-1}^{(k-1)}$ tends to zero, the quantities $e_n^{(k)}$ as computed by (5a) also have large relative errors. These errors in turn contaminate the quotients

$$\frac{e_n^{(k)}}{e_{n-1}^{(k)}}$$

required to evaluate (5b); hence, $q_n^{(k+1)}$ also will have large relative errors. It is clear that this instability becomes ever more pronounced as $k$ increases.

**DEMONSTRATION 3.3-4:** The loss of significant digits was illustrated in Example 3.3-2. By Theorem 3.3d, the column $e_n^{(2)}$ should theoretically consist of zeros. The fact that these elements are not zero, and even increase with increasing $n$, shows the growing influence of rounding errors. The example was artificially calculated using six decimals only. The reader is invited to verify that using more decimals will delay, but not ultimately prevent, the phenomenon of numerical instability.

The fact that the method of generating the $qd$ scheme described by the relations (5) is unstable does not, of course, prevent the convergence theorems stated above from being true. These theorems concern the mathematically exact, unrounded $qd$ scheme. They can be used numerically as soon as we succeed in generating the $qd$ scheme in a numerically stable manner. One method of avoiding rounding errors, applicable to polynomials with rational coefficients, would be to perform all operations in exact rational arithmetic. Experience shows that, in rational computations of this kind, the integer numerators and denominators grow very rapidly, putting severe demands on the speed and memory capacities of the computer. Fortunately, as will be shown below, there is a simple way of generating a stable $qd$ scheme also in conventional arithmetic.

## Progressive Form of the Algorithm

The $qd$ scheme can be generated in a stable manner if it is built up row by row instead of column by column, as described earlier. To see this, we solve each of the recurrence relations (5) for the south element of the rhombus involved:

$$q_{n+1}^{(k)} := (e_n^{(k)} - e_n^{(k-1)}) + q_n^{(k)}, \tag{17a}$$

$$e_{n+1}^{(k)} := \frac{q_{n+1}^{(k+1)}}{q_{n+1}^{(k)}} e_n^{(k)}. \tag{17b}$$

Let us assume that a row of $q$'s and a row of $e$'s is known, each affected by normal rounding errors of several digits in the last place. The new row of $q$'s, as calculated from (17a), will then have absolute errors of the same magnitude. The fact that the relative errors in the $e$'s are large (due to the smallness of the $e$'s) is not important now. Furthermore, the relative errors in the new row of $e$'s determined by (17b) are of the same order as in the old row of $e$'s. While a normal amount of error propagation also must be expected in the present mode of generating the scheme, it is bound to be much less serious than when the scheme is generated column by column.

If the scheme is to be generated row by row, a first couple of rows must somehow be secured. The following algorithm shows how this is accomplished. Let $a_0, a_1, \ldots, a_m$ be constants, all different from zero. Put

$$q_0^{(1)} := -\frac{a_1}{a_0}; \qquad q_0^{(k)} := 0, \qquad k = 2, 3, \ldots, m; \tag{18a}$$

$$e_0^{(k)} := \frac{a_{k+1}}{a_k}, \qquad k = 1, 2, \ldots, m-1. \tag{18b}$$

Consider the elements thus generated as the first two rows of a $qd$ scheme and generate further rows by means of (17), using the side conditions

$$e_n^{(0)} = e_n^{(m)} = 0, \qquad n = 1, 2, \ldots . \tag{19}$$

As before, there is a theoretical possibility of breakdown of the scheme because a denominator is zero. However, we have

**THEOREM 3.3e:** *If the scheme of the elements $q_n^{(k)}$ and $e_n^{(k)}$ defined by (18), (17), and (19) exists, it is (for $n \geq k$) identical with the qd scheme of the polynomial*

$$p(z) = a_0 z^m + a_1 z^{m-1} + \cdots + a_m,$$

*where the sequence $\{x_n\}$ is started in the manner (3).*

The proof of Theorem 3.3e requires some involved algebra, and it is omitted here. A necessary and sufficient condition for the existence of the extended $qd$ scheme is that the determinants (6) be different from zero also for $n > -k$. (Here, we have to interpret $x_n := 0$ for $n < 0$.)

**EXAMPLE 3.3-5:** For $m = 4$, the top rows of the progressive form of the $qd$ scheme look thus:

$$
\begin{array}{ccccccccc}
 & -\dfrac{a_1}{a_0} & & 0 & & 0 & & 0 & \\[2ex]
0 & & \dfrac{a_2}{a_1} & & \dfrac{a_3}{a_2} & & \dfrac{a_4}{a_3} & & 0 \\[2ex]
 & q_1^{(1)} & & q_1^{(2)} & & q_1^{(3)} & & q_1^{(4)} & \\[1ex]
0 & & e_1^{(1)} & & e_1^{(2)} & & e_1^{(3)} & & 0 \\[1ex]
 & q_2^{(1)} & & q_2^{(2)} & & q_2^{(3)} & & q_2^{(4)} & \\
\end{array}
$$

. . . . . . . . . . . . . . . . . . . . . . . . . . . . . . . . . . . . . . . . . .

**DEMONSTRATION 3.3-6:** Here we apply the progressive $qd$ algorithm to the polynomial

$$p(z) = z^4 - 16z^3 + 72z^2 - 96z + 24$$

considered in Demonstration 2.3-5. Since all zeros are distinct, real, and positive, the $qd$ scheme may be expected to converge in the manner indicated in Theorem 3.3a. Listing $q$ values only, the following scheme results.

| $n$ | $q_n^{(1)}$ | $q_n^{(2)}$ | $q_n^{(3)}$ | $q_n^{(4)}$ |
|---|---|---|---|---|
| 0 | 16.000000 | 0 | 0 | 0 |
| 1 | 11.500000 | 3.166667 | 1.083333 | 0.250000 |
| 2 | 10.260870 | 3.949657 | 1.481781 | 0.307692 |
| 3 | 9.783898 | 4.255499 | 1.640930 | 0.319672 |
| 4 | 9.576440 | 4.396970 | 1.704584 | 0.322006 |
| 5 | 9.481187 | 4.466641 | 1.729725 | 0.322447 |
| . . . | . . . | . . . | . . . | . . . |
| 14 | 9.395192 | 4.536502 | 1.745758 | 0.322548 |
| 15 | 9.395129 | 4.536563 | 1.745760 | 0.322548 |
| 16 | 9.395099 | 4.536593 | 1.745761 | 0.322548 |
| . . . | . . . | . . . | . . . | . . . |
| 25 | 9.395071 | 4.536620 | 1.745761 | 0.322548 |
| 26 | 9.395071 | 4.536620 | 1.745761 | 0.322548 |

In all $q$ columns, convergence within $10^{-6}$ has been achieved within 25 iterations. The limits agree with the zeros found in Demonstration 2.3-5.

**DEMONSTRATION 3.3-7:**  Here we consider

$$p(z) = z^4 - 8z^3 + 39z^2 - 62z + 50.$$

The complete $qd$ scheme, including the $e$ columns, is shown below.

| $q_n^{(1)}$ | $e_n^{(1)}$ | $q_n^{(2)}$ | $e_n^{(2)}$ | $q_n^{(3)}$ | $e_n^{(3)}$ | $q_n^{(4)}$ |
|---|---|---|---|---|---|---|
| 8.000000 | | 0.000000 | | 0.000000 | | 0.000000 |
| | −4.875000 | | −1.589744 | | −0.806452 | |
| 3.125000 | | 3.285256 | | 0.783292 | | 0.806452 |
| | −5.125000 | | −0.379037 | | −0.830296 | |
| −2.000000 | | 8.031220 | | 0.332033 | | 1.636748 |
| | 20.580000 | | −0.015670 | | −4.092923 | |
| 18.580000 | | −12.564451 | | −3.745220 | | 5.729671 |
| | −13.916921 | | −0.004671 | ǀ | 6.261609 | |
| 4.663079 | | 1.347799 | | 2.521060 | | −0.531938 |
| | −4.022497 | | −0.008737 | | −1.321186 | |
| 0.640581 | | 5.361559 | | 1.208612 | | 0.789248 |
| | −33.667611 | | −0.001970 | | −0.862761 | |
| −33.027030 | | 39.027201 | | 0.347820 | | 1.652009 |

We now have $e_n^{(2)} \to 0$, but $e_n^{(1)}$ and $e_n^{(3)}$ do not tend to zero. This indicates the presence of two pairs of complex conjugate zeros. To make use of Theorem 3.3c, we form the quantities

$$b_n^{(k)} := q_{n+1}^{(k+1)} + q_{n+1}^{(k+2)}, \qquad c_n^{(k)} := q_n^{(k+1)} q_{n+1}^{(k+2)}$$

for $k = 0$ and $k = 2$, obtaining the following values:

| $b_n^{(0)}$ | $c_n^{(0)}$ | $b_n^{(2)}$ | $c_n^{(2)}$ |
|---|---|---|---|
| 6.410256 | 26.282051 | 1.589744 | 0.000000 |
| 6.031220 | 25.097561 | 1.968780 | 1.282051 |
| 6.015549 | 25.128902 | 1.984451 | 1.902439 |
| 6.010878 | 25.042114 | 1.989122 | 1.992225 |
| 6.002141 | 25.001372 | 1.997859 | 1.989741 |
| 6.000171 | 25.000110 | 1.999829 | 1.996637 |

The limits are 6, 25, 2, 2, respectively, indicating that the polynomials

$$z^2 - 6z + 25 \quad \text{and} \quad z^2 - 2z + 2$$

are quadratic factors of the given polynomial. In fact,

$$(z^2 - 6z + 25)(z^2 - 2z + 2) = z^4 - 8z^3 + 39z^2 - 62z + 50.$$

We have yet to deal with the complication that arises if some of the coefficients $a_0, a_1, \ldots, a_m$ are zero. In that case, the extended $qd$ scheme defined by (18) clearly does not exist. A possible remedy is to introduce a new variable

$$z^* := z - a$$

and to consider the polynomial

$$p^*(z^*) := p(a + z^*)$$

$$= p(a) + \frac{1}{1!} p'(a)z^* + \cdots + \frac{1}{m!} p^{(m)}(a)z^{*m}.$$

Here, $a$ denotes a suitably chosen parameter. The coefficients of the polynomial $p^*$ are easily calculated by the Horner algorithm. (See §2.3.) It can be shown that if $p$ has some zero coefficients, then all coefficients of $p^*$ are different from zero for sufficiently small values of $a \neq 0$. If the zeros $z_k^*$ of $p^*$ have been computed, those of $p$ are given by the formula

$$z_k = z_k^* + a.$$

**DEMONSTRATION 3.3-8:**  Let

$$p(z) = 81z^4 - 108z^3 + 24z + 20.$$

Here, $a_2 = 0$, and the $qd$ algorithm cannot be started. We form $p^*$ with $a = 1$. The Horner scheme results in:

| 81 | −108 | 0 | 24 | 20 |
|----|------|-----|-----|-----|
| 81 | −27  | −27 | −3  | 17  |
| 81 | 54   | 27  | 24  |     |
| 81 | 135  | 162 |     |     |
| 81 | 216  |     |     |     |
| 81 |      |     |     |     |

The new polynomial

$$p^*(z^*) = 81z^{*4} + 216z^{*3} + 162z^{*2} + 24z^* + 17$$

has all its coefficients different from zero, and the first rows of its $qd$ scheme are as follows:

$$-\frac{216}{81} \qquad\qquad 0 \qquad\qquad 0 \qquad\qquad 0$$

$$0 \qquad\qquad \frac{162}{216} \qquad\qquad \frac{24}{162} \qquad\qquad \frac{17}{24} \qquad\qquad 0$$

As in Demonstration 3.3-7, continuation of the scheme shows two pairs of non-converging $q$ columns, separated by an $e$ column that does converge to 0. Forming the appropriate quantities $b_n^{(k)}$ and $c_n^{(k)}$, we find that $p^*$ has the two quadratic factors

$$z^{*2} + 2.666666667z^* + 1.888888889$$

and

$$z^{*2} + 0.111111111$$

with the zeros

$$z_{1,2}^* = -1.333333333 \pm i0.333333333$$

and

$$z_{3,4}^* = \pm i0.333333333.$$

Consequently $p$ has the zeros

$$z_{1,2} = -0.333333333 \pm i0.333333333,$$
$$z_{3,4} = \phantom{-}1.000000000 \pm i0.333333333.$$

## Computational Checks

Even if the progressive form of the $qd$ algorithm is used, excessively large (or small) elements may cause some loss of accuracy. The mathematical results given below may be used for checking purposes; failure of any of these checks indicates excessive rounding error.

  (i)   Relation (5a) implies that the sum of the $q$ values in any row of the scheme is constant. For the top row this sum is $-a_1/a_0$, which by Vieta's formula equals the algebraic sum of the zeros of the polynomial. Thus, we have

$$q_n^{(1)} + q_n^{(2)} + \cdots + q_n^{(m)} = -\frac{a_1}{a_0} \qquad\qquad (20)$$

for any $n$. Note that by Theorem 3.3a this relation confirms Vieta's formula for $n \to \infty$!

**(ii)** It can be shown that the product of all $q$ elements in any diagonal sloping downward also is independent of $n$ and equals the product of all the zeros of the polynomial $p$. Thus, again by Vieta,

$$q_n^{(1)} q_{n+1}^{(2)} q_{n+2}^{(3)} \cdots q_{n+m-1}^{(m)} = (-1)^m \frac{a_m}{a_0}. \tag{21}$$

It should be noted that while (20) checks only the additions and subtractions performed in constructing the scheme, equation (21) checks all operations.

**(iii)** If the $qd$ scheme is constructed by the progressive version of the algorithm, the quantities $x_n$ are not needed. However, we may calculate the $x_n$ from (2) and should find

$$\frac{x_{n+1}}{x_n} = q_n^{(1)}, \qquad n = 0, 1, 2, \ldots. \tag{22}$$

### $qd$ Versus Newton

In comparison with other methods for determining the zeros of a polynomial, the $qd$ algorithm enjoys the tremendous advantage of furnishing simultaneous approximations to all zeros of the polynomial. No information about the polynomial other than the values of its coefficients is required.

These advantages have to be paid for by the rather slow convergence of the algorithm. Since the $qd$ method contains the Bernoulli method as a special case, the convergence can be no better than that of Bernoulli's method. In fact it can be shown that under the hypotheses of Theorem 3.3b the errors $q_n^{(k)} - z_k$ tend to zero like the larger of the ratios

$$\left(\frac{z_k}{z_{k-1}}\right)^n \quad \text{and} \quad \left(\frac{z_{k+1}}{z_k}\right)^n.$$

Even if the figures in the $q$ columns eventually settle down, the accuracy of the zeros thus obtained is somewhat uncertain because the large number of arithmetic operations may have contaminated the scheme with rounding error.

For the above reasons, the $qd$ algorithm is not recommended for the purpose of determining the zeros with final accuracy. Instead, the following two-stage procedure is advocated:

**STAGE 1:** Use the $qd$ algorithm to obtain crude first approximations to the zeros, or to the quadratic factors containing complex zeros.

**STAGE 2:** Using these approximations as starting values, obtain the zeros accurately by Newton's or Bairstow's method.

This combination of several methods has the advantage that the final values of the zeros are obtained from the original, undisturbed polynomial, and thus are practically free of rounding error.

# PROBLEMS

**1** Let $x_n := n!$, $n = 0, 1, 2, \ldots$ . Determine the $qd$ scheme corresponding to the sequence $\{x_n\}$ (a) numerically, (b) analytically. (The sequence $\{x_n\}$ does not arise as solution of a difference equation in this case.)

**2** Give analytical formulas for the entries of the $qd$ scheme, if $x_n := 1 + q^n$, where $0 < |q| < 1$. Show that

$$\lim_{n \to \infty} q_n^{(1)} = 1, \qquad \lim_{n \to \infty} e_n^{(1)} = 0, \qquad \lim_{n \to \infty} q_n^{(2)} = q,$$

and that $e_n^{(2)} = 0$ for all $n$.

**3** Show that

$$q_n^{(1)} = \frac{H_{n+1}^{(1)}}{H_n^{(1)}}, \qquad e_n^{(1)} = \frac{H_n^{(2)}}{H_n^{(1)} H_{n+1}^{(1)}}, \qquad q_n^{(2)} = \frac{H_n^{(2)} H_{n-1}^{(1)}}{H_{n-1}^{(2)} H_n^{(1)}}.$$

**4** Show that the determinants $H_n^{(1)}$ and $H_n^{(2)}$ formed with the elements $x_n = 1 + q^n$, where $0 < |q| < 1$ are always different from zero.

**5** Prove Theorem 3.3a for a polynomial of degree $m = 2$ whose zeros satisfy $|z_1| > |z_2| > 0$.

**6** Use the $qd$ algorithm to determine simultaneously all zeros of the polynomial

$$p(x) = x^4 - 1{,}111x^3 + 112{,}110x^2 - 1{,}111{,}000x + 1{,}000{,}000.$$

[See Problem 8, §2.3. This is an extremely favorable case.]

**7** The polynomial

$$p(z) = 5z^4 + 4z^3 + 3z^2 + 2z + 1$$

has four complex zeros whose absolute values are almost equal. Direct application of the $qd$ algorithm thus converges slowly. By letting

$$z = -\tfrac{1}{2} + z^*,$$

obtain a polynomial $p^*(z^*)$ which has two pairs of complex zeros with distinctly different absolute values. Determine these by the $qd$ algorithm, and hence find the values of the zeros of $p$. (Compare Problem 2, §2.5.)

**8** Determine approximate values for the zeros of the polynomial

$$p(z) = 70z^4 - 140z^3 + 90z^2 - 20z + 1.$$

Then find more exact values by Newton's method.

**9\*** Solve the problem of Demonstration 3.2-5 by means of the $qd$ algorithm.

**10** Prove Theorem 3.3e for polynomials of degree $m = 2$. [It suffices to show that the formulas (17), (18) generate the correct values of $q_0^{(1)}$, $e_0^{(1)}$, and $q_1^{(2)}$.]

**11** In quantum mechanics one studies the Laguerre polynomials $L_n^{(\alpha)}$ defined by

$$L_n^{(\alpha)}(x) := \sum_{k=0}^{n} \binom{n+\alpha}{n-k} \frac{(-x)^k}{k!} ;$$

for $\alpha > -1$; their zeros all are simple, real, and positive. Use the $qd$ algorithm to determine the zeros of $L_4^{(\alpha)}$ for $\alpha = 0.0, 0.5, 1.0, 1.5, \ldots$, and use the results to draw a graph showing the zeros as a function of $\alpha$ in the $(x, \alpha)$-plane. [In this example, the ratios $a_{k+1}/a_k$ are simpler functions of $\alpha$ than the coefficients themselves.]

---

### Bibliographical Notes and Recommended Reading

The streamlined version of the Bernoulli method given in §3.2, including its use of balanced starting values, is unpublished. It is included because the method appears to be particularly suitable for small calculators. The original reference for the quotient-difference algorithm is Rutishauser [1954]. For a fuller treatment see Henrici [1974].

# CHAPTER 4

# Elimination

The key word, elimination, is intended to cover those algorithms which aim at the direct (non-iterative) *solution of systems of linear equations.* Some of these algorithms also can be used to solve some of the basic problems of linear *optimization theory.* Optimization theory in its modern form became generally known only after 1945. Since that time, it has developed into one of the most significant practical disciplines of *operations research.*

## §4.1  GAUSSIAN ELIMINATION

The solution of a system of linear equations is one of those basic problems of numerical mathematics that occur most frequently in practice. Most often the solution of the system is not an end in itself but occurs as part of a larger problem. Linear systems thus occur in the context of approximation problems (especially least-squares approximation), in the solution of boundary-value problems by the method of discretization, and also in vibration problems if not only the frequencies but also the modes of the vibration are sought. If a system of nonlinear equations is solved by Newton's method, then, at each step of the iteration, a system of linear equations must be solved.

We presuppose the basic theoretical facts on systems of linear equations. Thus, we know that a given system of $n$ linear equations with $n$ unknowns either has no solution, or has precisely one solution, or has an infinity of solutions. (Indeed, the three possibilities may already occur in the case $n = 1$!) Pure mathematicians usually focus their attention on *exceptional cases,* where either no or an infinity of solutions exist. In our numerical treatment we shall consider only the *normal case,* where the system possesses precisely one solution. As always in numerical mathematics, the accent here is not so much on the existence but rather on the algorithmic construction of the solution.

The notation used for the given linear system is as follows:

$$
\begin{aligned}
a_{11}x_1 + a_{12}x_2 + \cdots + a_{1n}x_n - a_{10} &= 0 \\
a_{21}x_1 + a_{22}x_2 + \cdots + a_{2n}x_n - a_{20} &= 0 \\
&\quad\cdots\cdots\cdots\cdots\cdots\cdots\cdots\cdots\cdots\cdots\cdots \\
a_{n1}x_1 + a_{n2}x_2 + \cdots + a_{nn}x_n - a_{n0} &= 0.
\end{aligned}
\tag{1}
$$

Here, the $a_{ik}$ ($i, k = 1, 2, \ldots, n$) are the **coefficients** of the system; the $x_i$ are the **unknowns**; and the $a_{i0}$ are the **constants** or **non-homogeneous terms**.

It is well known that the system (1) may be written more compactly by the use of *matrix notation*. Defining the matrix

$$\mathbf{A} := \begin{pmatrix} a_{11} & a_{12} & \cdots & a_{1n} \\ a_{21} & a_{22} & \cdots & a_{2n} \\ \multicolumn{4}{c}{\dotfill} \\ a_{n1} & a_{n2} & \cdots & a_{nn} \end{pmatrix}$$

and the vectors

$$\mathbf{v} := \begin{pmatrix} a_{10} \\ a_{20} \\ \vdots \\ a_{n0} \end{pmatrix}, \qquad \mathbf{x} := \begin{pmatrix} x_1 \\ x_2 \\ \vdots \\ x_n \end{pmatrix}, \qquad \mathbf{0} := \begin{pmatrix} 0 \\ 0 \\ \vdots \\ 0 \end{pmatrix},$$

the system may be written as

$$\mathbf{Ax} - \mathbf{v} = \mathbf{0}. \tag{2}$$

By matrix calculus, the solution is

$$\mathbf{x} = \mathbf{A}^{-1}\mathbf{v}. \tag{3}$$

The simple form of this solution frequently leads the uninitiated into believing that solutions of linear systems are found via the inverse matrix. Unfortunately this is not so, for the following reasons:

(i)   The determination of the inverse matrix requires considerably more work (more precisely: about three times as much work) than the direct solution of the system.

(ii)   Experience and theoretical analysis both show that the solution which is obtained directly by a judicious algorithm generally is less affected by rounding errors than the solution obtained via the inverse matrix.

There is a further fallacy against which the beginner should be warned. As shown in linear algebra, linear systems may be solved by *Cramer's rule*. It is then usually pointed out that this rule is not suitable for numerical computation because the evaluation of the determinants that are involved allegedly requires too many operations. Although the conclusion that Cramer's rule should not be used is correct, its justification is false. Although it is indeed costly to evaluate determinants by expansion by minors, there exists another algorithm (see §4.2) which is cheap and effective. Cramer's formula is merely an explicit form of using the inverse matrix, however, and thus is subject to the same problems of error propagation.

How, then, should one solve a given system of linear equations?

We arrange the *data* of the system, that is, the elements of the matrix $\mathbf{A}$ and of the vector $\mathbf{v}$, in a two-dimensional rectangular tableau, as follows:

$$
\begin{array}{cccc|cl}
x_1 & x_2 & x_n & 1 & & \\
\hline
a_{11} & a_{12} & a_{1n} & -a_{10} & & = 0 \\
a_{21} & a_{22} & a_{2n} & -a_{20} & & = 0 \\
\multicolumn{6}{c}{\dotfill} \\
a_{n1} & a_{n2} & a_{nn} & -a_{n0} & & = 0
\end{array}
\tag{4}
$$

At the top of the tableau we imagine the row vector $(x_1, x_2, \ldots, x_n, 1)$. The vector $(x_1, x_2, \ldots, x_n)$ evidently is a solution of the system (1) if, and only if, the scalar product of each row with the vector on top of the tableau is zero. This property is preserved if

· two rows of the tableau are interchanged;

· a row is multiplied by a constant;

· a row is added to another row.

Now let $(x_1, x_2, \ldots, x_n)$ be the unique solution of the system. (As pointed out earlier, it is assumed that such a solution exists.) Using the three operations just mentioned, we now shall try to transform the tableau (4) into one where the values of the solution are put in evidence.

Under our hypotheses there exists an $i$ so that $a_{i1} \neq 0$. (Otherwise, $x_1$ could be chosen arbitrarily, and thus the solution would not be unique.) Suppose, if necessary, that by means of an exchange of rows a row where $a_{i1} \neq 0$ has been put in the position of the first row. Let us divide by $a_{11}$. This yields the tableau

$$
\begin{array}{cccc|c}
x_1 & x_2 & x_n & 1 & \\
\hline
1 & c_{12} & c_{1n} & -c_{10} & \\
a_{21} & a_{22} & a_{2n} & -a_{20} & \\
\multicolumn{5}{c}{\dotfill} \\
a_{n1} & a_{n2} & a_{nn} & -a_{n0} &
\end{array}
\tag{5}
$$

where

$$
c_{1m} := \frac{a_{1m}}{a_{11}}, \qquad m = 1, 2, \ldots, n, 0.
\tag{6}
$$

Now we subtract for $k = 2, 3, \ldots, n$ the $a_{k1}$-fold of the new first row from the $k$-th row. This yields

| | $x_1$ | $x_2$ | $x_n$ | 1 | |
|---|---|---|---|---|---|
| 1 | | $c_{12}$ | $c_{1n}$ | $-c_{10}$ | |
| 0 | | $a_{22}^*$ | $a_{2n}^*$ | $-a_{20}^*$ | (7) |
| 0 | | $a_{n2}^*$ | $a_{nn}^*$ | $-a_{n0}^*$ | |

where

$$a_{km}^* := a_{km} - a_{k1} c_{1m}, \qquad k = 2, \dots, n; \qquad m = 2, \dots, n, 0. \tag{8}$$

The tableaus (4), (7) are equivalent in the sense that if $(x_1, \dots, x_n)$ satisfies one tableau it satisfies the other. The tableau (7) is divided into

(a)   an **end equation**, which may be written in the form

$$x_1 = c_{10} - \sum_{m=2}^{n} c_{1m} x_m; \tag{9}$$

(b)   a **reduced tableau** (whose entries are starred) describing $n - 1$ equations for the $n - 1$ unknowns $x_2, x_3, \dots, x_n$.

When the reduced system has been solved, the solution may be substituted into the end equation (9), which immediately yields the missing unknown $x_1$.

It is now possible to proceed inductively by treating the reduced system as we treated the original system. The first column of the reduced tableau must contain a coefficient $a_{k2}^* \neq 0$; if necessary, the corresponding row is brought into first position by an interchange of rows. We then have $a_{22}^* \neq 0$ and may divide the row by $a_{22}^*$. This yields another end equation, this time with coefficients

$$c_{2m} := \frac{a_{2m}^*}{a_{22}^*}, \qquad m = 2, 3, \dots, n, 0.$$

Subtracting appropriate multiples of this end equation from the remaining rows, we obtain a re-reduced array describing $n - 2$ equations in $n - 2$ unknowns, and so on. The final tableau consists of $n$ end equations:

| | $x_1$ | $x_2$ | $x_3$ | $x_4$ | $x_n$ | 1 | |
|---|---|---|---|---|---|---|---|
| 1 | | $c_{12}$ | $c_{13}$ | $c_{14}$ | $c_{1n}$ | $-c_{10}$ | |
| 0 | | 1 | $c_{23}$ | $c_{24}$ | $c_{2n}$ | $-c_{20}$ | |
| 0 | | 0 | 1 | $c_{34}$ | $c_{3n}$ | $-c_{30}$ | (10) |
| 0 | | 0 | 0 | 1 | $c_{4n}$ | $-c_{40}$ | |
| 0 | | 0 | 0 | 0 | 1 | $-c_{n0}$ | |

This tableau permits us to determine the unknowns one after another according to the formula

$$x_n = c_{n0},$$

$$x_{n-1} = c_{n-1,0} - c_{n-1,n}x_n,$$

and generally

$$x_k = c_{k0} - \sum_{m=k+1}^{n} c_{km}x_m, \qquad k = n, n-1, n-2, \ldots, 1. \tag{11}$$

Because the unknown $x_n$ is obtained first and then the unknowns $x_{n-1}$, $x_{n-2}$, ..., $x_1$, in that order, the process described by (11) is called **back substitution**. The process of setting up the end equations is called **forward reduction**. The entire algorithm is known as **Gaussian elimination**.

**DEMONSTRATION 4.1-1:** Here we seek to find a quadratic polynomial

$$p(x) = a_0 + a_1 x + a_2 x^2 \tag{12}$$

so that

$$\int_0^1 p(x)\, dx = 1,$$

$$\int_0^1 xp(x)\, dx = \tfrac{7}{12}, \tag{13}$$

$$\int_0^1 x^2 p(x)\, dx = \tfrac{13}{30}.$$

Substituting (12) and carrying out the integrations results in the linear system

$$a_0 + \tfrac{1}{2}a_1 + \tfrac{1}{3}a_2 = 1,$$

$$\tfrac{1}{2}a_0 + \tfrac{1}{3}a_1 + \tfrac{1}{4}a_2 = \tfrac{7}{12}, \tag{14}$$

$$\tfrac{1}{3}a_0 + \tfrac{1}{4}a_1 + \tfrac{1}{5}a_2 = \tfrac{13}{30},$$

for the three unknowns $a_0$, $a_1$, $a_2$. Written in the form of a tableau, the system appears in the form

$$
\begin{array}{cccc}
a_0 & a_1 & a_2 & 1 \\
\hline
1 & \tfrac{1}{2} & \tfrac{1}{3} & -1 \\
\tfrac{1}{2} & \tfrac{1}{3} & \tfrac{1}{4} & -\tfrac{7}{12} \\
\tfrac{1}{3} & \tfrac{1}{4} & \tfrac{1}{5} & -\tfrac{13}{30}
\end{array}
\tag{15}
$$

To begin with, we construct the sequence of tableaus generated by the Gaussian algorithm in rational arithmetic:

$$
\begin{array}{cccc}
1 & \frac{1}{2} & \frac{1}{3} & -1 \\
0 & \frac{1}{12} & \frac{1}{12} & -\frac{1}{12} \\
0 & \frac{1}{12} & \frac{4}{45} & -\frac{1}{10}
\end{array}
$$

$$
\begin{array}{cccc}
1 & \frac{1}{2} & \frac{1}{3} & -1 \\
0 & 1 & 1 & -1 \\
0 & 0 & \frac{1}{180} & -\frac{1}{60}
\end{array}
$$

$$
\begin{array}{cccc}
1 & \frac{1}{2} & \frac{1}{3} & -1 \\
0 & 1 & 1 & -1 \\
0 & 0 & 1 & -3
\end{array}
$$

The last line of the last tableau immediately yields

$$a_2 = 3;$$

back substitution produces

$$a_1 = 1 - a_2 \qquad = -2,$$
$$a_0 = 1 - \tfrac{1}{3}a_2 - \tfrac{1}{2}a_1 = 1.$$

The exact solution vector thus is

$$
\begin{pmatrix} a_0 \\ a_1 \\ a_2 \end{pmatrix} = \begin{pmatrix} 1 \\ -2 \\ 3 \end{pmatrix}. \tag{16}
$$

We now repeat this computation in floating decimal arithmetic. In order to illustrate the influence of the rounding errors, we simulate in our computation a machine working with a mantissa of four decimal digits. (On a pocket calculator, this simulation requires a short auxiliary program rounding the mantissa of every result to four digits.) This simulation means, first of all, that the data of the problem have to be represented by four-digit floating numbers. Thus, the problem actually solved on our hypothetical machine is represented by the tableau

| $a_0$ | $a_1$ | $a_2$ | 1 |
|--------|--------|--------|--------|
| 1.000 | 0.5000 | 0.3333 | −1.000 |
| 0.5000 | 0.3333 | 0.2500 | −0.5833 |
| 0.3333 | 0.2500 | 0.2000 | −0.4333 |

(17)

The majority of the entries of this tableau are identical with the corresponding entries of (15), and those that do not agree differ by a mere $\frac{1}{3} * 10^{-4}$. We thus would expect that the solution of (17) does not differ much from the solution (16). This, however, is not quite the case, for a careful computation (see Demonstration 4.2-3) shows that the solution of (17), correctly rounded to four decimal places, is

$$\begin{pmatrix} a_0 \\ a_1 \\ a_2 \end{pmatrix} = \begin{pmatrix} 1.005 \\ -2.023 \\ 3.021 \end{pmatrix}. \tag{18}$$

The changes by less than $4 * 10^{-5}$ in the data thus cause some components of the solution to change by more than 500 times—a typical example of an ill-conditioned problem.

The solution (18) is not the solution that is obtained by Gaussian elimination however. Always working with four-digit decimal arithmetic, we obtain the reduced tableaus

| 1.000 | 0.5000 | 0.3333 | −1.000 |
|-------|--------|--------|---------|
| 0 | 0.08330 | 0.08330 | −0.08330 |
| 0 | 0.08330 | 0.08890 | −0.1000 |

| 1.000 | 0.5000 | 0.3333 | −1.000 |
|-------|--------|--------|---------|
| 0 | 1.000 | 1.000 | −1.000 |
| 0 | 0 | 0.005600 | −0.01670 |

| 1.000 | 0.5000 | 0.3333 | −1.000 |
|-------|--------|--------|---------|
| 0 | 1.000 | 1.000 | −1.000 |
| 0 | 0 | 1.000 | −2.982 |

Back substitution yields the solution

$$a_2 = 2.982$$
$$a_1 = -1.982 \tag{19}$$
$$a_0 = 0.9971,$$

which differs considerably from the correct solution (18). We shall see later (Demonstration 4.2-3) how to improve the accuracy of the solution.

## Operations Count

Here we count the number of arithmetic operations that are required in the various steps of the Gaussian elimination algorithm.

(a)   The forward reduction requires in

| | | |
|---|---|---|
| Step 1: | $n\delta + (n-1)n$ | $(\alpha + \mu)$ |
| Step 2: | $(n-1)\delta + (n-1)(n-2)$ | $(\alpha + \mu)$ |
| . . . . . . . . . . . . . . . . . . . . . . . . . . . . . . . . . . . . . . . . . . . . . . . . . . . . . . | | |
| Step $n-1$: | $2\delta + 1 * 2$ | $(\alpha + \mu)$ |
| Step $n$: | $1\delta + 0$ | $(\alpha + \mu),$ |

thus a total of

$$[1 + 2 + \cdots + (n-1) + n]\delta = \tfrac{1}{2}n(n+1)\delta$$

plus

$$[1 * 2 + 2 * 3 + \cdots + (n-2)(n-1) + (n-1)n](\alpha + \mu)$$
$$= \tfrac{1}{3}(n-1)n(n+1) \quad (\alpha + \mu);$$

(b)   Back substitution requires in

| | | |
|---|---|---|
| Step 1: | 0 | $(\alpha + \mu)$ |
| Step 2: | 1 | $(\alpha + \mu)$ |
| . . . . . . . . . . . . . . . . . . . . . . . . . . . . . | | |
| Step $n$: | $(n-1)$ | $(\alpha + \mu),$ |

thus a total of

$$[1 + 2 + \cdots + (n-1)](\alpha + \mu) = \tfrac{1}{2}(n-1)n(\alpha + \mu).$$

The total arithmetic effort for one Gaussian elimination of a system of order $n$ thus amounts to

$$\tfrac{1}{2}n(n+1)\delta + \tfrac{1}{6}(n-1)n(2n+5)(\alpha + \mu)$$

which asymptotically behaves like

$$\tfrac{1}{2}n^2\delta + \tfrac{1}{3}n^3(\alpha + \mu).$$

We note that the main arithmetic effort is spent in the forward reduction process.

In theoretical studies of the complexity of algorithms, the count of operations is frequently restricted to the multiplications. Until a few years ago, the belief was generally held that the number of multiplications required to solve a system of $n$ linear equations with $n$ unknowns could not be improved beyond a number that for $n \to \infty$ grows like the $\tfrac{1}{3}n^3$ obtained above. However, V. Strassen showed in 1969 that there exists an algorithm for the solution of such a system that requires a number of multiplications growing asymptotically only like

$$Cn^{\log_2 7} = Cn^{2.807355\,\cdots},$$

and more recently this result has been further improved by A. Schönhage who found algorithms with a number of multiplications that grows only like $Cn^{2.49\,\cdots}$. Because the smaller exponent for sufficiently large $n$ offsets any large value which the constant $C$ may have, these algorithms indeed are asymptotically better than Gaussian elimination. Strassen's algorithm is based on the very simple observation that two $2 \times 2$ matrices can be multiplied in a mere 7 multiplications (in place of the usual 8) if the operations are arranged in a certain clever way. The algorithm for general $n$ is defined in a recursive manner that requires complicated bookkeeping operations, however. Even leaving aside questions of numerical stability, it is therefore questionable whether the algorithms of Strassen and Schönhage will ever be implemented so that they become competitive in the majority of problems that are encountered in daily life.

## Pivot Strategy

The **pivots** of the Gaussian algorithm are the divisors used in making the first coefficient in each end equation equal to one. We have already seen that, in general, the pivots are not uniquely determined. Indeed, the only condition that must be satisfied from the mathematical point of view is that the pivots are not zero. Apart from that, any row with a first element different from zero may be used as pivot row.

We now try to use the freedom which we have in the selection of pivots for the purpose of keeping rounding errors small. The construction of the totally reduced array consists mainly in forming differences. Thus, it is cancellation which must be avoided. Back substitution consists of the forming of increasingly large sums. Here, smearing may be dangerous. While it seems difficult to design a general strategy to avoid cancellation, smearing can be combatted by keeping the non-diagonal numbers in the totally reduced array small. Thus, it becomes clear that we should not only avoid pivots that are zero but also pivots that are small.

**DEMONSTRATION 4.1-2:** Consider the system described by the following tableau:

| $x_1$ | $x_2$ | 1 |
|---|---|---|
| 0.0003100 | 1.000 | $-3.000$ |
| 1.000 | 1.000 | $-4.000$ |

Consistently carrying four decimals, choosing $a_{11}$ for the first pivot yields

| 1.000 | 3226. | −9677. |
|-------|-------|--------|
| 0 | −3225. | 9670. |

| 1.000 | 3226. | −9677. |
|-------|-------|--------|
| 0 | 1.000 | −2.998 |

,

and by back substitution we get

$$x_2 = 2.998,$$

$$x_1 = 9677 - 3226 * 2.998 = 5.452;$$

however, these results are inaccurate due to smearing which here is identical with cancellation. A better solution is obtained by interchanging the two equations. The sequence of tableaus now becomes

| $x_1$ | $x_2$ | 1 |
|-------|-------|---|
| 1.000 | 1.000 | −7.000 |
| 0.0003100 | 1.000 | −3.000 |

| 1.000 | 1.000 | −7.000 |
|-------|-------|--------|
| 0 | 0.9997 | −2.998 |

| 1.000 | 1.000 | −7.000 |
|-------|-------|--------|
| 0 | 1.000 | −2.999 |

,

and back substitution yields the solution

$$x_2 = 2.999, \qquad x_1 = 4.001,$$

which is accurate to all digits given.

To keep the main part of an elimination program simple, one will usually want to place the pivot in the upper lefthand corner of the tableau under consideration. What options are at our disposal to keep the pivots large and to put them into the desired position?

(1)   An interchange of *rows*.

To keep the pivot as large as possible, one will select as pivot row the one with the largest (or a largest, if there are several) first element. Since the order of the equations is immaterial, no record of the interchanges must be kept unless the same matrix is used repeatedly. This strategy of choosing the pivot is called **partial pivoting strategy**. Although matrices can be constructed where the strategy fails, it is the strategy most commonly used today. More complicated strategies involve

(2)  An interchange of *columns*.

Here it is possible to choose as pivot the absolutely largest element in the currently remaining tableau. This is called the **complete pivoting strategy**. A record must now be kept of the interchanges, because back substitution now does not produce the unknowns in their natural (descending) order. A further option at our disposal is

(3)  Multiplication of the *equations* with constant factors.

This operation obviously does not change the solution of the system. It may be shown that it does not even change the numerical solution if, for the scale factors, powers of the base $b$ are used. It clearly changes the absolute values of the coefficients, however, and hence the choice of the pivots if either of the above pivoting strategies are used. In fact, every coefficient in the first column can be made pivot by appropriate scaling if the partial pivoting strategy is used. Thus, it is clear that partial pivoting cannot be universally valid. It will only be used after the equations have been scaled so that the coefficients in each equation have the same order of magnitude (for instance, by making the largest coefficient in each equation equal to 1). A further option that influences pivot selection is

(4)  Multiplication of the *unknowns* by constant factors.

This amounts to multiplying the columns of the system by the reciprocal factors. A record of the scale factors must be kept since the unknowns are changed. By a combination of (3) and (4), the complete pivoting strategy may be corrupted in an arbitrary fashion.

As these paradoxical results indicate, the design of a pivoting strategy that works in all cases is no simple matter. The problem is confounded by the fact that a strategy which appears to be locally best is not always globally best. No complete theoretical analysis of the problem appears to be possible at this time. However, a combination of scaling—or equilibrating—the equations and partial pivoting seems to work in most cases.

# PROBLEMS

1  The system

$$5x_1 + 5x_2 + 7x_3 + 6x_4 = 1$$
$$7x_1 + 7x_2 + 10x_3 + 8x_4 = 1$$
$$6x_1 + 9x_2 + 8x_3 + 10x_4 = 1$$
$$5x_1 + 10x_2 + 7x_3 + 9x_4 = 1$$

has the exact solution

$$x_1 = 20, \qquad x_2 = 3, \qquad x_3 = -12, \qquad x_4 = -5.$$

Determine the solution by the Gaussian algorithm, using simulated four-digit decimal arithmetic.

2   Let **A** and **B** be square matrices of order $n$, and let **v** be a vector with $n$ components. The product **ABv** can either be formed as **(AB)v** or as **A(Bv)**. Which way is more advantageous from the point of view of computational economy?

## §4.2   THE L-R DECOMPOSITION

Gaussian elimination was described in section 4.1 without reference to matrices. It possesses an interesting interpretation in terms of the matrix calculus, however.

We begin by describing a variant of the sequence of tableaus generated by the Gaussian algorithm. As described in §4.1, this algorithm generates zeros below the main diagonal of the tableau. On the main diagonal all entries eventually become $+1$. Naturally, there is no real need to store these zeros and ones, because they do not depend on the system of equations under consideration. Let us now change the scheme by storing, in place of these trivial elements, the *multipliers* of the scheme, that is, using the notation of §4.1, in the

| 1st column | 2nd column | 3rd column |
|---|---|---|
| $a_{11}$ | | |
| $a_{21}$ | $a_{22}^*$ | |
| $a_{31}$ | $a_{32}^*$ | $a_{33}^{**}$,  and so on |
| $\vdots$ | $\vdots$ | $\vdots$ |
| $a_{n1}$ | $a_{n2}^*$ | $a_{n3}^{**}$ |

It is then possible to read from the final tableau how it was generated. Denoting the multipliers uniformly by $b_{pq}$, the final tableau now looks as follows:

| $b_{11}$ | $c_{12}$ | $c_{13}$ | $\cdots$ | $c_{1n}$ | $-c_{10}$ |
| $b_{21}$ | $b_{22}$ | $c_{23}$ | $\cdots$ | $c_{2n}$ | $-c_{20}$ |
| $b_{31}$ | $b_{32}$ | $b_{33}$ | $\cdots$ | $c_{3n}$ | $-c_{30}$ |
| | | | | | |
| $b_{n1}$ | $b_{n2}$ | $b_{n3}$ | $\cdots$ | $b_{nn}$ | $-c_{n0}$ |

We now define two triangular matrices as follows:

$$
\mathbf{L} = (b_{ij}) := \begin{pmatrix}
b_{11} & 0 & 0 & \cdots & 0 \\
b_{21} & b_{22} & 0 & \cdots & 0 \\
b_{31} & b_{32} & b_{33} & \cdots & 0 \\
\multicolumn{5}{c}{\dotfill} \\
b_{n1} & b_{n2} & b_{n3} & \cdots & b_{nn}
\end{pmatrix},
$$

$$
\mathbf{R} = (c_{ij}) := \begin{pmatrix}
1 & c_{12} & c_{13} & \cdots & c_{1n} \\
0 & 1 & c_{23} & \cdots & c_{2n} \\
0 & 0 & 1 & \cdots & c_{3n} \\
\multicolumn{5}{c}{\dotfill} \\
0 & 0 & 0 & & 1
\end{pmatrix}.
$$

In using these notations, it is understood that

$$
b_{ij} = 0 \quad \text{for} \quad j > i \tag{1}
$$

and

$$
c_{ii} = 1, \qquad c_{ij} = 0 \quad \text{for} \quad j < i \tag{2}
$$

[A matrix is called **triangular**, if all elements above the main diagonal are zero or all elements below the main diagonal are zero. In the first case, the matrix is called **left** or **lower triangular**; in the second, it is called **right** or **upper triangular**.]

**THEOREM 4.2:** *If the matrix* **A** *is such that the pivots may be chosen in their natural order, then*

$$
\mathbf{A} = \mathbf{LR}. \tag{3}
$$

We have seen that by suitably permuting the rows any non-singular matrix can be transformed into one where the pivots may be selected in their natural order. The Gaussian algorithm thus generates a representation of any (suitably permuted) non-singular matrix as a product of a left and a right triangular matrix. This is called the **L-R decomposition** or the **triangular decomposition** of **A**.

*Proof of Theorem 4.2:* It is to be shown that the numbers $b_{ij}$ and $c_{ij}$ defined above satisfy

$$
\sum_{k=1}^{n} b_{ik} c_{kj} = a_{ij} \tag{4}
$$

for all $i, j$. To this end, we simply look at the history of the tableaus generated by the Gaussian algorithm.

**(i)** We have

$$b_{i1} = a_{i1}, \qquad\qquad i = 1, 2, 3, \ldots, n$$

$$b_{i2} = a_{i2} - \underbrace{b_{i1} c_{12}}_{a_{i2}^*}, \qquad\qquad i = \phantom{1,} 2, 3, \ldots, n$$

$$b_{i3} = \underbrace{\underbrace{a_{i3} - b_{i1} c_{13}}_{a_{i3}^*} - b_{i2} c_{23}}_{a_{i3}^{**}}, \qquad i = \phantom{1, 2,} 3, \ldots, n$$

and generally

$$b_{ij} = a_{ij} - \sum_{k=1}^{j-1} b_{ik} c_{kj}, \qquad i \geq j.$$

There follows

$$a_{ij} = \sum_{k=1}^{j-1} b_{ik} c_{kj} + b_{ij}, \qquad i \geq j,$$

which in view of (1) and (2) may be written

$$a_{ij} = \sum_{k=1}^{n} b_{ik} c_{kj}, \qquad i \geq j.$$

This proves (4) for the elements on and below the main diagonal.

**(ii)** To prove (4) for the elements above the main diagonal, we see how the $c_{ij}$ are generated. We have

$$c_{1j} = \frac{a_{1j}}{a_{11}} = \frac{a_{1j}}{b_{11}}, \qquad\qquad j > 1,$$

$$c_{2j} = \frac{a_{2j}^*}{a_{22}^*} = \frac{a_{2j} - b_{2j} c_{1j}}{b_{22}}, \qquad\qquad j > 2,$$

$$c_{3j} = \frac{a_{3j}^{**}}{a_{33}^{**}} = \frac{a_{3j} - b_{31} c_{1j} - b_{32} c_{2j}}{b_{33}}, \qquad j > 3,$$

and generally

$$c_{ij} = \frac{a_{ij} - b_{i1} c_{1j} - b_{i2} c_{2j} - \cdots - b_{i,i-1} c_{i-1,j}}{b_{ii}}, \qquad j > i. \qquad (5)$$

There follows

$$a_{ij} = \sum_{k=1}^{i-1} b_{ik} c_{kj} + b_{ii} c_{ij}, \qquad j > i,$$

which in view of (1) and (2) may be written

$$a_{ij} = \sum_{k=1}^{n} b_{ik} c_{kj}, \qquad j > i,$$

proving (4) for the elements above the main diagonal.

**DEMONSTRATION 4.2-1:** For the matrix of the problem considered in Demonstration 4.1-1 we get

$$L = \begin{pmatrix} 1 & 0 & 0 \\ \frac{1}{2} & \frac{1}{12} & 0 \\ \frac{1}{3} & \frac{1}{12} & \frac{1}{180} \end{pmatrix}, \qquad R = \begin{pmatrix} 1 & \frac{1}{2} & \frac{1}{3} \\ 0 & 1 & 1 \\ 0 & 0 & 1 \end{pmatrix},$$

and it may indeed be verified that $LR = A$. In four-digit decimal arithmetic we obtained

$$L = \begin{pmatrix} 1. & 0 & 0 \\ 0.5000 & 0.08330 & 0 \\ 0.3333 & 0.08335 & 0.005470 \end{pmatrix},$$

$$R = \begin{pmatrix} 1.000 & 0.5000 & 0.3333 \\ 0 & 1.000 & 1.001 \\ 0 & 0 & 1.000 \end{pmatrix},$$

and the product (computed without further rounding error) equals

$$\begin{pmatrix} 1.000 & 0.5000 & 0.3333 \\ 0.5000 & 0.3333 & 0.250033 \\ 0.3333 & 0.2500 & 0.1999224 \end{pmatrix}.$$

**Computation of Determinants**

For reasons mentioned earlier, this is not a problem of paramount importance in numerical linear algebra. However, if the determinant of $A$ should ever be required, it may conveniently be obtained from the L-R decomposition of $A$. Because the determinant of a product of square matrices equals the product of the determinants of the factors,

$$\det A = \det L \cdot \det R.$$

Because the determinant of a triangular matrix equals the product of its diagonal elements, $\det \mathbf{R} = 1^n = 1$ and

$$\det \mathbf{L} = b_{11} b_{22} \cdots b_{nn}.$$

There follows

$$\det \mathbf{A} = b_{11} b_{22} \cdots b_{nn}.$$

**DEMONSTRATION 4.2-2:** For the matrix considered in Demonstration 4.1-1 we find

$$\det \mathbf{A} = 1 \cdot \tfrac{1}{12} \cdot \tfrac{1}{180} = \tfrac{1}{2160} = 0.000462963 \ldots$$

or, for the rounded matrix and working in four-digit decimal arithmetic,

$$\det \mathbf{A} = 1.000 \cdot 0.08330 \cdot 0.005600$$

$$= 0.000466480.$$

Compared to the entries of $\mathbf{A}$, the determinant is rather small. The matrix $\mathbf{A}$ thus is nearly singular, which throws some light on the ill-conditioning of the problem considered in Demonstration 4.1-1.

**A New Look at Gaussian Elimination**

By means of the triangular decomposition of the matrix $\mathbf{A}$, the mechanism of Gaussian elimination can be represented in a compact way. Equation (5) also holds for $j = 0$; that is, for the column of constants. Thus,

$$c_{i0} = \frac{a_{i0} - \sum_{k=1}^{i-1} b_{ik} c_{k0}}{b_{ii}}, \qquad i = 1, 2, \ldots, n, 0$$

which in view of (1) implies

$$a_{i0} = \sum_{k=1}^{n} b_{ik} c_{k0}.$$

Introducing the vectors,

$$\mathbf{v} := \begin{pmatrix} a_{10} \\ a_{20} \\ \vdots \\ a_{n0} \end{pmatrix}, \qquad \mathbf{w} := \begin{pmatrix} c_{10} \\ c_{20} \\ \vdots \\ c_{n0} \end{pmatrix},$$

this may be written compactly as

$$\mathbf{Lw} = \mathbf{v}. \tag{6}$$

The end equations

$$x_k = c_{k0} - \sum_{m=k+1}^{n} c_{km} x_m, \qquad k = n, n-1, \ldots, 1,$$

also may be expressed in matrix notation. In view of (2), we have

$$\sum_{m=1}^{n} c_{km} x_m - c_{k0} = 0,$$

which is the same as

$$\mathbf{Rx} = \mathbf{w}. \tag{7}$$

Thus, if the matrices $\mathbf{L}$ and $\mathbf{R}$ in the $L$-$R$ decomposition of the matrix $\mathbf{A}$ are already known, $\mathbf{A} = \mathbf{LR}$, Gaussian elimination works as follows: Let the system to be solved be

$$\mathbf{Ax} = \mathbf{v}, \quad \text{that is,} \quad \mathbf{LRx} = \mathbf{v}.$$

We first determine an auxiliary vector $\mathbf{w}$ from (6), and then $\mathbf{x}$ from (7). Because $\mathbf{v} = \mathbf{Lw} = \mathbf{L(Rx)} = (\mathbf{LR})\mathbf{x} = \mathbf{Ax}$, the $\mathbf{x}$ thus determined solves the given system. In this way, the solution of $\mathbf{Ax} = \mathbf{v}$ is reduced to the solution of two systems whose matrices are triangular. Requiring only about $\frac{1}{2}n^2(\alpha + \mu)$, the solution of a triangular system is much easier than the solution of a full system. Thus, if the $L$-$R$ decomposition is known, the solution of a system with the same matrix but with a new vector of constants, may be accomplished in about $n^2(\alpha + \mu)$, which compares very favorably to the $\frac{1}{3}n^3(\alpha + \mu)$ that are required when the solution is started from scratch.

## Iterative Refinement

One application of the technique of using the $L$-$R$ decomposition repeatedly occurs in the **iterative refinement** of an inaccurate solution. Suppose the solution $\mathbf{x}$ of the system $\mathbf{Ax} = \mathbf{v}$ has been determined only approximately due to rounding errors. Calling the approximate solution $\mathbf{x}^{(0)}$, the **residual vector**

$$\mathbf{r}^{(0)} := \mathbf{Ax}^{(0)} - \mathbf{v} \tag{8}$$

will then be different from the zero vector. In order to improve the solution, we set

$$\mathbf{x} = \mathbf{x}^{(0)} + \mathbf{c},$$

where **c** is a correction to be determined. Substituting into the system yields

$$A(x^{(0)} + c) - v = 0,$$

which by virtue of $Ax^{(0)} - v = r^{(0)}$ simplifies to

$$Ac + r^{(0)} = 0. \qquad (9)$$

This system has the same coefficient matrix as the original system; only the vector of constants is different. Thus, the *L-R* decomposition that already has been obtained can be used again.

What has been gained by all this? If the original solution was inaccurate, why should the solution of (9) be more accurate? The answer lies in the difference between relative and absolute error. If the vector of constants in a system is scaled by a factor $m$, and if $m$ is a power of the number base of the computer, then not only the solution, but also its errors are scaled by the same factor $m$. The *relative* error of the solution thus remains constant. Thus, if, as is reasonable to expect, the components of $r^{(0)}$ are small, the absolute errors in the solution **c** of (9) can be expected to be small, although the relative errors will be of the same order as the relative errors in the initial approximation $x^{(0)}$. Thus, provided that the residual vector $r^{(0)}$ has been determined accurately (which in practice means using double precision), $x^{(1)} := x^{(0)} + c$ will indeed be a more precise solution than $x^{(0)}$. The process can even be repeated, if necessary, by computing the residual vector for $x^{(1)}$ and computing a corresponding correction $c^{(1)}$. Conditions for the theoretical convergence of the method are given in Forsythe & Moler [1967], section 22.

**DEMONSTRATION 4.2-3:**   We apply iterative refinement to the solution (19) of the moment problem studied in Demonstration 4.1-1. The first step consists of computing the residual vector of (19) in double precision. (Since single precision here means four decimals, double precision is simulated by full precision.) We find

$$r^{(0)} := Ax^{(0)} - v = \begin{pmatrix} 1.000 & 0.5000 & 0.3333 \\ 0.5000 & 0.3333 & 0.2500 \\ 0.3333 & 0.2500 & 0.2000 \end{pmatrix} \begin{pmatrix} 0.9971 \\ -1.982 \\ 2.982 \end{pmatrix}$$

$$- \begin{pmatrix} 1.000 \\ 0.5833 \\ 0.4333 \end{pmatrix} = \begin{pmatrix} 0.00000060 \\ 0.00014940 \\ 0.00006657 \end{pmatrix}.$$

The remainder of the iteration step is performed in four-digit precision. We first solve $Lw = -r^{(0)}$; that is,

$$\begin{pmatrix} 1.000 & 0 & 0 \\ 0.5000 & 0.08330 & 0 \\ 0.3333 & 0.08330 & 0.005600 \end{pmatrix} \begin{pmatrix} w_1 \\ w_2 \\ w_3 \end{pmatrix} = \begin{pmatrix} -0.00000060 \\ 0.00014940 \\ -0.00006657 \end{pmatrix},$$

and obtain

$$\begin{pmatrix} w_1 \\ w_2 \\ w_3 \end{pmatrix} = \begin{pmatrix} -0.00000060 \\ -0.001790 \\ 0.03855 \end{pmatrix}.$$

Next, we solve $\mathbf{Rc} = \mathbf{w}$; that is,

$$\begin{pmatrix} 1.000 & 0.5000 & 0.3333 \\ 0 & 1.000 & 1.000 \\ 0 & 0 & 1.000 \end{pmatrix} \begin{pmatrix} c_1 \\ c_2 \\ c_3 \end{pmatrix} = \begin{pmatrix} -0.00000060 \\ -0.001790 \\ 0.03855 \end{pmatrix},$$

with the result

$$\begin{pmatrix} c_1 \\ c_2 \\ c_3 \end{pmatrix} = \begin{pmatrix} 0.007319 \\ -0.04034 \\ 0.03855 \end{pmatrix}.$$

The more accurate solution $\mathbf{x}^{(1)} := \mathbf{x}^{(0)} + \mathbf{c}$ now naturally must be computed in double precision. We obtain

$$\mathbf{x}^{(1)} = \begin{pmatrix} 1.004419 \\ -2.022340 \\ 3.020550 \end{pmatrix}.$$

The residual vector (computed in double precision) now is

$$\mathbf{r}^{(1)} := \mathbf{Ax}^{(1)} - \mathbf{v} = \begin{pmatrix} -1.685 * 10^{-6} \\ 1.078 * 10^{-6} \\ -2.147 * 10^{-6} \end{pmatrix}.$$

Using it to refine once again, we find the second improved solution

$$\mathbf{x}^{(2)} = \begin{pmatrix} 1.0045366 \\ -2.0229892 \\ 3.0211761 \end{pmatrix}.$$

Its residual vector (as always computed in simulated double precision) is

$$\mathbf{r}^{(2)} := \mathbf{Ax}^{(2)} - \mathbf{v} = \begin{pmatrix} -6 * 10^{-9} \\ 25 * 10^{-9} \\ -31 * 10^{-9} \end{pmatrix};$$

however, these results are unreliable due to the heavy cancellation in forming the difference $\mathbf{Ax}^{(2)} - \mathbf{v}$. Moreover, from the fact that the maximum component of the residual vector is $10^{-8}$, it should not be concluded that the errors in the components of $x^{(2)}$ are less than $10^{-8}$. In fact, an independent calculation (see Problem 8) shows that the correct nine-digit representation of the exact solution of the rounded system is

$$\mathbf{x} = \begin{pmatrix} 1.00453849 \\ -2.02299985 \\ 3.02118642 \end{pmatrix}.$$

# PROBLEMS

1   What is the *L-R* decomposition of the matrix considered in Problem 1, §4.1? Use iterative refinement to improve the solution which you have found.

2   Let

$$\mathbf{A} := \begin{pmatrix} 1 & -1 & 1 \\ -1 & 3 & -3 \\ 1 & -2 & 5 \end{pmatrix}.$$

    **(a)**   Find the *L-R* decomposition of **A**.
    **(b)**   Find $\mathbf{A}^{-1}$ by solving the system $\mathbf{Ax} = \mathbf{e}_i$ for the three coordinate unit vectors $\mathbf{e}_i$.

3   For a general nonsingular matrix **A**, its inverse $\mathbf{A}^{-1}$ may be found as

$$\mathbf{A}^{-1} = (\mathbf{x}_1, \mathbf{x}_2, \ldots, \mathbf{x}_n),$$

where $\mathbf{x}_i$ is the solution of

$$\mathbf{Ax}_i = \mathbf{e}_i,$$

$\mathbf{e}_i$ denoting the *i*-th coordinate unit vector. Taking advantage of an *L-R* decomposition of **A** that has to be computed only once, how many arithmetic operations are required to construct $\mathbf{A}^{-1}$ by this method?

4   A matrix $\mathbf{A} = (a_{ij})$ is called **tridiagonal** if $a_{ij} = 0$ for $|i - j| > 1$.
    **(a)**   Design an algorithm, based on Gaussian elimination, for solving the tridiagonal system

$$\mathbf{Ax} = \mathbf{v},$$

where

$$
A = \begin{pmatrix}
b_1 & c_1 & & & & & & 0 \\
a_2 & b_2 & c_2 & & & & & \\
& a_3 & b_3 & c_3 & & & & \\
& & & \ddots & & & & \\
& & & & a_{n-1} & b_{n-1} & c_{n-1} & \\
0 & & & & & a_n & b_n &
\end{pmatrix}.
$$

Your algorithm should use simply indexed quantities only; let

$$
L = \begin{pmatrix}
e_1 & & & & 0 \\
d_2 & e_2 & & & \\
& d_3 & e_3 & & \\
& & & \ddots & \\
0 & & & d_n & e_n
\end{pmatrix}, \qquad
R = \begin{pmatrix}
1 & f_1 & & & 0 \\
& 1 & f_2 & & \\
& & \ddots & \ddots & \\
& & & 1 & f_{n-1} \\
0 & & & & 1
\end{pmatrix}.
$$

**(b)** Test your algorithm by solving the system $y_0 = y_{n+1} = 0$,

$$
-y_{k-1} + 2y_k - y_{k+1} = \frac{8}{(n+1)^2}, \qquad k = 1, 2, \ldots, n,
$$

for various values of $n$, and compare your solution with the mathematical solution

$$
y_k = 4\left[ \frac{k}{n+1} - \left(\frac{k}{n+1}\right)^2 \right].
$$

**5** For positive definite symmetric matrices **A**, the **Cholesky decomposition**

$$
A = LL^T \tag{10}
$$

($L$ = left triangular matrix, $L^T$ = transposed matrix $L$) is used in place of the $L$-$R$ decomposition.

**(a)** Devise an algorithm for constructing $L$, and in doing so show that there is precisely one $L$ with positive diagonal elements satisfying (10).

**(b)** The matrix **A** occurring in Demonstration 4.1-1 is positive definite. Construct its Cholesky decomposition, working consistently with four-digit decimal arithmetic.

**(c)** Show how to solve the system $Ax = v$, using the Cholesky decomposition, and do the problem of Demonstration 4.1-1 by this method.

**(d)** Count the number of operations required for the Cholesky decomposition and for solving the linear system. How does this compare to Gaussian elimination? [Count square roots as a separate category.]

**6** Find analytically the Cholesky decomposition $A = LL^T$ of the $n$-th order tridiagonal matrix,

$$A = \begin{pmatrix} a & -1 & & & 0 \\ -1 & a & -1 & & \\ & -1 & a & -1 & \\ & & & \ddots & \\ 0 & & & -1 & a \end{pmatrix}$$

where $a \geq 2$. What are the limiting values of the elements of $L$ as $n \to \infty$?

**7** The system

$$2x_1 + x_2 = 1$$
$$x_1 + 2x_2 = 0$$

has the exact solution $x_1 = \frac{2}{3}$, $x_2 = -\frac{1}{3}$. The exact factors of the $L$-$R$ decomposition of the matrix $A$ of the system are

$$L = \begin{pmatrix} 2 & 0 \\ 1 & 1.5 \end{pmatrix}, \qquad R = \begin{pmatrix} 1 & 0.5 \\ 0 & 1 \end{pmatrix}.$$

Experiment with approximate factors

$$L = \begin{pmatrix} a & 0 \\ b & c \end{pmatrix}, \qquad R = \begin{pmatrix} 1 & d \\ 0 & 1 \end{pmatrix}$$

and show that the method of iterative refinement converges to the exact solution even for matrices $L$ and $R$ that deviate considerably from the correct factors. *Assuming all operations exact, what are the mathematical limits for the convergence of the method in this case?

**8** *Perturbed systems.* If $A$ and $B$ are square matrices and $A$ is nonsingular, let $x(\varepsilon)$ denote the solution of

$$(A + \varepsilon B)x = v.$$

**(a)** By using a geometric series in the solution formula

$$x(\varepsilon) = (A + \varepsilon B)^{-1}v$$
$$= A^{-1}(I + \varepsilon A^{-1}B)v$$

show that for $|\varepsilon|$ sufficiently small

$$\mathbf{x}(\varepsilon) = \mathbf{A}^{-1} \sum_{n=0}^{\infty} (-1)^n \varepsilon^n (\mathbf{A}^{-1}\mathbf{B})^n \mathbf{v}. \tag{11}$$

**(b)** Show that the series (11) may be evaluated by the formula

$$\mathbf{x}(\varepsilon) = \sum_{n=0}^{\infty} (-1)^n \varepsilon^n \mathbf{x}_n,$$

where the vectors $\mathbf{x}_n$ are defined as the solutions of

$$\mathbf{A}\mathbf{x}_n = \mathbf{y}_n,$$

and the $\mathbf{y}_n$ are defined recursively by

$$\mathbf{y}_0 := \mathbf{v}, \qquad \mathbf{A}\mathbf{y}_n = \mathbf{B}\mathbf{y}_{n-1}, \qquad n = 1, 2, \ldots .$$

**(c)** Denoting by $\tilde{\mathbf{A}}$ the rounded representation of the matrix $\mathbf{A}$ and letting $\varepsilon\mathbf{B} := \tilde{\mathbf{A}} - \mathbf{A}$, use the foregoing series to compute the exact solution of $\tilde{\mathbf{A}}\mathbf{x} = \mathbf{v}$, where $\mathbf{A}\mathbf{x} = \mathbf{y}$ is the system considered in Demonstration 4.1-1.

---

## §4.3   THE METHOD OF LEAST SQUARES

Frequently, in applications, more than $m$ equations are given for $m$ unknowns. In this case, no exact solution of the system exists in general. What is wanted, then, is a solution which satisfies the given system as well as possible, in a sense to be specified.

**EXAMPLE 4.3-1:**   Let there be given $n > 3$ points $P_i = (x_i, y_i)$ in the $(x, y)$-plane, $i = 1, 2, \ldots, n$. We are looking for the equation of a circle that passes through the points $P_i$ as well as possible. The general equation of a circle is given by

$$x^2 + y^2 + ax + by + c = 0.$$

If this circle would pass through all points $P_i$, there would hold

$$x_i^2 + y_i^2 + ax_i + by_i + c = 0, \qquad i = 1, 2, \ldots, n.$$

Thus, the unknown coefficients $a$, $b$, $c$ would have to be determined so that the foregoing $n > 3$ equations are all satisfied. In general, this will not be possible; one will have to be satisfied with a (in some sense) best approximate solution of the problem.

If there are more equations than unknowns, the system is called **over-determined**. We assume the overdetermined system of $n$ equations for $m > n$ unknowns in the form

$$\sum_{j=1}^{m} a_{ij}x_j = v_j, \qquad j = 1, 2, \ldots, n.$$

Written in matrix form, this appears as

$$\mathbf{Ax = v}. \tag{1}$$

Here, $\mathbf{A}$ now is a rectangular matrix having more rows than columns. Represented schematically, (1) has the form

It will be assumed in the following that the matrix $\mathbf{A}$ has its maximum rank, that is, rank $m$. This assumption is not always satisfied in practice; we refer to the literature for the exceptional case where rank $\mathbf{A} < m$.

What should we now mean by the best possible solution of (1)? Should we mean, for instance, the solution of some $m$ of the $n$ given equations (circle passing through three of the given points)? Such a procedure would not be fair because it would not treat the given conditions in a symmetric manner. Since Gauss it has become customary to mean by best solution the solution determined according to a certain well-defined method, called the **method of least squares**.

Let us substitute an arbitrary vector $\mathbf{x}$ into the expression $\mathbf{Ax - v}$ which we wish to be zero. The result, in general, will not be $\mathbf{0}$, but a certain **residual vector** $\mathbf{r = r(x)}$ depending on $\mathbf{x}$,

$$\mathbf{r := Ax - v}. \tag{2}$$

The solution of (1) in the sense of the method of least squares is the (as will be seen, unique) vector $\mathbf{x}$ for which the residual vector becomes as small as possible, or equivalently, for which the real function

$$f(\mathbf{x}) := \|\mathbf{r}\|^2 = \|\mathbf{Ax - v}\|^2 \tag{3}$$

becomes a minimum. Here $\|\cdot\|$ denotes the **Euclidean norm** defined by

$$\|\mathbf{r}\|^2 := \sum_{i=1}^{n} r_i^2 .$$

The Euclidean norm, and not some other norm such as $\max |r_i|$, is used for theoretical as well as practical reasons. For this choice of the norm, the analytical as well as the numerical computations become most easily manageable; moreover, as Gauss has shown, under certain assumptions on the errors in the data the most probable values of the unknown quantities are those which minimize the Euclidean norm of the residual vector.

The problem now arises to characterize the $\mathbf{x}$ minimizing $f(\mathbf{x})$ so that it can be calculated. We shall solve this problem in three different ways, all leading to the same characterization.

## (a) Analytical Method

The problem can be solved by calculus. By matrix algebra, we have, denoting the transpose of a vector or matrix by the superscript $T$,

$$f(\mathbf{x}) = (\mathbf{Ax} - \mathbf{v})^T (\mathbf{Ax} - \mathbf{v})$$
$$= (\mathbf{Ax})^T \mathbf{Ax} - (\mathbf{Ax})^T \mathbf{v} - \mathbf{v}^T \mathbf{Ax} + \mathbf{v}^T \mathbf{v}$$
$$= \mathbf{x}^T \mathbf{A}^T \mathbf{Ax} - 2\mathbf{x}^T \mathbf{A}^T \mathbf{v} + \mathbf{v}^T \mathbf{v}$$

or, letting $\mathbf{B} := \mathbf{A}^T \mathbf{A} = (b_{ij})$,

$$f(\mathbf{x}) = \sum_{i,j=1}^{m} b_{ij} x_i x_j - 2 \sum_{j=1}^{m} \sum_{i=1}^{n} a_{ij} v_i x_j + \text{const.}$$

The $k$-th component of the gradient of $f$ at the point $\mathbf{x}$ is

$$\frac{\partial f}{\partial x_k} = 2 \sum_{j=1}^{m} b_{kj} x_j - 2 \sum_{i=1}^{n} a_{ik} v_i .$$

(In differentiating the first term, a contribution arises from the terms where $j = k$ as well as from those where $i = k$.) Because the gradient vanishes at a minimum, any $\mathbf{x}$ minimizing $f$ satisfies the equations,

$$\sum_{j=1}^{m} b_{kj} x_j = \sum_{i=1}^{n} a_{ik} v_i , \qquad k = 1, 2, \ldots, m$$

or

$$\mathbf{A}^T \mathbf{Ax} = \mathbf{A}^T \mathbf{v}. \tag{4}$$

Here, $A^T A$ is a square matrix of order $m$; thus (4) comprises as many equations as there are unknowns. We show that a unique solution exists by verifying that, under our hypotheses, $A^T A$ is nonsingular. If this matrix were singular, then $x \neq 0$ would exist so that $A^T A x = 0$, implying that $x^T A^T A x = 0$, implying that $(Ax)^T (Ax) = 0$, implying that $Ax = 0$, implying that the columns of $A$ are linearly dependent, contrary to our hypothesis that $A$ has full rank.

The equations (4) are called the **normal equations** of Gauss. We just have seen that the normal equations are *necessary* for $f$ to have a minimum at the point $x$. That the unique $x$ satisfying (4) really defines a minimum (and not a maximum or some other stationary point) follows from the fact that $A^T A$ is a positive definite matrix or, perhaps more simply, from the approaches (b) and (c) below.

## (b) Geometric Approach

The mapping

$$u \to y := Au$$

is from $R^m$ *into* $R^n$. The range of this mapping, that is, the set of all vectors $y = Au$, forms a linear subspace of dimension $m$ in $R^n$. Let $\mathscr{L}$ be this linear subspace. If $v$ is in $\mathscr{L}$, then an exact solution of (1) exists. If not, then we seek to determine $x_0$ so that the distance of $Ax_0$ from $v$ becomes a minimum. It is geometrically obvious that this minimum is achieved for the point $x_0$ for which the vector $r = Ax_0 - v$ is perpendicular to $\mathscr{L}$; that is, perpendicular to all vectors in $\mathscr{L}$. Thus, for any vector $y = Au$, we want

$$(Au)^T (Ax_0 - v) = 0; \qquad \text{that is,} \qquad u^T A^T (Ax_0 - v) = 0,$$

or, since this is to hold for all vectors $u$ and since $A$ has full rank,

$$A^T A x_0 - A^T v = 0.$$

Thus, $x_0$ satisfies the normal equations (4). The special case where $m = 1$ and $n = 2$ is shown in Figure 4.3a.

## (c) Algebraic Method

Let $x = x_0 + y$, where $x_0$ is the solution of the system of normal equations. Then,

$$r = Ax - v = A(x_0 + y) - v = r_0 + Ay,$$

where $r_0 := Ax_0 - v$ is the residual vector belonging to $x_0$. Thus,

$$\|r\|^2 = (r_0^T + y^T A^T)(r_0 + Ay)$$

$$= r_0^T r_0 + r_0^T Ay + y^T A^T r_0 + yA^T Ay.$$

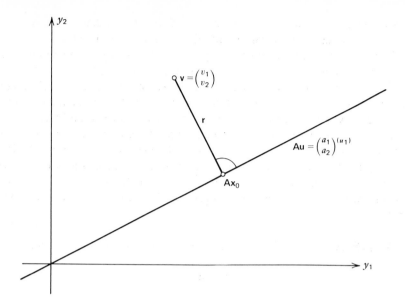

**Fig. 4.3a.** The method of least squares.

Using

$$\mathbf{x}_0 = (\mathbf{A}^T\mathbf{A})^{-1}\mathbf{A}^T\mathbf{v}, \qquad \mathbf{r}_0 = \mathbf{A}(\mathbf{A}^T\mathbf{A})^{-1}\mathbf{A}^T\mathbf{v} - \mathbf{v},$$

we see that the two middle terms are zero:

$$\mathbf{y}^T\mathbf{A}^T\mathbf{r}_0 = (\mathbf{r}_0^T\mathbf{A}\mathbf{y})^T = \mathbf{y}^T\mathbf{A}^T[\mathbf{A}(\mathbf{A}^T\mathbf{A})^{-1}\mathbf{A}\mathbf{v} - \mathbf{v}] = \mathbf{y}^T\mathbf{A}^T\mathbf{v} - \mathbf{y}^T\mathbf{A}^T\mathbf{v} = 0,$$

thus there remains

$$\|\mathbf{r}\|^2 = \|\mathbf{r}_0\|^2 + \|\mathbf{A}\mathbf{y}\|^2,$$

showing that $\|\mathbf{r}\|$ achieves its minimum for $\mathbf{y} = \mathbf{0}$ and (because $\mathbf{A}$ has full rank) only for $\mathbf{y} = \mathbf{0}$.

In summary, we have proved:

**THEOREM 4.3:** *Let* $\mathbf{A}$ *be a rectangular matrix with m columns and* $n > m$ *rows, and let* $\mathbf{A}$ *have rank m. Then, for each vector* $\mathbf{v}$, *the overdetermined linear system*

$$\mathbf{A}\mathbf{x} = \mathbf{v} \tag{1}$$

*has a unique solution in the sense of the method of least squares. This solution is identical with the solution of the nonsingular system*

$$\mathbf{A}^T\mathbf{A}\mathbf{x} = \mathbf{A}^T\mathbf{v}. \tag{4}$$

As a mnemonic device, it may be noted that the system of normal equations (4) is obtained from (1) simply by multiplying by $A^T$ from the left.

Being an ordinary nonsingular system with equally many unknowns as equations, the system of normal equations can, in principle, be solved by any method appropriate for such systems, for instance by Gaussian elimination. Experience shows, however, that if $m$ and $n$ are not very small, the normal equations are extremely sensitive to rounding errors. Special methods have been devised for dealing with large overdetermined systems which do not require setting up the normal equations. If the normal equations are used at all, double precision should be used in the computation of $A^TA$ and $A^Tv$ in order to counteract the loss of significant digits that is likely to occur in the forming of scalar products of large vectors. Some work may be saved by noting that the matrix $A^TA$ is symmetric.

**DEMONSTRATION 4.3-2:**   We wish to pass a circle through the 4 points

$$P_1 = (0, 0)$$
$$P_2 = (4, 0)$$
$$P_3 = (0, 4)$$
$$P_4 = (5, 6).$$

(See Example 4.3-1.) The equations for the coefficients $a$, $b$, $c$ of the general equation of a circle,

$$x^2 + y^2 + ax + by + c = 0,$$

may be written as

$$ax_i + by_i + c = -(x_i^2 + y_i^2),$$

yielding the four equations

$$0a + 0b + 1c = 0$$
$$4a + 0b + 1c = -16$$
$$0a + 4b + 1c = -16$$
$$5a + 6b + 1c = -61.$$

Thus, in the present case,

$$A = \begin{pmatrix} 0 & 0 & 1 \\ 4 & 0 & 1 \\ 0 & 4 & 1 \\ 5 & 6 & 1 \end{pmatrix}, \quad v = \begin{pmatrix} 0 \\ -16 \\ -16 \\ -61 \end{pmatrix}.$$

Computation yields

$$A^T A = \begin{pmatrix} 41 & 30 & 9 \\ 30 & 52 & 10 \\ 9 & 10 & 4 \end{pmatrix}, \qquad A^T v = \begin{pmatrix} -369 \\ -430 \\ -93 \end{pmatrix}.$$

Thus, the normal equations are

$$41a + 30b + 9c = -369$$
$$30a + 52b + 10c = -430;$$
$$9a + 10b + 4c = -93$$

they are symmetric, as expected. The Gaussian algorithm, carried through in three-digit floating arithmetic, yields the following sequence of arrays:

| $a$ | $b$ | $c$ | $1$ |
|---|---|---|---|
| 41.0 | 30.0 | 9.00 | 369 |
| 30.0 | 52.0 | 10.0 | 430 |
| 9.00 | 10.0 | 4.00 | 93.0 |

| | | | |
|---|---|---|---|
| 1.00 | 0.732 | 0.220 | 9.00 |
| 0 | 30.0 | 3.40 | 160 |
| 0 | 3.41 | 2.02 | 12.0 |

| | | |
|---|---|---|
| 1.00 | 0.113 | 5.33 |
| 0 | 1.64 | -6.20 |

| | |
|---|---|
| 1.00 | -3.78 |

Back substitution yields

$$c = 3.78$$
$$b = -5.76$$
$$c = -5.61.$$

The coordinates of the center of the desired circle thus are

$$x_0 = -\tfrac{1}{2}a = 2.81, \qquad y_0 = -\tfrac{1}{2}b = 2.88,$$

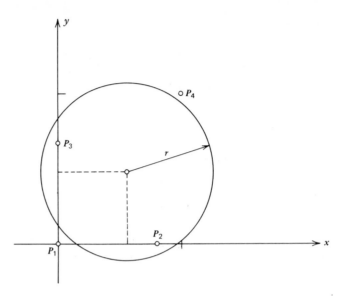

**Fig. 4.3***b***.**   Circle through four points.

and the radius is

$$r = \left[ \left( \frac{a}{2} \right)^2 + \left( \frac{b}{2} \right)^2 - c \right]^{1/2} = 3.52,$$

which is fortunately real. (See Figure 4.3b.)

### Linear Regression

Let $(x_i, y_i)$, $i = 1, 2, \ldots, n$ be a sequence of data points; for instance, the results of measuring the values of a function $y$ for selected values of the independent variable $x$. We wish to pass a straight line

$$y = ax + b$$

through these points as well as possible. Thus, we try to solve the over-determined system

$$ax_i + b = y_i, \qquad i = 1, 2, \ldots, n,$$

for the two unknowns $a$ and $b$. The application of the method of least squares to this problem is called **linear regression** in statistics. The matrix of the system is

$$\mathbf{A} = \begin{pmatrix} x_1 & 1 \\ x_2 & 1 \\ \cdots\cdots \\ x_n & 1 \end{pmatrix}.$$

For the normal equations we require

$$\mathbf{A}^T\mathbf{A} = \begin{pmatrix} \sum x_i^2 & \sum x_i \\ \sum x_i & n \end{pmatrix}, \qquad \mathbf{A}^T\mathbf{v} = \begin{pmatrix} \sum x_i y_i \\ \sum y_i \end{pmatrix}$$

(all sums from $i = 1$ to $i = n$). The normal equations thus are

$$\sum x_i^2 a + \sum x_i b = \sum x_i y_i,$$
$$\sum x_i a + nb = \sum y_i. \tag{5}$$

The algebraic solution of the system (5) is

$$a = \frac{\sum x_i y_i - \dfrac{1}{n}\sum x_i \sum y_i}{\sum x_i^2 - \dfrac{1}{n}(\sum x_i)^2}, \tag{6a}$$

$$b = \frac{1}{n}\sum y_i - a\frac{1}{n}\sum x_i, \tag{6b}$$

and this form of the solution is also frequently used numerically, especially on pocket calculators, where its evaluation is facilitated by the $\sum+$ instruction.

Unfortunately, the formulas (6) are subject to cancellation, and hence unstable. This is particularly so if the $x_i$ are close together but far removed from the origin. For instance, for the data

| $i$ | $x_i$ | $y_i$ |
|---|---|---|
| 1 | 1.000001 | 2.000005 |
| 2 | 1.000002 | 2.000007 |

which are perfectly fitted by the linear function $y = 2x + 3 * 10^{-6}$, (6a) cannot be evaluated on some calculators because the denominator numerically equals zero.

In order to compute the linear regression function in a stable manner, we introduce the running means

$$\mu := \frac{1}{n}\sum_{i=1}^{n} x_i,$$

$$v := \frac{1}{n}\sum_{i=1}^{n} y_i.$$

Because the slope of the regression line is not affected by a shift of the origin of the $x$ and of the $y$ axis, we also have

$$a = \frac{\sum (x_i - \mu)(y_i - v)}{\sum (x_i - \mu)^2},$$ 

(7a)

which is a stable formula. As before,

$$b = v - \mu a.$$ 

(7b)

A problem arises if we wish to compute the regression line dynamically; that is, if we wish to allow for the possibility of adding or removing data points. If the unstable formula (6a) is used, this is easily accomplished by adding or subtracting the corresponding terms from the sums $\sum x_i, \sum y_i, \sum x_i^2, \sum x_i y_i$. The use of formula (7a), if this formula is taken literally, would seem to require recomputing the means $\mu$ and $v$, and then recomputing from scratch the sums occurring in (7a), using the new values of $\mu$ and $v$. In order to avoid this, we find recurrence relations for the sums

$$q_n := \sum_{i=1}^{n} (x_i - \mu_n)^2,$$

$$r_n := \sum_{i=1}^{n} (x_i - \mu_n)(y_i - v_n).$$

Simple algebra shows that

$$q_{n+1} = q_n + \frac{n+1}{n}(x_{n+1} - \mu_{n+1})^2,$$

$$r_{n+1} = r_n + \frac{n+1}{n}(x_{n+1} - \mu_{n+1})(y_{n+1} - v_{n+1}).$$

Similarly, if the point $(x_n, y_n)$ is removed,

$$q_{n-1} = q_n - \frac{n-1}{n}(x_n - \mu_{n-1})^2,$$

$$r_{n-1} = r_n - \frac{n-1}{n}(x_n - \mu_{n-1})(y_n - v_{n-1}).$$

Both sets of formulas may be combined into one set by letting $q := q_n$, $r := r_n$,

$\mu := \mu_n$, $v := v_n$ and marking the corresponding values after adding or removing a point $(x, y)$ by a prime. Thus,

$$n' := \begin{vmatrix} n + 1, \text{ if the point is added,} \\ n - 1, \text{ if the point is removed,} \end{vmatrix}$$

so that, for instance, $q' = q_{n'}$. We then have

$$q' = q + (n' - n)\frac{n'}{n}(x - \mu')^2$$

$$r' = r + (n' - n)\frac{n'}{n}(x - \mu')(y - v'). \tag{8}$$

The means $\mu$, $v$ are computed from the sums

$$s := s_n := \sum_{i=1}^{n} x_i,$$

$$t := t_n := \sum_{i=1}^{n} y_i \tag{9}$$

by forming

$$\mu = \frac{1}{n}s, \qquad v = \frac{1}{n}t; \tag{10}$$

the sums $s$, $t$ are calculated recursively by

$$s' = s + (n' - n)x, \qquad t' = t + (n' - n)y,$$

where the data point $(x, y)$ is either added or removed. After the quantities $q'$, $r'$, $\mu'$, $v'$ are computed, the coefficients of the new regression line are simply

$$a = \frac{r'}{q'}, \qquad b = v' - a\mu'. \tag{11}$$

**DEMONSTRATION 4.3-3:** This demonstration compares the stability properties of the formulas (6) and (11). The values of $x_i$ are $\pi$, $e$, $e^{-\pi}$ plus $10^{2m}$ for $m = 0, 1, 2, 3$. The values of $y_i$ are the (machine computed) values of $\sqrt{2}x_i + \frac{1}{3}$. The exact regression function thus is $y = \sqrt{2}x + \frac{1}{3}$ in all cases.

| $x_k$ | Equations (6) | | Equations (11) | |
|---|---|---|---|---|
| | $a$ | $b$ | $a$ | $b$ |
| $\pi$ | — | — | — | — |
| $e$ | 1.14121344 | 0.33333369 | 1.41421357 | 0.33333332 |
| $e^{-\pi}$ | 1.41421356 | 0.33333334 | 1.41421356 | 0.33333333 |
| $\pi + 10^2$ | — | — | — | — |
| $e + 10^2$ | 1.41433196 | 0.32114680 | 1.41421319 | 0.33337140 |
| $e^{-\pi} + 10^2$ | 1.41421068 | 0.33362730 | 1.41421359 | 0.33333010 |
| $\pi + 10^4$ | — | — | — | — |
| $e + 10^4$ | 1.00000000 | 4143.68257 | 1.41420202 | 0.44876000 |
| $e^{-\pi} + 10^4$ | 1.38596491 | 282.875420 | 1.41421557 | 0.31325000 |
| $\pi + 10^6$ | — | — | — | — |
| $e + 10^6$ | Error | — | 1.41509434 | −880.447000 |
| $e^{-\pi} + 10^6$ | — | — | 1.41404298 | 170.919000 |

## PROBLEMS

**1**  The decay of a radioactive substance is described by the formula

$$y(t) = y(0)e^{-ct},$$

where $y(t)$ denotes the amount of substance at time $t$, and where $c > 0$ is a constant typical for the substance. The *halftime* of the substance is the time $T$ for which $y(T) = \frac{1}{2}y(0)$. For a sample of the recently discovered substance Paulium the following amounts $y_i$ have been measured at the times $t_i$:

| $t_i$ | 1.5 | 4.0 | 6.5 | 9.0 | 11.5, |
|---|---|---|---|---|---|
| $y_i$ | 25.9 | 22.7 | 20.3 | 17.7 | 15.6. |

Compute the halftime $T_p$ of Paulium.
   [Theoretically the equations

$$y_i = ae^{-ct_i}, \qquad i = 1, 2, \ldots, 5,$$

should hold, with unknown values of $a$ and $c$. The equations can be made linear by taking logarithms. The resulting overdetermined system is to be solved by the method of least squares. Knowing $c$ it is easy to find $T_p$.]

2  A satellite moves in a strongly excentric orbit around the earth. The following observations of its position have been made at various tracking stations in a certain system of polar coordinates $(r, \phi)$:

| $\phi$ | 48° | 88° | 150° | 221° | 247° | 311° | 359° |
|--------|------|------|------|------|------|------|------|
| $r$ | 4.32 | 2.05 | 1.18 | 1.26 | 1.52 | 4.25 | 9.98 |

By Kepler's law the coordinates should satisfy

$$r = \frac{p}{1 - e \cos \phi},$$

where $p$ is a parameter and $e$ is the excentricity. Estimate $p$ and $e$ from the given observations by the method of least squares.

[The parameters must appear linearly for the method to be applicable. Hence, remove the denominator.]

3  The weight $w$ in pounds of an infant at the age of $t$ weeks has been recorded as follows:

| $t$ | 0 | 4 | 9 | 13 | 18 | 22 | 27 |
|-----|------|------|-------|-------|-------|-------|-------|
| $w$ | 8.62 | 9.02 | 11.07 | 12.10 | 14.52 | 16.15 | 17.67 |

(a)  Fit a linear function $y = at + b$ through these points. How much weight did the infant gain per week on the average?

(b)  Fit a linear function through any four consecutive data points. How does the average gain per week change as a function of time?

---

## §4.4  THE EXCHANGE ALGORITHM

Here we discuss another elimination algorithm of numerical linear algebra called the **exchange algorithm**. This algorithm is more general than Gaussian elimination; it solves the following problems:

(1)  matrix inversion;

(2)  solution of linear systems of equations;

(3)  linear optimization problems (also called linear programming problems, see §4.5).

Let $t$ and $s$ be positive integers, not necessarily equal. We consider a linear map $R^t \to R^s$. Denoting the coordinates in $R^t$ by $(x_1, x_2, \ldots, x_t)$ and the coordinates in $R^s$ by $(y_1, y_2, \ldots, y_s)$, the map shall be given by the formulas

$$y_i = \sum_{j=1}^{t} a_{ij} x_j, \qquad i = 1, 2, \ldots, s. \tag{1}$$

As in the case of linear systems, the map is fully determined by its *tableau*,

$$
\begin{array}{c|cccc}
 & x_1 & x_2 & & x_t \\
\hline
y_1 = & a_{11} & a_{12} & \cdots & a_{1t} \\
y_2 = & a_{21} & a_{22} & \cdots & a_{2t} \\
\cdots & \cdots & \cdots & & \cdots \\
y_s = & a_{s1} & a_{s2} & \cdots & a_{st}
\end{array}
$$

which is to be read in the usual manner by forming scalar products with the variables indicated at the head of the tableau.

We now shall describe the basic operation of the exchange algorithm, called an **exchange step** ($E$ step for brevity). Mathematically, this simply consists of selecting a pair of indices $(p, q)$ so that $a_{pq} \neq 0$ and interchanging the roles of $y_p$ and $x_q$. That is, we wish to express the variables

$$y_1, \ldots, y_{p-1}, x_q, y_{p+1}, \ldots, y_s$$

in terms of

$$x_1, \ldots, x_{q-1}, y_p, x_{q+1}, \ldots, x_t$$

while at the same time maintaining the linear relations (1). The element $a_{pq}$ is called the **pivot** of the exchange step. Obviously, if the tableau is square, $s = t$, all $x_i's$ may be exchanged against all $y_j's$ by successive $E$ steps, thus producing the inverse map of the map (1). Because the matrix of the inverse map is the inverse of the matrix of the given map, it is clear that the exchange algorithm can be used to compute inverses of matrices.

To produce the algebraic formulas necessary to carry out an $E$ step algorithmically, we solve the $p$-th equation (1) for $x_q$,

$$x_q = \frac{1}{a_{pq}} \left( y_p - \sum_{\substack{j=1 \\ j \neq q}}^{t} a_{pj} x_j \right),$$

and substitute the result into the $i$-th equation $(i \neq q)$:

$$y_i = \sum_{j=1}^{t} a_{ij} x_j$$

$$= \sum_{\substack{j=1 \\ j \neq q}}^{t} a_{ij} x_j + a_{iq} x_q$$

$$= \sum_{\substack{j=1 \\ j \neq q}}^{t} a_{ij} x_j + \frac{a_{iq}}{a_{pq}} \left( y_p - \sum_{\substack{j=1 \\ j \neq q}}^{t} a_{pj} x_j \right)$$

or

$$y_i = \sum_{\substack{j=1 \\ j \neq q}}^{t} \left( a_{ij} - \frac{a_{iq}}{a_{pq}} a_{pj} \right) x_j + \frac{a_{iq}}{a_{pq}} y_p .$$

Describing the linear relations (1) after the exchange step again by a tableau, appropriately marked,

|         | $x_1$      | $x_{q-1}$ | $y_p$      | $x_{q-1}$ | $x_t$      |
|---------|------------|-----------|------------|-----------|------------|
| $y_1$   | $a'_{11}$  |           | $a'_{1q}$  |           | $a'_{1t}$  |
| $y_{p-1}$|           |           |            |           |            |
| $x_q$   | $a'_{p1}$  |           | $a'_{pq}$  |           | $a'_{pt}$  |
| $y_{p+1}$|           |           |            |           |            |
| $y_s$   | $a'_{s1}$  |           | $a'_{sq}$  |           | $a'_{st}$  |

the new elements $a'_{ij}$ are given in terms of the old ones as follows:

(a)  for $i = p$ (pivot row)

$$a'_{pq} = \frac{1}{a_{pq}}$$

$$a'_{pj} = -\frac{a_{pj}}{a_{pq}}, \qquad j \neq q$$

(b)  for $i \neq p$ (elements not in pivot row)

$$a'_{iq} = \frac{a_{iq}}{a_{pq}}$$

$$a'_{ij} = a_{ij} - \frac{a_{iq}}{a_{pq}} a_{pj}, \qquad j \neq q.$$

In machine computation one will write the new tableau $(a'_{ij})$ over the old tableau $(a_{ij})$. This is possible without intermediate storage if the operations are arranged as follows:

$$a'_{pq} = \frac{1}{a_{pq}} \qquad\qquad\qquad \text{(pivot element)}$$

$$a'_{pj} = -a'_{pq}a_{pj}, \qquad j \neq q \qquad \text{(remaining elements in pivot row)}$$

$$a'_{ij} = a_{ij} + a'_{pj}a_{iq}, \quad j \neq q, \;\; i \neq p \qquad \text{(elements neither in pivot row nor in pivot column)}$$

$$a'_{iq} = a'_{pq}a_{iq}, \qquad\qquad\qquad i \neq p \qquad \text{(remaining elements in pivot column)}$$

(2)

These are the basic formulas that will be used in all the examples in this and the following section.

### Matrix Inversion

As indicated, the exchange algorithm may be used to compute the inverse of a square matrix $(s = t^* = n)$ by exchanging all variables. In selecting the pivots, care should be taken not to undo exchanges that already have been made. Unless all pivots are chosen in the main diagonal, the variables $x_i$ and $y_j$ will not be in their natural order when all the exchanges have been made. Thus in order to obtain the inverse matrix, the rows and columns of the final tableau must be permuted so that the variables appear in their natural order.

**DEMONSTRATION 4.4-1:** Here we invert the matrix of Demonstration 4.1-1, using rational arithmetic. The starting tableau is

|       | $x_1$ | $x_2$         | $x_3$         |
|-------|-------|---------------|---------------|
| $y_1$ | $1$   | $\frac{1}{2}$ | $\boxed{\frac{1}{3}}$ |
| $y_2$ | $\frac{1}{2}$ | $\frac{1}{3}$ | $\frac{1}{4}$ |
| $y_3$ | $\frac{1}{3}$ | $\frac{1}{4}$ | $\frac{1}{5}$ |

Although it would be feasible in this example to select the pivots in their natural order, we avoid doing so for the sake of illustration. Thus selecting the (1, 3) element as first pivot*, we obtain

|       | $x_1$          | $x_2$           | $y_1$         |
|-------|----------------|-----------------|---------------|
| $x_3$ | $-3$           | $-\frac{3}{2}$  | $3$           |
| $y_2$ | $-\frac{1}{4}$ | $-\frac{1}{24}$ | $\frac{3}{4}$ |
| $y_3$ | $-\frac{4}{15}$ | $\boxed{-\frac{1}{30}}$ | $\frac{3}{5}$ |

---

* Pivot elements are framed.

The choice of $(3, 2)$ as second pivot yields

|       | $x_1$              | $y_3$           | $y_1$          |
|-------|--------------------|-----------------|----------------|
| $x_3$ | 5                  | 30              | $-15$          |
| $y_2$ | $-\frac{1}{36}$    | $\frac{5}{6}$   | $\frac{1}{4}$  |
| $x_2$ | $-\frac{16}{3}$    | $-20$           | 12             |

In the last step, $(2, 1)$ necessarily must be pivot because the variables $x_1$ and $y_2$ are the only ones that have not yet been exchanged. This yields

|       | $y_2$   | $y_3$   | $y_1$  |
|-------|---------|---------|--------|
| $x_3$ | $-180$  | 180     | 30     |
| $x_1$ | $-36$   | 30      | 9      |
| $x_3$ | 192     | $-180$  | $-36$  |

This tableau does not yet represent the inverse matrix because the marks $x_i$ and $y_j$ are not yet in their natural order. Interchanging the rows in order to put the $x_i$ in their natural order yields

|       | $y_2$   | $y_3$   | $y_1$  |
|-------|---------|---------|--------|
| $x_1$ | $-36$   | 30      | 9      |
| $x_2$ | 192     | $-180$  | $-36$  |
| $x_3$ | $-180$  | 180     | 30     |

and interchanging the columns accordingly yields

|       | $y_1$   | $y_2$   | $y_3$  |
|-------|---------|---------|--------|
| $x_1$ | 9       | $-36$   | 30     |
| $x_2$ | $-36$   | 192     | $-180$ |
| $x_3$ | 30      | $-180$  | 180    |

Thus, the given matrix has the inverse

$$\mathbf{A}^{-1} = 3 \begin{pmatrix} 3 & -12 & 10 \\ -12 & 64 & -60 \\ 10 & -60 & 60 \end{pmatrix}.$$

Being the inverse of a symmetric matrix, $A^{-1}$ is again symmetric. Its large elements point to the ill-conditioned nature of the problem considered in Demonstration 4.1-1.

To count the number of operations required to invert a matrix by the exchange algorithm, we note that each $E$ step requires the same number of operations, namely [if performed according to (2)]

$$\delta + [(n-1) + (n-1)^2 + (n-1)]\mu + (n-1)^2\alpha = \delta + (n^2-1)\mu + (n-1)^2\alpha.$$

Thus, for $n$ $E$ steps, the operations count is

$$n\delta + n(n^2-1)\mu + n(n-1)^2\alpha,$$

which approximately equals the count for three Gaussian eliminations.

Of course, the inverse matrix of $A$ also can be computed by solving the $n$ linear systems $Ax = e_i$, where the $e_i$ are the $n$ coordinate unit vectors. Advantage may be taken of the fact that all these systems have the same matrix, and that therefore the $L$-$R$ decomposition must be computed only once (see §4.2). Some further savings result from the special form of the vectors $e_i$. The total operations count again results in approximately $n^3(\mu + \alpha)$. However, the program for computing the inverse by the exchange algorithm is more straightforward.

**Gauss-Jordan Algorithm**

Let a system of $n$ equations with $n$ unknowns be written in the form

$$\sum_{j=1}^{n} a_{ij}x_j + a_{i0} = 0, \qquad i = 1, 2, \ldots, n. \tag{3}$$

This system may also be solved via the exchange algorithm, but without computing the inverse matrix. The result is called the **Gauss-Jordan algorithm**. Consider the system of linear forms described by the following tableau:

|       | $x_1$    | $x_2$    | $\cdots$ | $x_n$    | $x_0$    |
|-------|----------|----------|----------|----------|----------|
| $y_1$ | $a_{11}$ | $a_{12}$ |          | $a_{1n}$ | $a_{10}$ |
| $y_2$ | $a_{21}$ | $a_{22}$ |          | $a_{2n}$ | $a_{20}$ |
| $\vdots$ |       |          |          |          |          |
| $y_n$ | $a_{n1}$ | $a_{n2}$ |          | $a_{nn}$ | $a_{n0}$ |

In terms of this tableau, the problem of solving the system (3) may be formulated as follows: Put $x_0 := 1$, and determine $x_1, \ldots, x_n$ so that all $y_j = 0$. This may be accomplished by the exchange algorithm by exchanging $x_1, \ldots, x_n$ against $y_1, \ldots,$

$y_n$, leaving $x_0$ in its place. The choice of the pivots is arbitrary. Thus, after all exchanges have taken place, the tableau may look as follows:

|  | $y_3$ | $y_7$ | $\cdots$ | $y_5$ | $x_0$ |
|---|---|---|---|---|---|
| $x_2$ | $a'_{11}$ | $a'_{12}$ |  | $a'_{1n}$ | $a'_{10}$ |
| $x_8$ | $a'_{21}$ | $a'_{22}$ |  | $a'_{2n}$ | $a'_{20}$ |
| $x_1$ | $a'_{n1}$ | $a'_{n2}$ |  | $a'_{nn}$ | $a'_{n0}$ |

The tableau means, for instance, that

$$x_2 = a'_{11}y_3 + a'_{12}y_7 + \cdots + a'_{1n}y_5 + a'_{10},$$
$$x_8 = a'_{21}y_3 + \ldots, \qquad \text{and so on.}$$

If all $y_j = 0$, there follows

$$x_2 = a'_{10},$$
$$x_8 = a'_{20},$$

and so on, and we recognize that *the values in the $x_0$ column of the final tableau represent the solution of the given system* (3).

The computational work may be reduced substantially by the following observation. Each $E$ step produces a new column which is marked with some $y_i$. Because all $y_i$ are set equal to zero, the values of the elements in these columns do not affect the final result and thus may be replaced by zeros. If performed in this manner, the solution of the system (3) by the exchange algorithm requires approximately $n\delta + \frac{1}{2}n^3(\mu + \alpha)$. This is somewhat more than the $n\delta + \frac{1}{3}n^3(\mu + \alpha)$ required by Gaussian elimination, but the program is more compact.

**DEMONSTRATION 4.4-2:** We solve the system of Demonstration 4.1-1 by the Gauss-Jordan algorithm, again using four-digit decimal arithmetic. The initial tableau is

|  | $x_1$ | $x_2$ | $x_3$ | $x_0$ |
|---|---|---|---|---|
| $y_1$ | 1.000 | 0.5000 | 0.3333 | $-1.000$ |
| $y_2$ | 0.5000 | 0.3333 | 0.2500 | $-0.5833$ |
| $y_3$ | 0.3333 | 0.2500 | 0.2000 | $-0.4333$ |

Selecting the pivots in their natural order, the following sequence of tableaus results (omitting entries that are not needed in the following):

|       | $y_1$     | $x_2$     | $x_3$     | $x_0$      |
|-------|-----------|-----------|-----------|------------|
| $x_1$ | (1.000)   | −0.5000   | −0.3333   | 1.000      |
| $y_2$ | —         | 0.08330   | 0.08330   | −0.08330   |
| $y_3$ | —         | 0.08330   | 0.08890   | −0.1000    |

|       | $y_1$ | $y_2$    | $x_3$     | $x_0$      |
|-------|-------|----------|-----------|------------|
| $x_1$ | —     | —        | 0.1665    | 0.5002     |
| $x_2$ | —     | (12.00)  | −0.9996   | 0.9996     |
| $y_3$ | —     | —        | 0.005630  | −0.01673   |

|       | $y_1$ | $y_2$ | $y_3$    | $x_0$    |
|-------|-------|-------|----------|----------|
| $x_1$ | —     | —     | —        | 0.9949   |
| $x_2$ | —     | —     | —        | −1.970   |
| $x_3$ | —     | —     | (177.6)  | 2.971    |

In the $x_0$-column, we read the solution

$$x_1 = 0.9949$$
$$x_2 = -1.970$$
$$x_3 = 2.971.$$

The solution is somewhat worse than that obtained by Gaussian elimination. The solution could be improved, of course, by computing the residues in double precision and using iterative refinement as described in §4.2. No advantage can now be taken of an $L$-$R$ decomposition that already exists, however, and a full new system must now be solved in order to compute the corrections.

## §4.5   LINEAR INEQUALITIES; OPTIMIZATION

Here we present an application of the exchange algorithm in optimization. The general problem to be considered is best illustrated by the following special *example*:

A grocer has at his disposal

|      |                          |
|------|--------------------------|
| 5000 | quarts of orange juice,  |
| 1500 | quarts of apricot juice. |

He can manufacture and sell the following mixes:

Mix 1:  Pure orange juice, at a price of $0.90 per quart;
Mix 2:  80% orange juice, 20% apricot juice, at a price of $0.96 per quart;
Mix 3:  60% orange juice, 40% apricot juice, at a price of $0.99 per quart.

How many quarts of each mix should he manufacture in order to maximize his profit?

Let $x_1$, $x_2$, $x_3$ be the amounts of the three mixes manufactured. The fact that the basic materials are available only in limited amounts yields the inequalities,

$$\text{Amount of orange juice} = x_1 + 0.8x_2 + 0.6x_3 \leq 5000$$
$$\text{Amount of apricot juice} = \phantom{x_1 + } 0.2x_2 + 0.4x_3 \leq 1500. \tag{1}$$

The problem is to maximize the profit (in $)

$$z := 0.90x_1 + 0.96x_2 + 0.99x_3. \tag{2}$$

Mathematically, the problem to be solved is an extremal problem with side conditions. If the side conditions (or constraints) have the form of equations, such problems are considered in analysis and are treated by the method of Lagrange multipliers. In the present situation, however, the constraints have the form of inequalities. Two kinds of constraints appear:

(a)  there are *upper bounds* on the total amounts of the two fruit juices used;

(b)  there are *lower bounds* on the amounts of the mixes that are manufactured. Clearly, it is not possible to manufacture a negative amount of a mix. Thus, we have the additional constraints

$$x_1 \geq 0, \qquad x_2 \geq 0, \qquad x_3 \geq 0. \tag{3}$$

For the mathematical treatment of the problem, it is convenient to reformulate it so that all constraints have the form $x \geq 0$. We therefore introduce auxiliary variables indicating the amount of orange juice and of apricot juice not used:

$$y_1 := \text{orange juice not used} = 5000 - x_1 - 0.8x_2 - 0.6x_3 \geq 0,$$
$$y_2 := \text{apricot juice not used} = 1500 - \phantom{x_1 - } 0.2x_2 - 0.4x_3 \geq 0.$$

Now the constraints (1) may be expressed as

$$y_1 \geq 0, \qquad y_2 \geq 0.$$

The variables $y_1$, $y_2$ are called **slack variables**. The function $z$ to be maximized is the **objective function**.

As in preceding sections, it is again possible to arrange the data of the problem in a rectangular tableau, which is to be read in the usual manner if names of variables are attached to the rows and columns.

|       | $x_1$   | $x_2$   | $x_3$   | 1     |
|-------|---------|---------|---------|-------|
| $y_1$ | $-1.00$ | $-0.80$ | $-0.60$ | 5000  |
| $y_2$ | 0       | $-0.20$ | $-0.40$ | 1500  |
| $z$   | 0.90    | 0.96    | 0.99    | 0     |

The problem to be solved now may be formulated thus: Maximize the linear function $z$ defined by the table under the constraint that all $x_i \geq 0$ and that the linear functions $y_i$ defined by the table are also non-negative.

This type of problem has turned out to be extremely important in operations research and mathematical economics. In real life problems there may be several hundreds of variables and constraints. Here, we merely wish to show how to solve the basic problem of the above type by means of the exchange algorithm.

The *basic idea* is to try to achieve, by suitable exchange steps, a tableau where all coefficients in the $z$-row with the exception of the last are *negative*. Suppose, for concreteness, that after a certain number of $E$ steps the following array has been obtained:

|       | $x_1$   | $y_1$   | $x_3$   | 1     |
|-------|---------|---------|---------|-------|
| $x_2$ | *       | *       | *       | *     |
| $y_2$ | *       | *       | *       | *     |
| $z$   | $-0.92$ | $-0.27$ | $-1.08$ | 6500  |

Because the side conditions $x_i \geq 0$, $y_j \geq 0$ always hold, the function $z$ assumes its maximum value for $x_1 = y_1 = x_3 = 0$; any other admissible system of values $x_1$, $y_1$, $x_3$ would yield a smaller value. The maximum value of $z$ equals 6500. If $x_1 = y_1 = x_3 = 0$, then the variables $x_2$ and $y_2$ have the values indicated in the "1" column. It follows that these values must not be negative.

There emerges the following strategy for the solution of our extremal problem: The given tableau should be transformed, by means of $E$ steps, into one where

(a)  the coefficients of the $z$-row (with the exception of the last) are negative;

(b)  the coefficients of the "1" column remain positive.

It is obvious that this strategic goal, in general, cannot be achieved in a single $E$ step. What *tactics* should be used to achieve it in several steps?

The answer is found by considering the general formulas for an $E$ step (equations (2) of §4.4). Let $a_{pq}$ be the chosen pivot. Then, the element in the "1" position of the *pivot row* changes according to the formula

$$a'_{pt} = -a_{pt}a'_{pq}, \qquad \text{where} \quad a'_{pq} = \frac{1}{a_{pq}}. \tag{4}$$

Because $a_{pt} > 0$ and $a'_{pt}$ should also be $>0$, there follows

**RULE 1:** *The pivot must be negative.*

The element in the $z$ position of the *pivot column* changes according to

$$a'_{sq} = a_{sq}a'_{pq}.$$

Because we want $a'_{sq} < 0$ and, as a consequence of Rule 1, have $a'_{pq} < 0$, we find that we need $a_{sq} > 0$. There follows

**RULE 2:** *The pivot must be selected in a column where the z element is positive.*

Finally, we take into consideration the fact that *all* elements in the "1" column (not only the one in the pivot row already accounted for) should remain positive. The new elements in the "1" column are for $i \neq p$ given by

$$a'_{it} = a_{it} + a_{iq}a'_{pt}$$

$$= a_{it} - a_{iq}\frac{a_{pt}}{a_{pq}}$$

$$= \frac{a_{it}a_{pt}}{a_{pq}}\left(\frac{a_{pq}}{a_{pt}} - \frac{a_{iq}}{a_{it}}\right).$$

According to Rules 1 and 2, we have $a_{it} > 0$, $a_{pt} > 0$, $a_{pq} < 0$, hence

$$\frac{a_{it}a_{pt}}{a_{pq}} < 0.$$

We want $a'_{it} > 0$, thus we must have

$$\frac{a_{pq}}{a_{pt}} - \frac{a_{iq}}{a_{it}} < 0 \qquad \text{for all } i \neq p$$

or

$$\frac{a_{pq}}{a_{pt}} < \frac{a_{iq}}{a_{it}} \qquad \text{for all } i \neq p.$$

There follows

**RULE 3:** *After the pivot column (index q) has been selected, the pivot row (index p) must be chosen such that*

$$\frac{a_{pq}}{a_{pt}} = \min_{1 \leq i < s} \frac{a_{iq}}{a_{it}}. \tag{5}$$

Thus, after selecting a pivot column satisfying Rules 1 and 2 (that is, having a positive element in the $z$ position and having some negative elements as potential pivots), one must form the so-called *characteristic quotients*

$$c_i := \frac{a_{iq}}{a_{it}}, \qquad i = 1, 2, \ldots, s - 1. \tag{6}$$

The correct index $p$ of the pivot row then is that value of $i$ for which $c_i$ is a minimum, that is, most negative.

If the extremum problem has any solution at all, it must be possible to find it in a finite number of $E$ steps according to the foregoing rules. Of course, it may happen that the problem has no solution, either because there exists no system of values satisfying all constraints, or because the function $z$ is unbounded on the set of all points satisfying the constraints. In most applications, it will be clear from the context whether a solution can be expected to exist.

The whole algorithm for solving an extremal problem of the above type by means of the exchange algorithm is called the **simplex algorithm** in view of a geometric interpretation which cannot be discussed here. The algorithm was developed in the late 1940s by G. B. Dantzig. Somewhat earlier, the algorithm had been found by L. V. Kantorovich in the U.S.S.R. who later was awarded the Nobel prize in economics essentially for this contribution.

We prove one general fact about the algorithm.

**THEOREM 4.5:** *In the simplex algorithm, the values $a_{st}$ increase monotonically.*

*Proof:* Let $a_{pq}$ be the pivot element. Evidently $p \neq s$, $q \neq t$, because the variable $z$ and the constant 1 are never exchanged. According to the general formulas for an $E$ step,

$$a'_{st} = a_{st} - a_{sq} \frac{a_{pt}}{a_{pq}}.$$

According to Rule 2, $a_{sq} > 0$, and according to Rule 1, $a_{pq} < 0$; furthermore $a_{pt} > 0$ always. There follows

$$a'_{st} - a_{st} = -a_{sq}\frac{a_{pt}}{a_{pq}} > 0,$$

as was to be shown.

**DEMONSTRATION 4.5-1:** We return to our problem about fruit juices. This problem was described by the tableau

|       | $x_1$   | $x_2$   | $x_3$   | 1    |
|-------|---------|---------|---------|------|
| $y_1$ | $-1.00$ | $-0.80$ | $-0.60$ | 5000 |
| $y_2$ | 0       | $-0.20$ | $-0.40$ | 1500 |
| $z$   | 0.90    | 0.96    | 0.99    | 0    |

A pivot for the first exchange must now be selected. Rule 1 can be satisfied in every column, and Rule 2 will be satisfied in all cases. We arbitrarily select the first column as our first pivot column; since $a_{11}$ is the only negative element there, it is the only possible choice for a pivot. (Mechanical application of Rule 3 would yield the same result.) Carrying out the $E$ step in three-decimal arithmetic yields

|       | $y_1$   | $x_2$   | $x_3$   | 1    |
|-------|---------|---------|---------|------|
| $x_1$ | $-1.00$ | $-0.80$ | $-0.60$ | 5000 |
| $y_2$ | 0       | $-0.20$ | $-0.40$ | 1500 |
| $z$   | $-0.90$ | 0.24    | 0.45    | 4500 |

The bottom element in the first column is negative, as expected, but the remaining bottom elements are still positive. We try an $E$ step with the pivot in the second column, which satisfies Rules 1 and 2. The characteristic quotients are

$$c_1 = \frac{a_{12}}{a_{14}} = \frac{-0.80}{5000} = -1.6 * 10^{-4},$$

$$c_2 = \frac{a_{22}}{a_{24}} = \frac{-0.20}{1500} = -1.33 * 10^{-4}.$$

Since $c_1 < c_2$, Rule 3 requires that the first row be selected as pivot row. Carrying out the $E$ step yields

|       | $y_1$   | $x_1$   | $x_3$   | 1    |
|-------|---------|---------|---------|------|
| $x_2$ | $-1.25$ | $-1.25$ | $-0.75$ | 6250 |
| $y_2$ | 0.25    | 0.25    | $-0.25$ | 250  |
| $z$   | $-1.20$ | $-0.30$ | 0.27    | 6000 |

One more $E$ step must be carried out, because $a_{33} > 0$ still. The characteristic quotients of the third column are

$$c_1 = \frac{a_{13}}{a_{14}} = \frac{-0.75}{6250} = -1.20 * 10^{-4},$$

$$c_2 = \frac{a_{23}}{a_{24}} = \frac{-0.25}{250} = -1.00 * 10^{-3},$$

and it is obvious that $a_{23}$ must be selected as pivot element. The $E$ step yields

|       | $y_1$   | $x_1$   | $y_2$   | 1    |
|-------|---------|---------|---------|------|
| $x_2$ | $-2.00$ | $-2.00$ | 3.00    | 5500 |
| $x_3$ | 1.00    | 1.00    | $-4.00$ | 1000 |
| $z$   | $-0.93$ | $-0.03$ | $-1.08$ | 6270 |

The desired strategic goal has now been reached: All coefficients of the objective function (with the exception of the last) are negative, while the "1" column has remained positive. Thus, the maximum of

$$z = 6270 - 0.93y_1 - 0.03x_1 - 1.08y_2$$

under the constraint that all variables are non-negative is obtained for

$$y_1 = x_1 = y_2 = 0.$$

This means that

$$x_2 = 5000 \text{ quarts of the } 80/20 \text{ mix}$$

and

$$x_3 = 1000 \text{ quarts of the } 60/40 \text{ mix}$$

should be produced, and that no pure orange juice should be put on the shelves. This will use all resources ($y_1 = y_2 = 0$), and the profit will amount to \$6270.

**Minimum Problems**

Here we consider the variant of the foregoing problem where the aim is to *minimize* the objective function. Of course, this may be transformed formally into a maximum problem simply by changing the sign of the objective function. A complication nevertheless arises, however, as may be seen from the following example.

Let us again consider the situation described in our problem about fruit juices, but let us now take the point of view of the *customer* who wishes to buy the equivalent of 20 quarts of orange juice and 10 quarts of apricot juice as cheaply as possible by buying suitable amounts of the grocer's mixes. If the amounts of the mixes are again denoted by $x_1$, $x_2$, $x_3$, the following conditions are now to be satisfied:

$$\text{amount of orange juice} = x_1 + 0.8x_2 + 0.6x_3 \geq 20;$$
$$\text{amount of apricot juice} = \phantom{x_1 + {}} 0.2x_2 + 0.4x_3 \geq 10. \tag{7}$$

The objective is to minimize the total expenses

$$0.90x_1 + 0.96x_2 + 0.99x_3,$$

and, of course, there are the constraints

$$x_1 \geq 0, \qquad x_2 \geq 0, \qquad x_3 \geq 0,$$

because it is not possible to buy negative amounts of any mix.

We transform this into a problem of the type considered earlier by considering the negative expenses

$$z := -0.90x_1 - 0.96x_2 - 0.99x_3,$$

which are to be *maximized*, and by formulating the constraints (7) in terms of the non-negative slack variables

$$y_1 := \text{surplus of orange juice} = x_1 + 0.8x_2 + 0.6x_3 - 20,$$
$$y_2 := \text{surplus of apricot juice} = \phantom{x_1 + {}} 0.2x_2 + 0.4x_3 - 10.$$

The problem is represented by the following tableau:

|       | $x_1$   | $x_2$   | $x_3$   | 1     |
|-------|---------|---------|---------|-------|
| $y_1$ | 1       | 0.8     | 0.6     | $-20$ |
| $y_2$ | 0       | 0.2     | 0.4     | $-10$ |
|       |         |         |         |       |
| $z$   | $-0.90$ | $-0.96$ | $-0.99$ | 0     |

The complication mentioned above now consists of the fact that the coefficients in the "1" column are negative. Therefore, although all coefficients in the $z$ row with the exception of the last are negative, $x_1 = x_2 = x_3 = 0$ does not constitute a solution to our problem, since this would yield negative and, hence, inadmissible values of the $y_j$. It thus becomes necessary to transform the tableau by means of a *preparatory step* into one where the elements in the "1" column have the correct sign for the application of the simplex algorithm.

To find the preparatory step, consider the formulas for the new elements in the "1" column. We have

$$a'_{pt} = -a_{pt}\frac{1}{a_{pq}}, \tag{8}$$

$$a'_{it} = a_{it}\frac{a_{pt}}{a_{pq}}\left(\frac{a_{pq}}{a_{pt}} - \frac{a_{iq}}{a_{it}}\right), \qquad i \neq p. \tag{9}$$

We assume for simplicity that all $a_{it} < 0$, and that therefore the signs of all $a_{it}$ are to be changed. Then (8) implies that we must have $a_{pq} > 0$. In this case,

$$a_{it}\frac{a_{pt}}{a_{pq}} > 0,$$

and the expression in parentheses must be $>0$. Therefore, the index $p$ must be chosen so that

$$\frac{a_{pq}}{a_{pt}} > \frac{a_{iq}}{a_{it}} \qquad \text{for all } i \neq p.$$

All this together yields the following:

**RULE for the preparatory step:** *Select a pivot column (index q) containing positive elements. Then select the pivot row (index p) such that*

$$\frac{a_{pq}}{a_{pt}} = \max_{1 \leq i < s} \frac{a_{iq}}{a_{it}}. \tag{10}$$

We recognize that the same characteristic quotients must be formed as before, but that now the choice of the index of the pivot row must be made so that the characteristic quotient is *maximized*.

**DEMONSTRATION 4.5-2:** In the foregoing fruit juice problem, either column 2 or 3 may be selected for the preparatory step because all their elements are positive. We arbitrarily select the second column. The characteristic quotients are

$$c_1 = \frac{a_{12}}{a_{14}} = \frac{0.8}{-20} = -0.04,$$

$$c_2 = \frac{a_{22}}{a_{24}} = \frac{0.2}{-10} = -0.02.$$

Because $c_2 > c_1$, $a_{22}$ must be selected as pivot for the preparatory step. The $E$ step yields

|       | $x_1$   | $y_2$  | $x_3$  | 1     |
|-------|---------|--------|--------|-------|
| $y_1$ | 1.00    | 4.00   | $\boxed{-1.00}$ | 20    |
| $x_2$ | 0       | 5.00   | $-2.00$ | 50    |
| $z$   | $-0.90$ | $-4.80$ | 0.93  | $-48$ |

The elements in the "1" column (with the exception of the last) are positive, as expected. The simplex algorithm now can be continued in the regular fashion. Evidently, the next pivot must be selected in the $x_3$ column, and, by forming the characteristic quotients, it is seen that $a_{13}$ is the correct pivot. Carrying out the $E$ step furnishes the tableau

|       | $x_1$   | $y_2$   | $y_1$   | 1        |
|-------|---------|---------|---------|----------|
| $x_3$ | 1.00    | 4.00    | $-1.00$ | 20       |
| $x_2$ | $-2.00$ | $-3.00$ | 2.00    | 10       |
| $z$   | 0.030   | $-1.08$ | $-0.93$ | $-29.40$ |

Unfortunately, $a_{31}$ has turned positive, so that an additional $E$ step with pivot in the first column is necessary. The correct pivot is found to be $a_{21}$, and performing the $E$ step we get the tableau

|       | $x_2$    | $y_2$   | $y_1$   | 1        |
|-------|----------|---------|---------|----------|
| $x_3$ | $-0.50$  | 2.50    | 0       | 25       |
| $x_1$ | $-0.50$  | $-1.50$ | 1.00    | 5        |
| $z$   | $-0.015$ | $-1.13$ | $-0.90$ | $-29.25$ |

All coefficients in the $z$ row (with the exception of the last) now are negative; the problem is solved. The customer minimizes expenses by buying 25 quarts of the 60/40 mix and 5 quarts of pure orange juice. Her expenses will be $29.25.

In some problems of practical interest not all elements $a_{it}$, $i = 1, 2, \ldots$, $s - 1$, in the "1" column are negative. It then may happen that a single preparatory step is not sufficient to make all $a_{it}$ positive. Several such steps should then be carried out. The pivot is always to be selected in those rows where $a_{jt} < 0$.

# PROBLEMS

1  Mr. Tischmacher, a cabinetmaker recently immigrated from Germany, prepares for the Christmas business. He makes tables and chairs. A table requires one sheet of plywood and 25 feet of 2 × 2 beam, and it can be manufactured in one day. A chair requires only 1/3 sheet of plywood and 15 feet of beam, but it requires two days of work. At Mr. Tischmacher's disposal are 50 working days, $16\frac{2}{3}$ sheets of plywood, and 600 feet of beam. He can sell a table for $300 and a chair for $150.

    **(a)**  How many tables and chairs should he make in order to maximize his profit?

    **(b)**  What will be his profit?

    **(c)**  Which of the available resources is not used up in the optimal solution?

2  A Swiss banker has at his disposal

> 4000 kilograms (kg) of pure silver,
> 7000 kg of gold.

He can manufacture and sell the following alloys:

    (i)  50% gold, 50% silver at a price of SFr. 9600 per kg;

    (ii)  75% gold, 25% silver at a price of SFr. 18000 per kg;

    (iii)  pure gold at a price of SFr. 15000 per kg.

    **(a)**  How much of each alloy should he manufacture in order to maximize his profit?

    **(b)**  A customer of the banker wishes to buy 2 kg of gold and 1 kg of silver as cheaply as possible by buying suitable amounts of the banker's alloys. How much of each alloy should he buy?

3  Two types of buildings are to be erected on a given piece of land. A type "A" building houses 45 people, a type "B" building houses 18 people. The area, cost, and labor required for one unit of each type are as follows:

| | Work (man-months) | Land (squ. feet) | Cost ($10^6$ $) |
|---|---|---|---|
| Type "A" | 180 | 8000 | 0.9 |
| Type "B" | 90 | 6000 | 0.3 |

The available resources are 6750 man-months of labor, 420,000 square feet of land, and $27 * 10^6$ \$.

    **(a)** How many buildings of each type are to be erected in order to accommodate as many people as possible?

    **(b)** How many people can be accommodated?

    **(c)** Which of the available resources is not exhausted?

**4** René Chiroc, an ambitious student at the elitist École Normale de Paris, tries to pass the comprehensive exams in the topics A, B, C simultaneously. Each exam is graded on a scale from 0 to 100. Chiroc has 60 days to study. If he allocates $x_T$ days to the preparation of topic $T$, the respective grades $g_T$ can be expected to be as follows:

$$g_A = 3x_A;$$

$g_B = 10 + 0.5x_A + 2x_B$ (topic $A$ helps to understand
      topic $B$, and the professor gives at least 10 points)

$g_C = -0.2x_A + 2x_C$ (topic $A$ confuses understanding of topic $C$)

$g_C \leq 90$ (professor never gives maximum number of points)

    **(a)** How much time should Mr. Chiroc allocate to the study of each topic in order to maximize the sum of the grades?

    **(b)** Suppose Chiroc suddenly loses his ambition and is content with an average grade of 40 which is sufficient for passing. How much time should he allocate to each topic in order to minimize the total preparation time?

    Point out any unusual features in this linear programming problem and comment on them.

---

## Bibliographical Notes and Recommended Reading

Concerning the theoretical background required for this chapter, Strang [1976] provides a balanced account with motivating applications. Concerning Gaussian elimination and choice of pivots, see Forsythe and Moler [1967]. Lawson and Hanson [1974] deal with the numerical solution of large-scale least squares problems. The exchange algorithm was pioneered by Stiefel [1963]. On linear optimization, see Gass [1969].

    Important problems of numerical linear algebra which are not dealt with here include the algebraic eigenvalue problem, for which Wilkinson [1965] still is the classical reference. For the theory of iterative methods for large systems, see Varga [1962] and Young [1971].

    Because the problems of numerical linear algebra are easily standardized, there exist comprehensive collections of practically tested algorithms (intended primarily for large computers). See, in particular, Wilkinson and Reinsch [1971] and Dongarra, *et al.* [1979].

# CHAPTER 5

# Approximation

In Chapter 2 we considered the approximation of *numbers*, such as fixed points or solutions of equations. In Chapters 5 and 6 we shall study the approximation of *functions* or, more generally, the approximation of numbers that depend on all or on infinitely many values of a function. Examples of numbers depending on functions, or **functionals**, are the definite integral of a function over a fixed interval or the value of the derivative at a fixed point.

To approximate a function means in practice to replace the function by a simpler function. Why should one want to do this? For the sake of illustration, consider the problem of computing values of the function sin $x$. Except for very special values of $x$, these values cannot be calculated exactly; in fact, it is shown in number theory that the number sin $x$ is transcendental for all rational $x \neq 0$. To calculate sin $x$ from its Maclaurin series,

$$\sin x = x - \frac{x^3}{3!} + \frac{x^5}{5!} - \frac{x^7}{7!} + \dots,$$

would require summing infinitely many terms of a series, which cannot be done. If values of sin $x$ are desired, this function must be approximated (with an error that is as small as possible) by a function which can actually be calculated.

But which functions can actually be calculated? For simplicity, let us disregard the fact that the number system of any computer is discrete. We thus hypothesize the existence of a computer with a mantissa having infinitely many decimal places or, more realistically, of a sequence of computers whose word length tends to infinity. Even such an idealized computer will be restricted to the execution of programs consisting of a finite number of instructions. Each instruction tells the computer to carry out one of the four basic arithmetical operations, subject perhaps to a condition which is expressed in terms of inequalities between real numbers. Any function that can be calculated by means of a finite number of additions, subtractions, multiplications, and divisions is *rational*. Thus, it is clear that the only functions that can actually be computed are the **piecewise rational functions**. The word, piecewise, refers to the fact that it is possible to partition the interval of definition into subintervals so that, in each

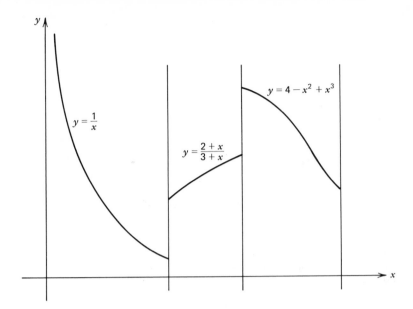

**Fig. 5.0.** Piecewise rational function.

subinterval, the function is represented by a rational function. These rational functions may be different in different subintervals, however. (See Figure 5.0.)

Thus, we see that the only functions that can be used for purposes of approximation are the piecewise rational functions. In this expository presentation, we shall go one step further and restrict ourselves to the approximation by *polynomials* or by *piecewise polynomial functions*.

## §5.1 TWO APPROXIMATION THEOREMS FROM CLASSICAL ANALYSIS

Here we quote, without proofs, two classical theorems concerning approximation. Very loosely speaking, the first theorem says that it is almost always possible to approximate by polynomials; the second says how to do it in a very special case.

**THEOREM 5.1a (the Weierstrass approximation theorem):** *Let $[a, b]$ be a finite, closed interval, let $f$ be continuous on $[a, b]$, and let $\varepsilon > 0$ be arbitrary. Then, there exists a polynomial $p$ so that*

$$| f(x) - p(x) | < \varepsilon \qquad \text{for all} \quad x \quad \text{in} \quad [a, b].$$

Thus, for instance, it is possible to approximate the function $f(x) := |x|$ on the interval $[-1, 1]$ with an error $< 10^{-12}$ by a single polynomial without

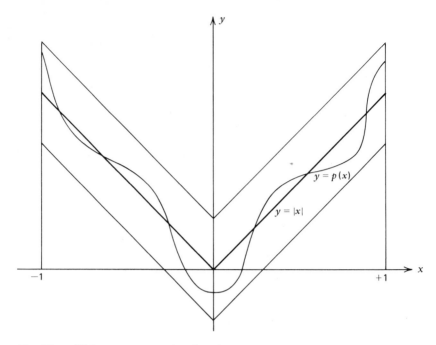

**Fig. 5.1a.** Weierstrass approximation theorem.

distinguishing the two cases $x \leq 0$ and $x \geq 0$. (See Figure 5.1a.) If the distinction were permitted, the function could be approximated with zero error by the two polynomials $p(x) = -x$ on $[-1, 0]$ and $p(x) = x$ on $[0, 1]$.

The Weierstrass approximation theorem (published by Weierstrass in 1886) can be proved in many different ways. Some proofs even contain an algorithm for the construction of the approximating polynomial. The approximating polynomials constructed in this manner are not usually suitable for purposes of numerical computation, however, because their degree is much higher than necessary.

The value of the Weierstrass theorem lies in the fact that nothing beyond continuity is assumed about the function $f$ to be approximated. If $f$ is assumed to have many derivatives, then it is not only possible to give a simple explicit formula for some approximating polynomial, but also to estimate the error committed in the approximation.

**THEOREM 5.1b (Taylor's theorem):** *Let $[a, b]$ be a finite or infinite interval, let $f$ be $n + 1$ times continuously differentiable on $[a, b]$, and let $t \in [a, b]$. If*

$$p(x) := f(t) + \frac{f'(t)}{1!}(x - t) + \cdots + \frac{f^{(n)}(t)}{n!}(x - t)^n,$$

*then for each* $x \in [a, b]$ *there exists a number* $t_x$ *lying strictly between t and* $x$ *so that*

$$f(x) - p(x) = \frac{f^{(n+1)}(t_x)}{(n+1)!}(x-t)^{n+1}. \tag{1}$$

The polynomial $p$ is called the **Taylor polynomial** of degree $n$ of $f$ at the point $t$. It is characterized by the fact that its value and the values of its first $n$ derivatives coincide with the corresponding values for $f$ at the point $x = t$.

**EXAMPLE 5.1-1:** We wish to approximate the function

$$f(x) := \sin x$$

on the interval $[0, \pi/2]$ with a *relative* error $< 10^{-14}$. Because $f(0) = 0$, the approximation must be exact at $x = 0$. We thus try to use the Taylor polynomial at $t = 0$. The Taylor polynomial of degree $2n - 1$ is

$$p(x) = x - \frac{x^3}{3!} + \frac{x^5}{5!} - \cdots + (-1)^{n-1}\frac{x^{2n-1}}{(2n-1)!}.$$

Equation (1) yields for the error

$$\sin x - p(x) = \pm \frac{x^{2n}}{(2n)!}\sin(t_x) \tag{2}$$

where $0 < t_x < x$. Now the *relative* error is to be bounded, which means that $n$ is to be chosen so that

$$\left|\frac{\sin x - p(x)}{\sin x}\right| < 10^{-14}, \qquad 0 < x \le \frac{\pi}{2}.$$

By (2), this condition is satisfied if

$$\left|\frac{x^{2n}}{(2n)!}\frac{\sin(t_x)}{\sin x}\right| < 10^{-14}, \qquad 0 < x \le \frac{\pi}{2}.$$

Because $0 < t_x < x$ and $\sin x$ is monotonically increasing in $[0, \pi/2]$, the last condition will be satisfied if

$$\left|\frac{x^{2n}}{(2n)!}\right| < 10^{-14} \qquad \text{for all} \quad x \quad \text{in} \quad \left[0, \frac{\pi}{2}\right],$$

that is, if

$$\frac{(\pi/2)^{2n}}{(2n)!} < 10^{-14}.$$

By trial and error, we find that $n = 10$ is the smallest integer for which this condition is satisfied. Thus, the Taylor polynomial of degree $2n - 1 = 19$ furnishes the desired approximation.

Approximation by the Taylor polynomial has the following *disadvantages*.
(i)  As may be seen from the error formula (1), the approximation generally will be good where $|x - t|$ is small and not so good where $|x - t|$ is large. Thus, in a practical sense, the approximation will not be uniformly good in the interval where an approximation is desired.
(ii)  In this age of portable software it is natural to ask for approximation algorithms that furnish arbitrarily good approximations; that is, approximations where the maximum error in a given interval can be made arbitrarily small. If the Taylor polynomial is used, this will be the case only if the remainder, that is, the expression on the right of (1), tends to zero for $n \to \infty$. For this to be the case it is not only necessary that all derivatives of $f$ exist—which is a very strong requirement in itself—but that $x$ lies in the interval of convergence of the *Taylor series* of $f$ about the point $x = t$. Thus, in this sense, approximation by the Taylor polynomial is applicable only in a very special situation.

## PROBLEMS

1  Here we consider various ways to compute the number log 2 by series expansions.
(a)  Estimate the number of terms required to obtain the value in question with an error $< 5 * 10^{-11}$ by putting $x = 1$ in the series

$$\log(1 + x) = x - \frac{x^2}{2} + \frac{x^3}{3} - \frac{x^4}{4} + \cdots.$$

(b)  How many terms are required to obtain the same accuracy by using

$$\log 2 = -\log \tfrac{1}{2} = -\log(1 - \tfrac{1}{2}) = x + \frac{x^2}{2} + \frac{x^3}{3} + \cdots$$

where $x = \tfrac{1}{2}$?
(c)  How many terms are required by using

$$\log 2 = \log \frac{1 + \tfrac{1}{3}}{1 - \tfrac{1}{3}} = 2\left(x + \frac{x^3}{3} + \frac{x^5}{5} + \cdots\right) \quad \text{where} \quad x = \tfrac{1}{3}?$$

(d)  Try to construct logarithms of other prime numbers in a similarly economic fashion, for example,

$$\log 17 = \log 16(1 + \tfrac{1}{16}) = 4 \log 2 + \log(1 + \tfrac{1}{16}),$$

and so on.

**2** Use the series*

$$(1 + x)^{-1/2} = \sum_{n=0}^{\infty} \binom{-1/2}{n} x^n$$

to obtain an accurate decimal value of $\sqrt{2}$ as follows:

$$\sqrt{2} = \sqrt{\tfrac{49}{25}(1 + \tfrac{1}{49})} = \tfrac{7}{5}\sqrt{\tfrac{50}{49}} = \tfrac{7}{5}(1 - \tfrac{2}{100})^{-1/2}.$$

Show, however, that Newton's method is still faster, even when only a moderate accuracy is required.

**3** Let $f$ be defined on $[0, 1]$. The function

$$p_n(x) := \sum_{k=0}^{n} f\left(\frac{k}{n}\right)\binom{n}{k} x^k(1 - x)^{n-k}$$

is called the $n$-th **Bernstein polynomial** associated with $f$. If $f$ is continuous, the sequence $\{p_n\}$ may be shown to converge to $f$ uniformly on $[0, 1]$, which is a constructive proof of the Weierstrass approximation theorem.

  **(a)** What is the condition of $p_n$ with regard to changes in the function $f$?

  **(b)** Devise an algorithm for evaluating $p_n$, and try it on the functions $f(x) := 1$, $x$, $x^2$, $|2x - 1|$. On the basis of this experimental evidence, how would you judge the convergence properties of the sequence $\{p_n\}$? How high would $n$ have to be chosen to approximate $f(x) := x^2$ with an error $< 10^{-6}$?

---

## §5.2 THE INTERPOLATING POLYNOMIAL

The *Taylor polynomial* of degree $n$ agrees with a given function $f$ and with its first $n$ derivatives at one point. An *interpolating polynomial* agrees with the given function, but not necessarily with any of its derivatives, at $n + 1$ points. (See Figure 5.2a.)

---

*Here, as elsewhere, $\binom{z}{n}$ denotes a binomial coefficient,

$$\binom{\alpha}{n} := \begin{cases} 1, & \text{if } n = 0 \\[2mm] \dfrac{\alpha(\alpha - 1)(\alpha - 2)\cdots(\alpha - n + 1)}{1 \cdot 2 \cdots n}, & \text{if } n = 1, 2, \ldots. \end{cases}$$

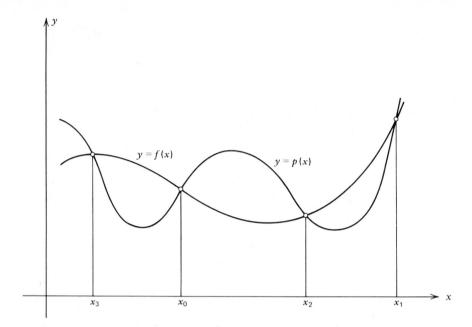

**Fig. 5.2*a*.** Interpolating polynomial.

Let $f$ be defined on some interval $I$, and let $x_0, x_1, \ldots, x_n$ be $n + 1$ distinct points of $I$. We do not assume that the $x_i$ are equidistant, nor need they be numbered in their natural order. For brevity, we write

$$f_k := f(x_k), \qquad k = 0, 1, \ldots, n.$$

Our problem is to find a polynomial, $p$, of degree not exceeding $n$, so that

$$p(x_k) = f_k, \qquad k = 0, 1, \ldots, n. \tag{1}$$

The points $x_k$ in this context are called **interpolating points**, and the numbers $f_k$ are the **interpolated values**. A polynomial $p$ satisfying (1) is called an **interpolating polynomial**.

**THEOREM 5.2:** *There exists precisely one polynomial of degree $\leq n$ which at $n + 1$ distinct points $x_k$ assumes given values $f_k$ ($k = 0, 1, \ldots, n$).*

*Proof:* As is customary in mathematics, existence and uniqueness of the desired mathematical object are proved by different methods.

(a) *Existence.* This is proved by exhibiting a *formula* for the interpolating polynomial. To obtain the desired formula, we first consider a simple special case. Let all interpolated values be equal to zero with the exception of one, say $f_m$, which is equal to 1. We thus seek a polynomial $l_m$, of degree not exceeding $n$, so that

$$l_m(x_k) = \begin{cases} 1, & k = m; \\ 0, & k \neq m. \end{cases}$$

Such a polynomial is easy to find. Because it must vanish at the points $x_0$, $x_1, \ldots, x_{m-1}, x_{m+1}, \ldots, x_n$, it must contain the factors $x - x_k$ $(k \neq m)$. Because there are $n$ such factors, $l_m$ must have the form

$$l_m(x) = c_m \prod_{\substack{k=0 \\ k \neq m}}^{n} (x - x_k),$$

where $c_m$ is a constant. The value of this constant may be found from the condition that $l_m(x_m) = 1$. This yields

$$1 = c_m \prod_{\substack{k=0 \\ k \neq m}}^{n} (x_m - x_k),$$

thus,

$$c_m = \frac{1}{\displaystyle\prod_{\substack{k=0 \\ k \neq m}}^{n} (x_m - x_k)}.$$

It follows that the desired polynomial $l_m$ is given by

$$l_m(x) = \prod_{\substack{k=0 \\ k \neq m}}^{n} \frac{x - x_k}{x_m - x_k}; \tag{2}$$

evidently, $l_m$ has the precise degree $n$.

We now assert that, for arbitrary $f_k$, the function

$$p(x) := \sum_{m=0}^{n} l_m(x) f_m \tag{3}$$

is a polynomial of degree $\leq n$ which at the points $x_k$ assumes the values $f_k$. As a sum of polynomials of degree $\leq n$, $p$ clearly is a polynomial of degree $\leq n$ (the degree may actually be $< n$, because the coefficients of the highest powers may cancel each other), and on setting $x := x_k$ we get

$$p(x_k) = \sum_{m=0}^{n} l_m(x_k) f_m.$$

All terms in the sum are zero with the exception of the term $m = k$, which has the value $f_k$. There follows

$$p(x_k) = f_k, \qquad k = 0, 1, \ldots, n,$$

completing the proof of the *existence* of the desired interpolating polynomial.

(b)   *Uniqueness.* Following the usual pattern of a uniqueness proof, we assume the existence of two objects with the desired properties. Thus, let us assume that there exist two interpolating polynomials satisfying the conditions of Theorem 5.2, say a polynomial $q$ in addition to the polynomial $p$ just constructed. We then have

$$p(x_k) = f_k,$$
$$q(x_k) = f_k,$$

$k = 0, 1, \ldots, n$. Now let $d := p - q$. This again is a polynomial of degree $\leq n$. Moreover $d$ satisfies

$$d(x_k) = f_k - f_k = 0, \qquad k = 0, 1, \ldots, n.$$

Thus, $d$ is a polynomial of degree $\leq n$ which has $n + 1$ distinct zeros. This is possible only if $d$ is the zero polynomial. It follows that $p = q$.

The special interpolating polynomials $l_m$ defined by (2) are called **Lagrangian interpolating coefficients**. The polynomial $p$ itself, as given by (3), is sometimes called the **Lagrangian interpolating polynomial**. It would be more correct, however, to call (3) the Lagrangian *representation* of the interpolating polynomial, because the function $p$, as we have seen, is uniquely determined. There are many representations of that function by formulas or algorithms, and (3) is just one of these many representations. By far, it is not the most efficient from a practical point of view.

# PROBLEMS

---

1   For a function $f$ defined at the integers, let $p_n$ denote the interpolating polynomial using the interpolating points $x_k = k$, $k = 0, 1, \ldots, n$.

(a)   Using the Lagrangian form of $p$, find an expression for

$$d_n(x) := p_n(x) - p_{n-1}(x).$$

**(b)**   Show that there results a representation for $p_n$ of the form

$$p_n(x) = \sum_{k=0}^{n} \binom{x}{k} a_k, \qquad (4)$$

where the $a_k$ depend only on $f$.

**(c)**   Show how to express the $a_k$ in terms of $f_0$, $\Delta f_0$, $\Delta^2 f_0$, ..., where $\Delta$ is the difference operator defined in §2.2.

**(d)**   Using a recurrence relation for the binomial coefficients $\binom{x}{k}$, show how to evaluate the sum (4) economically by a Horner-like algorithm.

[(4) is known as **Newton's form** of the interpolating polynomial.]

---

# §5.3   THE ERROR OF THE INTERPOLATING POLYNOMIAL

How accurate is the approximation of a function $f$ by some interpolating polynomial $p$? Lacking further information, nothing at all can be said about the quality of this approximation because between any two interpolating points the function $f$ may be changed at will without any influence on the interpolating polynomial. (See Figure 5.3a.)

A more precise statement about the error is possible, however, if we presuppose at least a qualitative knowledge of some high derivative of $f$.

**THEOREM 5.3:**   *Let $f$ be $n + 1$ times continuously differentiable on the interval $I$, and let $p$ be the polynomial of degree $\leq n$ which interpolates $f$ at $n + 1$ distinct points $x_k$ $(k = 0, 1, \ldots, n)$ of $I$. Let $x$ be any point in $I$. Then, the smallest interval containing $x$ and all the points $x_k$ contains a point $t_x$ in its interior so that*

$$f(x) - p(x) = \frac{1}{(n+1)!} l(x) f^{(n+1)}(t_x), \qquad (1)$$

*where*

$$l(x) := (x - x_0)(x - x_1) \cdots (x - x_n). \qquad (2)$$

*Proof:*   Nothing is to be proved if $x = x_k$ for some $k$ because then the expressions on either side of (1) are zero for any $t_x$. Thus, let us suppose that $x \neq x_k$ for all $k$; this value of $x$ will be fixed throughout the proof. To prove the existence of a point $t_x$ so that (1) holds, consider the auxiliary function of $t$,

$$g(t) := f(t) - p(t) - cl(t), \qquad (3)$$

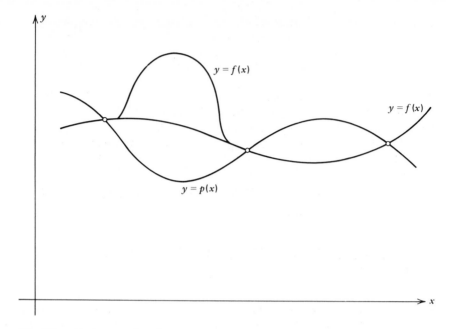

**Fig. 5.3a.** Bad approximation.

where $c$ denotes a constant yet to be chosen. For any choice of $c$,

$$g(x_k) = f_k - f_k - 0 = 0,$$

$k = 0, 1, \ldots, n$. We now choose $c$ so that $g$ vanishes also for $t = x$, which yields

$$0 = f(x) - p(x) - cl(x)$$

or

$$c = \frac{f(x) - p(x)}{l(x)}.$$

With this choice of $c$, the auxiliary function $g$ has $n + 2$ zeros $x_0, x_1, \ldots,$ $x_n$, and $x$, all contained in $I$. By Rolle's theorem, $g'$ has at least $n + 1$ zeros, $g''$ has at least $n$ zeros, and so on, until we find that $g^{(n+1)}$ has at least one zero. All these zeros lie in the interior of the smallest interval containing all the zeros of $g$; that is, in the smallest interval containing all the points $x_k$ as well as $x$. Let $t_x$ be such a zero of $g^{(n+1)}$. Differentiating (3) $n + 1$ times, we obtain

$$g^{(n+1)}(t) = f^{(n+1)}(t) - c(n+1)!;$$

here we have used the fact that the $(n + 1)$st derivative of a polynomial of degree $n$ is zero, while the $(n + 1)$st derivative of a polynomial of degree $n + 1$ with leading coefficient 1 has the value $(n + 1)!$. Letting $t = t_x$ yields, in view of $g^{(n+1)}(t_x) = 0$,

$$f^{(n+1)}(t_x) = c \cdot (n + 1)!,$$

which, on remembering the definition of $c$, yields the desired relation (1).

**EXAMPLE 5.3-1:** Let $n = 0$ (*one* interpolating point). Here $p$, the interpolating polynomial of degree 0, reduces to the constant $f(x_0)$. Relation (1) thus becomes

$$f(x) - f(x_0) = (x - x_0)f'(t_x),$$

where $t_x$ lies between $x_0$ and $x$. This is the analytical content of what is known as the *mean value theorem*.

**EXAMPLE 5.3-2:** Now let $n = 1$ (*two* interpolating points; *linear* interpolation, see Figure 5.3b). According to (1),

$$f(x) - p(x) = \frac{(x - x_0)(x - x_1)}{2} f''(t_x).$$

What is the largest possible error if $x$ lies between $x_0$ and $x_1$, and it is known that $|f''(x)| \le M_2$? The maximum of

$$\left| \frac{(x - x_0)(x - x_1)}{2} \right| \qquad \text{for } x \in [x_0, x_1]$$

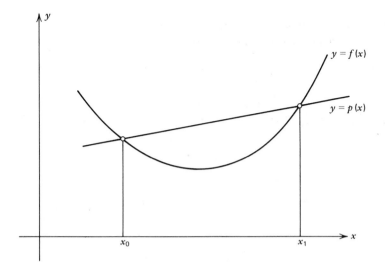

**Fig. 5.3b.** Linear interpolation.

is taken at the center of this interval and has the value

$$\frac{(x_1 - x_0)^2}{8}.$$

Thus, if linear interpolation is used, the error is bounded by

$$|f(x) - p(x)| \leq \frac{(x_1 - x_0)^2}{8} M_2; \tag{4}$$

here, $M_2$ denotes a bound for the second derivative, and it is assumed that $x$ lies between the two interpolating points.

As a numerical example, consider $f(x) := \sin x$. We wish to construct a six-digit table of this function with the property that the error $\varepsilon$ due to linear interpolation does not exceed the maximum rounding error of the entries; that is,

$$\varepsilon \leq 0.5 * 10^{-6}.$$

Evidently, $M_2 = 1$. We thus see from (4) that the desired property can be guaranteed if the step of the table, $h$ say, satisfies

$$\tfrac{1}{8}h^2 \leq 0.5 * 10^{-6}.$$

This condition is fulfilled for $h = 2 * 10^{-3}$. It is thus sufficient to tabulate $\sin x$ for $x = 0.000, 0.002, 0.004, \ldots$ in order to be sure that the total error (rounding error of entries plus interpolating error) will not exceed one unit of the last digit given, provided that the interpolating polynomial is evaluated exactly.

**EXAMPLE 5.3-3:** What can be gained by using cubic interpolation ($n = 3$)? We assume that the interpolating points are equidistant,

$$x_k = x_0 + k \cdot h, \qquad k = 1, 2, 3,$$

and that $x$ is chosen in the middle interval $[x_1, x_2]$, where the error can be expected to be smallest. Let $|f^{(4)}(x)| \leq M_4$. Then, according to Theorem 5.3,

$$|f(x) - p(x)| \leq M_4 \frac{1}{4!} \max_{x_1 \leq x \leq x_2} |l(x)|.$$

The maximum is taken at the midpoint of $[x_1, x_2]$, and its value is

$$\tfrac{3}{2}h \cdot \tfrac{1}{2}h \cdot \tfrac{1}{2}h \cdot \tfrac{3}{2}h = \tfrac{9}{16}h^4.$$

Hence,

$$|f(x) - p(x)| \leq M_4 \tfrac{9}{16} \tfrac{1}{24}h^4 = \tfrac{3}{128}M_4 h^4.$$

Considering again the example of the six-digit sine table, we have $M_4 = 1$. The step $h$ therefore now must satisfy

$$\tfrac{3}{128}h^4 \le 0.5 * 10^{-6},$$

which is the case if

$$h \le 0.068.$$

Thus, it suffices to tabulate the sine function for $x = 0.00$, $0.05$, $0.10$, $0.15$, ... to achieve the desired accuracy, which means that the step chosen may be 25 times as large as when linear interpolation is used.

# PROBLEMS

**1**  A table is called *well-suited for linear interpolation* if the error due to interpolation is not larger than the rounding error of the tabular entries. The function $f(x) := \sqrt{x}$ is tabulated at $x = 0, 1, 2, 3, \dots$. The table shows four digits after the decimal point. From which value of $x$ is this table well-suited for linear interpolation?

**2**  A table of the function $f(x) := \log_{10}(\sin x)$ shows five digits after the decimal point. The step of the table is $1/60$ of $1°$. From which point is this table well-suited for linear interpolation?

**3**  *Missing entry in a table.* The function $f$ is defined and infinitely differentiable for all real $x$, and its derivatives satisfy

$$|f^{(m)}(x)| \le 1, \qquad -\infty < x < \infty, \qquad m = 1, 2, \dots.$$

(Example: $f(x) := \sin x$.) Let $h > 0$, and let $p_n$ denote the polynomial of degree $<2n$ which interpolates $f$ at the $2n$ points $-nh$, $-nh + h$, ..., $-h$, $h$, $2h$, ..., $nh$. For which values of $h$ does there hold

$$\lim_{n \to \infty} p_n(0) = f(0)?$$

**4**  Can the function,

$$f(x) := \cos \frac{\pi x}{2}, \qquad -1 \le x \le 1,$$

be calculated as limit of the sequence of interpolating polynomials $p_n$ of degree $4n$ using merely the values of the function $f$ at the points $x = k = -2n$, $-2n + 1$, ..., $2n$, which are either 0 or $\pm 1$?

## §5.4 NUMERICAL EVALUATION OF THE INTERPOLATING POLYNOMIAL

We have seen above that, given a function $f$ and $n + 1$ interpolating points $x_k$, there exists precisely one polynomial of degree $\leq n$ which interpolates $f$ at these given points. Thus, as a *function*, the interpolating polynomial is uniquely determined. On the other hand, there exist many different *algorithms* for *evaluating* the interpolating polynomial for a given value of $x$. We review some of these from the numerical point of view.

### (a) Direct Use of the Lagrangian Formula (3), §5.2

This was a preferred method in the days of hand computation; extensive tables of the Lagrangian interpolating coefficients $l_m(x)$ were constructed for that purpose. The literal execution of the formula on the computer, however, requires $O(n^2)$ operations ($O(n)$ for each coefficient), which renders the formula uneconomical in comparison with other methods that will be discussed.

### (b) Finite Difference Formulas

If the interpolating points are equidistant, it is possible to represent the interpolating polynomial in terms of finite differences of the interpolated values $f_k$. Many such formulas exist, and a veritable calculus of finite differences has been developed for their manipulation. (See, for example, Henrici [1964], chapter 11.) It seems that these formulas were useful mainly in the days of computation by hand, possibly because their judicious application requires human judgment. Except for special purposes, such as the numerical integration of ordinary differential equations, these formulas are not much used today.

### (c) Aitken-Neville Algorithm

(See Henrici [1964], chapter 10.) This is a simple, recursive, numerically stable algorithm for the evaluation of the interpolating polynomial. It requires $O(n^2)$ operations per evaluation.

### (d) Representation in Normal Form

Being a polynomial of degree $n$, the interpolating polynomial may be brought into the form

$$p(x) = a_0 + a_1 x + a_2 x^2 + \cdots + a_n x^n.$$

Once it is in this form, only $O(n)$ operations are required for its evaluation. This would seem a feasible method if the same interpolating polynomial has to be evaluated many times. However, the dependence of the $a_i$ on the $f_k$ is complicated and, in most cases, ill-conditioned. (An exception occurs when interpolation is done in the complex plane, and the interpolating points are chosen as roots of unity.) Thus, the normal form of the interpolating polynomial normally should not be used.

## (e)  Barycentric representation

Here, we wish to discuss a representation of the interpolating polynomial which, although it is eminently suitable for machine computation, is seldom discussed in the literature. We obtain it by a simple transformation of the Lagrangian formula. We begin by simplifying the Lagrangian interpolation coefficients

$$l_m(x) := \frac{\prod_{k \neq m} (x - x_k)}{\prod_{k \neq m} (x_m - x_k)}, \qquad m = 0, 1, \ldots, n.$$

Let

$$l(x) := \prod_{k=0}^{n} (x - x_k) = (x - x_0)(x - x_1) \cdots (x - x_n),$$

as before. Then, clearly the numerator of $l_m(x)$ for $x \neq x_m$ is

$$\frac{l(x)}{x - x_m}.$$

For the denominator, we introduce the abbreviation

$$w_m := \frac{1}{\prod_{k \neq m} (x_m - x_k)}, \qquad m = 0, 1, \ldots, k. \tag{1}$$

The **weights** $w_m$ do not depend on $x$ or on the function $f$ to be interpolated; they are numbers that depend only on the interpolating points, and that, once these interpolating points are fixed, can be computed once for all. With this notation we have

$$l_m(x) = l(x) \frac{w_m}{x - x_m},$$

and hence,

$$p(x) = l(x) \sum_{m=0}^{n} \frac{w_m f_m}{x - x_m}. \tag{2}$$

This formula can be useful for purposes of numerical computation. It is capable of further simplification by the following simple observation: If the function $f$ to be interpolated itself is a polynomial of degree $\leq n$, it follows from the error formula or from the uniqueness statement of Theorem 5.2 that

$$p(x) = f(x)$$

for all $x$. The constant function $f(x) := 1$ surely is a polynomial of degree $\leq n$. Because $f_m = 1$ for all $m$, equation (2) shows that

$$1 = l(x) \sum_{m=0}^{n} \frac{w_m}{x - x_m},$$

$x \neq x_0, x_1, \ldots, x_n$. Thus, for such values of $x$,

$$l(x) = \frac{1}{\displaystyle\sum_{m=0}^{n} \frac{w_m}{x - x_m}},$$

and we obtain

$$p(x) = \frac{\displaystyle\sum_{m=0}^{n} \frac{w_m f_m}{x - x_m}}{\displaystyle\sum_{m=0}^{n} \frac{w_m}{x - x_m}}. \tag{3}$$

Formula (3) expresses the value of $p(x)$ as a weighted average of the interpolated values $f_m$ and is therefore called a **barycentric formula**.

One advantage of the barycentric formula (3) is its numerical stability. If $x$ is close to some $x_m$, the corresponding terms, both in the numerator and in the denominator, get a heavy weight, as they should. In addition, the formula is very economical, especially if we wish to compute interpolating polynomials which use the same interpolating points but different interpolated values. In this case, the numbers $w_m$ need to be evaluated only once, which in general requires $(n + 1)n\mu + \delta$; in many cases, the $w_m$ may even be calculated analytically. Once the $w_m$ are available, we require $(n + 1)(\delta + \alpha)$ for the calculation of the numbers

$$\frac{w_m}{x - x_m}$$

and $(n + 1)\mu$ for the calculation of the numbers

$$\frac{w_m f_m}{x - x_m} = \frac{w_m}{x - x_m} \cdot f_m.$$

No storage is required, because the sums in the numerator and denominator of

(3) may be accumulated simultaneously. A final $\delta$ is required for forming the quotient in (3). Thus, if the $w_m$ are known, only

$$(3n + 1)\alpha + (n + 1)\mu + (n + 2)\delta = O(n)$$

operations are required to evaluate the polynomial.

The barycentric formula takes a very simple form for equidistant interpolating points. Let

$$x_k = x_0 + kh, \qquad k = 0, 1, \ldots, n.$$

Then,

$$w_m^{-1} = \prod_{\substack{k=0 \\ k \neq m}}^{n} (x_m - x_k) = (-1)^{n-m} h^n m! (n - m)!.$$

Now obviously the value resulting from (3) is unchanged if the weights $w_m$ are replaced by weights $w_m^* := c w_m$ where $c \neq 0$ is a constant. We choose $c$ so that $w_0^* = 1$; that is,

$$c = (-1)^n h^n n!,$$

which yields the modified weights

$$w_m^* = (-1)^m \binom{n}{m}, \qquad m = 0, 1, \ldots, n,$$

that is, just the binomial coefficients with alternating signs. Thus, in the case of equidistant $x_k$, the interpolating polynomial is given by

$$p(x) = \frac{\displaystyle\sum_{m=0}^{n} (-1)^m \binom{n}{m} \frac{f_m}{x - x_m}}{\displaystyle\sum_{m=0}^{n} (-1)^m \binom{n}{m} \frac{1}{x - x_m}}. \tag{4}$$

For instance, if $n = 3$,

$$p(x) = \frac{\dfrac{f_0}{x - x_0} - 3\dfrac{f_1}{x - x_1} + 3\dfrac{f_2}{x - x_2} - \dfrac{f_3}{x - x_3}}{\dfrac{1}{x - x_0} - 3\dfrac{1}{x - x_1} + 3\dfrac{1}{x - x_2} - \dfrac{1}{x - x_3}} \tag{5}$$

It is remarkable that, in this representation of the interpolating polynomial, the step $h$ does not enter explicitly, and the weights are simple integers.

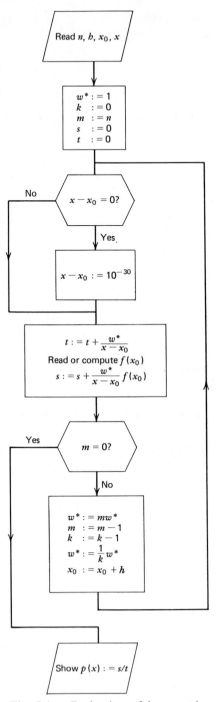

**Fig. 5.4a.** Evaluation of barycentric formula.

A flow diagram for the evaluation of (4) for arbitrary $n$, $h$, $x_0$, and $x$ is given in Figure 5.4a. To avoid division by 0 if $x = x_k$ for some $k$, the program checks whether $x - x_k = 0$ and, if so, replaces 0 by $10^{-30}$. If all $|f_k| < 10^{20}$ and if $h$ is not exceedingly small, this will result in the correct display of $p(x_k) = f_k$. The weights $w_k^*$ are calculated recursively by

$$w_0^* = 1, \quad w_{k+1}^* = \frac{n-k}{-1-k} w_k^*, \quad k = 0, 1, \ldots, n-1.$$

We write $w_k^* = w^*$ and denote the numerator and denominator of (4) by $s$ and $t$, respectively.

**DEMONSTRATION 5.4-1:**  Referring to Example 5.3-3, we use a cubic interpolating polynomial to interpolate the function $\sin x$ (with $x$ given in degrees) in the interval [44, 46], using the interpolating points 42, 44, 46, 48. Because $2°$ is less than 0.05 radians, the theory predicts that at least six decimals will be correct. The barycentric formula (5) yields the following results.

| $x$ (degrees) | $p(x)$ | $p(x) - \sin x$ |
|---|---|---|
| 44.2 | 0.697165094 | −0.000000009 |
| 44.4 | 0.699663326 | −0.000000015 |
| 44.6 | 0.702153033 | −0.000000020 |
| 44.8 | 0.704634187 | −0.000000024 |
| 45.0 | 0.707106757 | −0.000000025 |
| 45.2 | 0.709570713 | −0.000000024 |
| 45.4 | 0.712026026 | −0.000000020 |
| 45.6 | 0.714472665 | −0.000000015 |
| 45.8 | 0.716910599 | −0.000000009 |

It has been observed that the barycentric formula is basically stable. Equation (4) points to the limits of that stability in the case of equidistant interpolating points. If $n$ is large, the binomial coefficients $\binom{n}{m}$ vary widely in magnitude. Hence, the sums in the numerator and in the denominator, which if $f$ has constant sign are alternating, contain terms of widely varying magnitude and thus potentially are subject to smearing.

**DEMONSTRATION 5.4-2:**  Here, we interpolate the function $f(x) = x$ by a polynomial of degree $n$, using the interpolating points $x_k := k$, $k = 0, 1, \ldots, n$. Because $f$ itself is a polynomial of degree 1, the exact values should result for all $n \geq 1$

and for all $x$. Here are the numerical results produced by (4) for $x = 0.5$ and for selected values of $n$:

| $n$ | $p(0.5)$ |
|---|---|
| 1 | 0.500000000 |
| 2 | 0.500000000 |
| 4 | 0.500000000 |
| 8 | 0.499999997 |
| 16 | 0.500000294 |
| 32 | 0.497819013 |
| 64 | $-368.3070433$ |

The bad results of Demonstration 5.4-2 may be explained by the fact that we tried to interpolate near the end of the interval spanned by the interpolating points. Here the values of the sums in the numerator and the denominator of (4) are small compared to the largest terms in the sum. Better results may be expected if the interpolating polynomial is evaluated near the center of the interval spanned by the points $x_k$ because then the values of these sums are of the same order as their largest terms.

**DEMONSTRATION 5.4-3:**  We compute

$$\sin(\pi/4) = 2^{-1/2} = 0.707106781$$

by interpolating the sine function with the step $h := \frac{1}{2}\pi$, using an even number of equally spaced interpolating points located symmetrically with respect to $\frac{1}{4}\pi$. Thus, the interpolated values all are either 0 or 1 or $-1$. Using the error formula of Theorem 5.3 it may be shown that the values computed by interpolation (theoretically) converge as the number of interpolating points tends to infinity, in spite of the fact that the step of the interpolation remains fixed. The following values result from (4):

| number of points | $p(\pi/4)$ |
|---|---|
| 2 | 0.500000000 |
| 4 | 0.625000000 |
| 8 | 0.691406250 |
| 16 | 0.706377029 |
| 32 | 0.707104704 |
| 64 | 0.707106782 |
| 128 | 0.707106781 |

We see that under these circumstances (4) is stable even for large values of $n$.

## A formula for $w_m$

The weights $w_m$ occurring in the general barycentric formula (3) were originally defined by (1). We have seen that, for equidistant interpolating points, the weights may be expressed simply in terms of binomial coefficients. We now derive a simple formula for $w_m$ that holds generally. If $l(x)$ has the same meaning as before, then clearly

$$
\begin{aligned}
w_m^{-1} &= \prod_{k \neq m} (x_m - x_k) \\
&= \lim_{x \to x_m} \prod_{k \neq m} (x - x_k) \\
&= \lim_{x \to x_m} \frac{l(x)}{x - x_m} \\
&= \lim_{x \to x_m} \frac{l(x) - l(x_m)}{x - x_m} \quad (\text{because} \quad l(x_m) = 0) \\
&= l'(x_m),
\end{aligned}
$$

by the definition of the derivative. Hence, we obtain

$$
w_m = \frac{1}{l'(x_m)}, \qquad m = 0, 1, \ldots, n. \tag{6}
$$

This formula is useful whenever the polynomial $l(x)$ can be expressed in some closed form.

# PROBLEMS

1  The function

$$
f(x) := \frac{1}{1 + 25x^2}
$$

is approximated on $[-1, 1]$ by the interpolating polynomial of degree $2n$, using the nodes $x_k := -1 + k/n$, $k = 0, 1, \ldots, 2n$.

(a)  Using the barycentric formula (4), evaluate the interpolating polynomial for $n = 10$ at points halfway between the interpolating points, and draw a rough graph of both $f$ and $p$.

(b)  Does the approximation become better or worse when $n$ is increased?

**2** Develop the barycentric formula for interpolation at the points

$$x_k := q^k, \qquad k = 0, 1, 2, \ldots, n,$$

where $q$ is fixed, $0 < q < 1$.

  **(a)** Construct an algorithm for evaluating the resulting expression at $x = 0$. Normalize the weights $w_k$ so that $w_0 = 1$, and find a simple recurrence relation expressing $w_k$ in terms of $w_{k-1}$.

  **(b)** Experiment with some functions, such as $f(x) := e^x$ or

$$f(x) := 1 + x + x^2 + x^4 + x^8 + x^{16}.$$

**3** Apply the algorithm of the preceding problem to extrapolation to infinity, using the variable $t := 1/x$. Examples:

  **(a)** Summation of infinite series $\sum a_k$ as limits of their partial sums $s_1, s_2, s_4, s_8, \ldots$. Try

$$\sum_{k=1}^{\infty} k^{-2}.$$

  **(b)** Computation of

$$s := \lim_{n \to \infty} s_n$$

by extrapolating from $s_1, s_2, s_4, s_8, \ldots$. Try

$$\lim_{n \to \infty} \left(1 + \frac{1}{n}\right)^n.$$

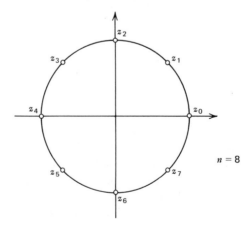

**Fig. 5.4b.**  Interpolation at roots of unity.

**4** Polynomial interpolation also is possible in the complex plane, and all *formal* results also hold for complex interpolation. [The error formula (Theorem 5.3) takes a different form, because Rolle's theorem does not hold for functions $\mathbb{C} \to \mathbb{C}$.] Derive the barycentric formula for the polynomial of degree $<n$, interpolating a given function $f$ at the $n$-th roots of unity. (See Figure 5.4a.)

**5** Show, theoretically and experimentally, that

$$\cos(\pi x/2) = \lim_{n \to \infty} \frac{\sum\limits_{k=-n}^{n} (-1)^k \binom{4n}{2n+2k} \dfrac{1}{x-2k}}{\sum\limits_{k=-2n}^{2n} (-1)^k \binom{4n}{2n+k} \dfrac{1}{x-k}}.$$

---

## §5.5 THE CONDITION OF POLYNOMIAL INTERPOLATION

In the preceding section, we were concerned with the *numerical evaluation* of the interpolating polynomial. An algorithm has been exhibited which, while satisfactory for moderate values of $n$, can become unstable for larger $n$ if the interpolating points are equally spaced.

We now consider the *condition* of the polynomial interpolation problem; that is, the question of how much the values of the mathematical interpolating polynomial change (independently of the algorithm used for its evaluation) if the data, the interpolated values, are changed.

Our first result will be that the condition of the polynomial interpolation problem becomes *extremely bad* as $n \to \infty$ if the interpolating points are *equally spaced*. We show this by the following mental experiment. Let $f$ be defined on $[-1, 1]$, and let us try to interpolate $f$ by the polynomial of degree $\leq 2n$, using the equally spaced interpolating points

$$x_k := -1 + \frac{k}{n}, \qquad k = 0, 1, \ldots, 2n.$$

Let us now assume that the interpolated value $f_n = f(0)$, and only this value, is changed by $\varepsilon$ as a consequence of rounding. (In reality, of course, all $f_k$ normally will be affected by rounding, but here we wish to consider the effect of a single error only.) By the Lagrangian formula, the effect of $f_n$ on the interpolating polynomial at the point $x$ is given by

$$f_n l_n(x),$$

where $l_n$ is the interpolating coefficient belonging to $x_n = 0$,

$$l_n(x) = \frac{(x - x_0) \cdots (x - x_{n-1})(x - x_{n+1}) \cdots (x - x_{2n})}{(0 - x_0) \cdots (0 - x_{n-1})(0 - x_{n+1}) \cdots (0 - x_{2n})}$$

$$= \frac{(x + 1)(x + 1 - h) \cdots (x + h)(x - h) \cdots (x - 1)}{(-1)^n h^{2n}(n!)^2}$$

$(h := 1/n)$. Using $f_n + \varepsilon$ in place of $f_n$ therefore changes the interpolating polynomial by

$$l_n(x)\varepsilon.$$

Let us evaluate this change at the point $x := 1 - h/2$, which is contained in the interval spanned by the interpolating points. We have

$$l_n\left(1 - \frac{h}{2}\right) = \pm \frac{\left(2 - \frac{h}{2}\right)\left(2 - 3\frac{h}{2}\right) \cdots \left(1 + \frac{h}{2}\right)\left(1 - 3\frac{h}{2}\right) \cdots \left(\frac{h}{2}\right)\left(-\frac{h}{2}\right)}{h^{2n}(n!)^2}$$

$$= \pm \frac{(4n - 1)(4n - 3) \cdots 3 \cdot 1(-1)h^{2n}}{2^{2n}h^{2n}(n!)^2(2n - 1)}$$

$$= \pm \frac{(4n)!}{2^{4n}(2n)!\,(n!)^2} \frac{1}{2n - 1}.$$

The order of magnitude of this expression can be made apparent by Stirling's formula (see §6.2), according to which $n!$ for large $n$ may be approximated (with a relative error that tends to zero for $n \to \infty$) as follows:

$$n! \sim \sqrt{2\pi n}\left(\frac{n}{e}\right)^n.$$

Thus, the last expression approximately equals

$$\frac{\sqrt{2\pi \cdot 4n}\left(\dfrac{4n}{e}\right)^{4n}}{2^{4n}\sqrt{2\pi \cdot 2n}\left(\dfrac{2n}{e}\right)^{2n} 2\pi n\left(\dfrac{n}{e}\right)^{2n}} \frac{1}{2n} = \frac{2^{2n}}{2^{3/2}\pi n^2}.$$

It follows that an error $\varepsilon$ at $x_n = 0$ changes the interpolating polynomial at $x = 1 - 1/2n$ approximately by

$$\pm \frac{2^{2n}}{2^{3/2}\pi n^2}\varepsilon.$$

It is evident that this expression grows extremely rapidly as $n \to \infty$. For $n = 10$, it already exceeds $10^3 \, \varepsilon$; for $n = 100$ it exceeds $10^{55} \, \varepsilon$. Thus, for large $n$, a small change in the data is magnified enormously even within the interval spanned by the equidistant interpolating points, proving our assertion that this kind of interpolation is extremely ill-conditioned.

The reason for the bad condition of this kind of interpolation clearly is the fact that the relative maxima of some interpolating coefficient, which occur between any two consecutive interpolating points, differ widely in magnitude. Is it possible to correct this deficiency by choosing the location of the interpolating points more judiciously?

Instead of looking at individual interpolating coefficients, we may as well look at the function

$$l(x) := (x - x_0)(x - x_1) \cdots (x - x_n)$$

(we return to writing $n$ in place of $2n$), which differs from the interpolating coefficients only by a linear factor. If the $x_k$ are equally spaced, the graph of $l(x)$ typically looks as shown in Figure 5.5a.

Clearly, the maximum value of $|l(x)|$ in the interval $[-1, 1]$, which we wish to make as small as possible, will be influenced by our choice of the $x_k$. Thus, the problem is to choose the $x_k$ so that the maximum becomes as small as possible. This is a famous extremum problem of classical analysis which was solved completely by P. Chebyshev. It turns out that the maximum becomes smallest if all *relative* maxima of $|l(x)|$ have the same value. (For a proof, see Henrici [1964], §9.4.) This, in turn, will be the case if $l(x)$ is chosen as a constant multiple of an appropriate *Chebyshev polynomial*.

The **Chebyshev polynomials** by definition are the functions

$$T_n(x) := \cos(n \arccos x), \qquad n = 0, 1, 2, \ldots . \tag{1}$$

We have

$$T_0(x) = 1$$
$$T_1(x) = x$$
$$T_2(x) = 2x^2 - 1$$
$$T_3(x) = 4x^3 - 3x, \ldots,$$

and, using standard trigonometric identities, we get the recurrence relation

$$T_{n+1}(x) = 2xT_n(x) - T_{n-1}(x), \qquad n = 1, 2, \ldots,$$

showing that $T_n$ indeed is a polynomial of degree $n$. Because the leading coefficient of $T_n$ is 1 for $n = 1$ and is thereafter doubled at each step, the leading coefficient of $T_n$ is $2^{n-1}$. Because $l(x)$ is a polynomial of degree $n + 1$ with leading

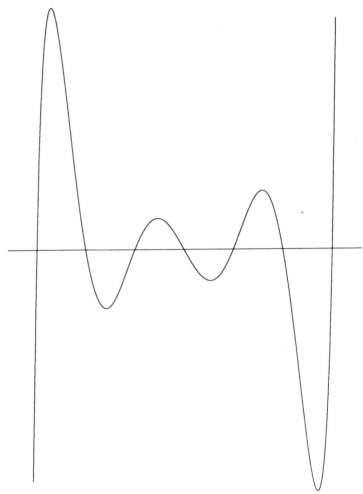

**Figure 5.5a.**   The function $l(x)$ for $n = 6$.

coefficient 1, it follows that, in order to make the maximum of $|l(x)|$ in $[-1, 1]$ as small as possible, we must choose

$$l(x) = 2^{-n}T_{n+1}(x).$$    (2)

(The maximum then has the value $2^{-n}$.) The interpolating points then are the zeros of $T_{n+1}$, which are calculated to be

$$x_k = \cos \phi_k, \qquad \phi_k := \frac{2k+1}{n+1}\frac{\pi}{2}, \qquad k = 0, 1, \dots, n.$$    (3)

These zeros are also called the **Chebyshev points** of order $n$.

Let us repeat our mental experiment if the Chebyshev points are chosen as interpolating points! Because $2n + 1$ interpolating points have been used, we must replace $n$ by $2n$ in the above formulas. Then,

$$l(x) = 2^{-2n} T_{2n+1}(x),$$

and the interpolating coefficient belonging to $x_n = 0$ becomes

$$l_n(x) = \frac{l(x)}{xl'(0)} = (-1)^n \frac{T_{2n+1}(x)}{(2n+1)x}.$$

Because evidently $|T_{2n+1}(x)| \leq 1$ for $-1 \leq x \leq 1$, the influence of an error $\varepsilon$ in $f_n = f(0)$ at every $x$ so that $1/(2n+1) \leq |x| \leq 1$ does not exceed $\varepsilon$. As may be shown by a slightly more elaborate analysis, the influence is $\leq \varepsilon$ also in the neighborhood of $x = 0$.

Our experiment admittedly concerns a very special situation. However, the following theorem of Powell (quoted after Dahlquist and Björck [1974], p. 107) shows that if it is at all reasonable to represent a function by a single polynomial on the interval $[-1, 1]$, then the choice of the Chebyshev points as interpolating points will always produce good results.

**THEOREM 5.5:**  *If the Chebyshev points are selected in order to approximate a continuous function on the interval $[-1, 1]$ by an interpolating polynomial of degree $n$, then for $n \leq 20$ the maximum error is at most four times as large, and for $n \leq 100$ at most five times as large, than the error in the approximation of that function by* **any** *polynomial of degree $\leq n$.*

It remains to discuss the numerical evaluation of the interpolating polynomial using the Chebyshev points. If the barycentric formula is used, it is necessary to evaluate the weights $w_m$. Using equation (6), §5.4, we find

$$w_k^{-1} = l'(x_k) = 2^n(n+1) \frac{(-1)^k}{\sin \phi_k},$$

$k = 0, 1, \ldots, n$. Because the value of the barycentric formula is unchanged if all weights are multiplied by the same nonzero constant, the modified weights

$$w_k^* := (-1)^k \sin \phi_k$$

serve equally well. With these weights, the barycentric formula becomes

$$p(x) = \frac{\displaystyle\sum_{k=0}^{n} (-1)^k \sin \phi_k \frac{f_k}{x - x_k}}{\displaystyle\sum_{k=0}^{n} (-1)^k \sin \phi_k \frac{1}{x - x_k}}, \tag{4}$$

where the $\phi_k$ and the $x_k$ are defined by (3). This formula may be programmed and evaluated much the same way as the corresponding formula for equidistant interpolating points. Concerning the numerical stability of the formula, we note that the absolute values of the weights now range between

$$w_0^* = \sin \phi_0 = \sin \frac{\pi}{2n+2} \sim \frac{\pi}{2n+2}$$

and 1. Thus, for large $n$, the ratio between the maximum and the minimum weight now is $\sim (2/\pi)n$ as opposed to

$$\binom{n}{n/2} \sim \left(\frac{2}{\pi n}\right)^{1/2} 2^n$$

in the case of equidistant points. Thus, the representation (4) may be expected to enjoy a much better numerical stability.

**DEMONSTRATION 5.5-1:** We interpolate the function $f(x) := x$, using the Chebyshev points. The following values result if the interpolating polynomial of degree $n$ is evaluated at $x = 1$:

| $n$ | $p(1)$ |
|-----|--------|
| 1 | 0.999999999 |
| 2 | 1.000000000 |
| 4 | 1.000000000 |
| 8 | 0.999999998 |
| 16 | 1.000000002 |
| 32 | 0.999999996 |
| 64 | 1.000000006 |
| 128 | 0.999999983 |

The theoretical answer, of course, is $p(1) = 1.000000000$ because the interpolated function is a polynomial of degree 1, and should be reproduced exactly. Thus, the errors which appear are exclusively due to rounding and to possible inaccuracies in the trigonometric function routines. If compared to Demonstration 5.4-2, however, the stability properties of the algorithm are excellent.

**DEMONSTRATION 5.5-2:** Here we show that interpolation at the Chebyshev points can even be used to approximate non-differentiable functions. Let $f(x) := |x|$. We interpolate this function by polynomials of odd degree; the number of interpolating points will then be even, and $x = 0$ is never an interpolating point. Evaluating the polynomials at $x = 0$, that is, at the point where the derivative does not exist, yields the following values:

| $n$ | $p(0)$ | Computing time on typical calculator |
|---|---|---|
| 1 | 0.707107 | 8 sec |
| 3 | 0.270598 | 15 |
| 7 | 0.127449 | 28 |
| 15 | 0.062802 | 55 |
| 31 | 0.031288 | 110 |
| 63 | 0.015630 | 217 |

We see that the computing times increase linearly with $n$, as they should for the barycentric formula.

## PROBLEMS

1  The function $f(x) := x^9$ is approximated on $[-1, 1]$ by an (interpolating) polynomial of degree 8. What is the minimum of the maximum error that can be achieved?

2  Let $f$ be approximated on $[-1, 1]$ by the Lagrangian interpolating polynomial $p$ of degree $\leq 2n$, using the $2n + 1$ equidistant interpolating points

$$x_k := -1 + \frac{k}{n}, \qquad k = 0, 1, \ldots, 2n,$$

with the interpolated values $f_k := f(x_k)$. We have seen in the main text that the condition of the value of $p$ at

$$x = 1 - \frac{1}{2n} \quad \text{with respect to the datum} \quad f_n = f(0)$$

is extremely bad. What is the condition of the value of $p$ at

$$x = \frac{1}{2n} \quad \text{with respect to the datum} \quad f_0 = f(-1)?$$

**3** *Salzer's formula.* Show that for interpolation at the Chebyshev points of the second kind,

$$x_k := \cos \frac{k\pi}{n}, \qquad k = 0, 1, \ldots, n,$$

the barycentric formula takes the especially simple form

$$p(x) = \frac{\sum_{k=0}^{n} (-1)^k \frac{\varepsilon_k}{x - x_k} f_k}{\sum_{k=0}^{n} (-1)^k \frac{\varepsilon_k}{x - x_k}}, \tag{5}$$

where $\varepsilon_0 = \varepsilon_n = \frac{1}{2}$, all other $\varepsilon_k = 1$. (H. B. Salzer [1972].)

**4** If $M_{n+1}$ is a bound on $|f^{(n+1)}(x)|$ in $[-1, 1]$, what is the maximum of the error of Salzer's formula (5)?

**5** Repeat Problem 1 of §5.4 using interpolation at the Chebyshev points, and compare the performance of the two kinds of interpolating polynomials.

## §5.6 HERMITE INTERPOLATION

The Lagrangian interpolation problem considered in the previous section is a mere special case of the following more general problem: Given a sufficiently differentiable function $f$ on some interval $I$, given $n + 1$ points $x_0, x_1, \ldots, x_n$ in $I$, and given $n + 1$ nonnegative integers $m_0, m_1, \ldots, m_n$, one wishes to determine a polynomial $p(x)$ of minimum degree which at each point $x_k$ satisfies

$$p^{(i)}(x_k) = f^{(i)}(x_k), \qquad i = 0, 1, \ldots, m_k.$$

That is, not only should, at each interpolating point, the value of $p$ agree with the value of $f$, but there should also be agreement of the derivatives of $p$ and $f$ up to a certain prescribed order (which in general may vary from point to point). Here we are concerned with the case where all $m_k = 1$. Writing

$$f_k := f(x_k), \qquad f'_k := f'(x_k), \qquad k = 0, 1, \ldots, n$$

for brevity, we thus wish to construct a polynomial $p$ of minimum degree so that

$$p(x_k) = f_k, \qquad p'(x_k) = f'_k, \qquad k = 0, 1, \ldots, n. \tag{1}$$

Such polynomials are required for the theory of spline interpolation. (To be discussed in §5.8.) Moreover, they lead to an efficient method for solving nonlinear equations. (See §5.7.)

Because the number of conditions to be satisfied now equals $2n + 2$, it is reasonable to expect the existence of a polynomial of degree $\leq 2n + 1$, satisfying the conditions (1). In fact, it may be shown as in §5.3 that there exists *at most one* polynomial $p$ of degree $\leq 2n + 1$ satisfying (1). To show the existence of such a polynomial, we again use the method of explicit construction. To this end, we seek to represent $p$ in the form

$$p(x) = \sum_{m=0}^{n} h_m(x)f_m + \sum_{m=0}^{n} \hat{h}_m(x)f'_m, \tag{2}$$

where the functions $h_m$ and $\hat{h}_m$, called the **Hermite interpolating coefficients**, themselves are polynomials of degree $\leq 2n + 1$. The polynomial $p$ will satisfy (1) for arbitrarily prescribed $f_m$ and $f'_m$ if, and only if, the Hermite interpolation coefficients satisfy the following set of conditions:

$$h_m(x_k) = h'_m(x_k) = 0, \qquad k \neq m; \tag{3a}$$

$$h_m(x_m) = 1; \tag{3b}$$

$$h'_m(x_m) = 0; \tag{3c}$$

and

$$\hat{h}_m(x_k) = \hat{h}'_m(x_k) = 0, \qquad k \neq m; \tag{4a}$$

$$\hat{h}_m(x_m) = 0; \tag{4b}$$

$$\hat{h}'_m(x_m) = 1. \tag{4c}$$

To construct $h_m$, we note that (3a) requires that $h_m$ contains twice each factor $(x - x_0), \ldots, (x - x_{m-1}), (x - x_{m+1}), \ldots, (x - x_n)$. If

$$l(x) := \prod_{k=0}^{n} (x - x_k)$$

as in preceding sections, $h_m$ must have the form

$$\left(\frac{l(x)}{x - x_m}\right)^2 \quad \text{times a linear polynomial.}$$

We represent the linear polynomial in the form $a(x - x_m) + b$ and face the task of determining $a$ and $b$ so that (3b) and (3c) are satisfied. Using the result,

$$\lim_{x \to x_m} \frac{l(x)}{x - x_m} = l'(x_m) = \frac{1}{w_m},$$

used in the derivation of equation (6) of §5.4, condition (3b) yields

$$h_m(x_m) = \frac{b}{w_m^2} = 1,$$

hence, $b = w_m^2$. To satisfy (3c) we evaluate

$$h_m'(x_m) = \lim_{x \to x_m} \frac{h_m(x) - h_m(x_m)}{x - x_m}$$

$$= \lim_{x \to x_m} \left\{ \frac{|l(x)|^2}{(x - x_m)^3} [a(x - x_m) + w_m^2] - \frac{1}{x - x_m} \right\}$$

$$= a \lim_{x \to x_m} \frac{[l(x)]^2}{(x - x_m)^2} + \lim_{x \to x_m} \frac{w_m^2 [l(x)]^2 - (x - x_m)^2}{(x - x_m)^3}.$$

The first limit equals $aw_m^{-2}$; the second may be evaluated by L'Hospital's rule*
(differentiating numerator and denominator three times) with the result
$w_m l''(x_m)$. Setting

$$u_m := l''(x_m), \qquad m = 0, 1, \ldots, n, \tag{5}$$

$h'(x_m) = 0$ now implies $a = -w_m^3 u_m$, and we find

$$h_m(x) = \left| \frac{l(x)}{x - x_m} \right|^2 [w_m^2 - w_m^3 u_m(x - x_m)].$$

To construct the polynomials $\hat{h}_m(x)$, we note that by (4a) they must contain
twice each factor $(x - x_k)$, where $k \neq m$, and by (4b) they must contain the factor
$x - x_m$. They thus have the form

$$\hat{h}_m(x) = c \frac{[l(x)]^2}{x - x_m},$$

where $c$ is to be determined so that (4c) holds. From

$$\hat{h}_m'(x_m) = \lim_{x \to x_m} \frac{\hat{h}_m(x) - \hat{h}_m(x_m)}{x - x_m}$$

$$= c \lim_{x \to x_m} \frac{[l(x)]^2}{(x - x_m)^2}$$

$$= cw_m^{-2},$$

_____

* See Salas and Hille [1978], p. 496; Anton [1980], p. 545.

there follows $c = w_m^2$, and we have

$$\hat{h}_m(x) = w_m^2 \frac{[l(x)]^2}{x - x_m}.$$

From (2) the **Hermite interpolating polynomial** thus is given by

$$p(x) = [l(x)]^2 \sum_{m=0}^{n} \frac{w_m^2}{x - x_m} \left\{ \left( \frac{1}{x - x_m} - u_m w_m \right) f_m + f'_m \right\}, \tag{6}$$

where $w_m := [l'(x_m)]^{-1}$ and $u_m := l''(x_m)$. A barycentric version of this formula is obtained by noting that, for $f(x) = 1$, the identity

$$1 = [l(x)]^2 \sum_{m=0}^{n} \frac{w_m^2}{x - x_m} \left\{ \frac{1}{x - x_m} - u_m w_m \right\}$$

results. On dividing this into (6) we obtain

$$p(x) = \frac{\displaystyle\sum_{m=0}^{n} \frac{w_m^2}{x - x_m} \left\{ \left( \frac{1}{x - x_m} - u_m w_m \right) f_m + f'_m \right\}}{\displaystyle\sum_{m=0}^{n} \frac{w_m^2}{x - x_m} \left( \frac{1}{x - x_m} - u_m w_m \right)}. \tag{7}$$

**The Error of the Hermite Interpolating Polynomial**

If $f$ has $2n + 2$ continuous derivatives on $I$, the error committed in approximating $f$ by $p$ may be estimated by the following result, which is established by the method used in proving Theorem 5.3: For each $x \in I$, there exists $t_x$ in the smallest interval containing $x$ and all interpolating points $x_k$ so that

$$f(x) - p(x) = \frac{[l(x)]^2}{(2n + 2)!} f^{(2n + 2)}(t_x). \tag{8}$$

**Two Interpolating Points**

In the following, Hermite interpolation with just two interpolating points, $x_0$ and $x_1$, will be required. We then have

$$l(x) = (x - x_0)(x - x_1)$$

and, setting

$$\Delta := x_1 - x_0,$$

there holds

$$w_0 = \frac{1}{l'(x_0)} = -\frac{1}{\Delta}, \qquad w_1 = \frac{1}{l'(x_1)} = \frac{1}{\Delta};$$

furthermore

$$u_1 = u_2 = l''(x) = 2.$$

Thus, (6) becomes

$$p(x) = (x - x_0)^2(x - x_1)^2 \left\{ \frac{1}{\Delta^2(x - x_0)} \left[ \left( \frac{1}{x - x_0} + \frac{2}{\Delta} \right) f_0 + f'_0 \right] \right.$$

$$\left. + \frac{1}{\Delta^2(x - x_1)} \left[ \left( \frac{1}{x - x_1} - \frac{2}{\Delta} \right) f_1 + f'_1 \right] \right\},$$

or on simplification

$$p(x) = \frac{1}{\Delta^2} \left\{ (x - x_1)^2 f_0 + (x - x_0)^2 f_1 \right.$$

$$+ (x - x_0)(x - x_1) \left[ (x - x_1) \left( f'_0 + \frac{2}{\Delta} f_0 \right) \right.$$

$$\left. + (x - x_0) \left( f'_1 - \frac{2}{\Delta} f_1 \right) \right] \right\} \tag{9}$$

**DEMONSTRATION 5.6-1:** We use (9) to evaluate $f(x) := \sin x$ (with $x$ measured in degrees) in the interval $[44, 46]$, using the interpolating points $x_0 = 44$ and $x_1 = 46$. It is to be noted that because $x$ is given in degrees, we have $f'(x) = \pi/180 \cos x$.

| $x$ | $p(x)$ | $p(x) - \sin x$ |
|-----|--------|-----------------|
| 44.2 | 0.697165103 | $-0.000000001$ |
| 44.4 | 0.699663340 | $-0.000000001$ |
| 44.6 | 0.702153051 | $-0.000000002$ |
| 44.8 | 0.704634208 | $-0.000000003$ |
| 45.0 | 0.707106779 | $-0.000000003$ |
| 45.2 | 0.709570734 | $-0.000000003$ |
| 45.4 | 0.712026044 | $-0.000000002$ |
| 45.6 | 0.714472679 | $-0.000000001$ |
| 45.8 | 0.716910607 | $-0.000000000$ |

## PROBLEMS

**1** Show that if the interpolating points are the Chebyshev points,

$$x_k = \cos \phi_k, \qquad \phi_k := \frac{2k + 1}{2n + 2} \pi, \qquad k = 0, 1, \ldots, n,$$

then

$$u_m w_m = \frac{\cos \phi_m}{(\sin \phi_m)^2},$$

and consequently (7) may be written

$$p(x) = \frac{\displaystyle\sum_{k=0}^{n} \left| \frac{1 - x \cos \phi_k}{(x - \cos \phi_k)^2} f_k + \frac{(\sin \phi_k)^2}{x - \cos \phi_k} f'_k \right|}{\displaystyle\sum_{k=0}^{n} \frac{1 - x \cos \phi_k}{(x - \cos \phi_k)^2}}.$$

**2** Show: For arbitrary interpolating points $x_0, x_1, \ldots, x_n$, the quantities

$$u_m w_m := \frac{l''(x_m)}{l'(x_m)}$$

occurring in the Hermite interpolation formula (7) are given by

$$u_m w_m = 2 \sum_{\substack{k=0 \\ k \neq m}}^{n} \frac{1}{x_m - x_k}.$$

**3** Can the function

$$f(x) := \sin x$$

be calculated in the interval $(-\pi/2, \pi/2)$ by using the values of $f$ and of $f'$ solely at the points $x = k\pi$, $k = -n, -n + 1, \ldots, -1, 0, 1, \ldots, n$, if $n$ is sufficiently large? Analyze and experiment!

## §5.7  INVERSE INTERPOLATION

Two-point Hermite interpolation offers an elegant approach to the problem of solving (scalar) non-linear equations to which we return here briefly. (See §2.3.)

Let $f$ be continuous and continuously differentiable on the interval $[a, b]$; let the signs of $f(x)$ at $x = a$ and $x = b$ be different; and let $f$ be *monotone* on $[a, b]$. Under these hypotheses, we are assured that the equation

$$f(x) = 0 \qquad (1)$$

has precisely one solution, say $x = s$, in $[a, b]$.

From the point of view of pure mathematics, finding $s$ is trivial. In view of the presumed monotonicity, the inverse function of $f$, that is, the uniquely determined function $f^{[-1]}$ satisfying

$$f^{[-1]}(f(x)) = x \quad \text{for all} \quad x \in [a, b]$$

exists. Since $f(a)$ and $f(b)$ have different signs, 0 belongs to the domain of definition of $f^{[-1]}$, and clearly

$$s = f^{[-1]}(0).$$

This insight is of no direct numerical use, since the determination of the value of $f^{[-1]}$ at any point $y$ of its domain of definition requires solving the equation $f(x) = y$ for $x$, which is precisely the problem which we are trying to deal with algorithmically. It is easy, however, to generate values of $f^{[-1]}$ at points $y$ that are not specified in advance, for if $x_0$ is any point of $[a, b]$ and $y_0 := f(x_0)$, then $x_0$ is the value of $f^{[-1]}$ at $y_0$. More generally, if $x_0, x_1, \ldots, x_n$ is any set of points in $[a, b]$ with corresponding function values $y_0, y_1, \ldots, y_n$, then $x_0, \ldots, x_n$ are the values of $f^{[-1]}$ at $y_0, \ldots, y_n$.

The method of inverse interpolation for solving (1) now simply consists of interpolating the inverse function $f^{[-1]}$, using the interpolating points $y_0, \ldots, y_n$ and corresponding function values $x_0, \ldots, x_n$, and in evaluating the resulting interpolating polynomial $p(y)$ at $y = 0$. If

$$x_{n+1} := p(0)$$

is such that $y_{n+1} := f(x_{n+1}) = 0$, then $s = x_{n+1}$ is the desired solution. If $y_{n+1} \neq 0$, then the pair $(y_{n+1}, x_{n+1})$ may be added to the data used for constructing $p$, and the process may be repeated. If it is desired to keep the degree of $p$ constant, one of the data points $(y_k, x_k)(k = 0, 1, \ldots, n)$ may be removed (for instance, the one with the largest $|y_k|$).

If linear interpolating polynomials are used, this process is known as the **regula falsi** or the **rule of false positions**. If the required methods for constructing and evaluating $p$ are available, it is easy to use the process with interpolating polynomials of arbitrary degree. If the derivative of $f$ is easily evaluated, the derivative of $f^{[-1]}$ at the points $y_k$ is

$$f^{[-1]'}(y_k) = \frac{1}{f'(x_k)} = \frac{1}{f'_k},$$

and therefore may be regarded as known, and the data for approximating $f^{[-1]}$ by Hermite interpolation are available. If two interpolating points are used, equation (9), §5.6, after interchanging $x_k$ and $y_k$ and replacing $f'_k$ by $1/f'_k$ yields

$$
p(y) = \frac{1}{(y_1 - y_0)^2} \Big\{ (y - y_1)^2 x_0 + (y - y_0) x_1
$$

$$
+ (y - y_0)(y - y_1) \Big[ \Big( \frac{1}{f'_0} + \frac{2x_0}{y_1 - y_0} \Big)(y - y_1)
$$

$$
+ \Big( \frac{1}{f'_1} - \frac{2x_1}{y_1 - y_0} \Big)(y - y_0) \Big] \Big\}.
$$

Evaluating this at $y = 0$, we obtain as approximation for the desired zero

$$
x_2 := \frac{1}{(y_1 - y_0)^2} \Big\{ x_0 y_1^2 + x_1 y_0^2 - y_0 y_1
$$

$$
\times \Big[ \Big( \frac{1}{f'_0} + \frac{2x_0}{y_1 - y_0} \Big) y_1 + \Big( \frac{1}{f'_1} - \frac{2x_1}{y_1 - y_0} \Big) y_0 \Big] \Big\}. \tag{2}
$$

Choosing two starting values $x_0$, $x_1$ (preferably so that the corresponding values of $f$ have opposing signs); evaluating the corresponding values $y_k = f_k$ and $f'_k$; calculating $x_2$ from (2); substituting $x_0 := x_1$, $x_1 := x_2$ and repeating constitutes the algorithm of **iterated inverse two-point Hermite interpolation**. Newton's method in this light may be looked at as iterated inverse one-point Hermite interpolation. Since the data used for the two-point process include those used for the one-point process, the convergence of the two-point process may be expected to be better. An analysis of the behavior of the error shows that the theoretical order of the process, that is, the supremum of all exponents $\alpha$ so that the sequence

$$
\frac{|x_{k+1} - s|}{|x_k - s|^\alpha}
$$

is bounded, equals $1 + \sqrt{3} = 2.7321$, the larger zero of the equation $x^2 = 2x + 2$.

**DEMONSTRATION 5.7-1:**  Let $f(x) := x^2 - 4$. Here $1/f'(x) = 1/2x$. With the starting values, $x_0 = 1$, $x_1 = 10$, we get

| $k$ | $x_k$ |
|---|---|
| 2 | 2.430353117 |
| 3 | 2.029524000 |
| 4 | 2.000008608 |
| 5 | 2.000000000 |

## PROBLEMS

**1** Find the solution of $f(x) := \cos x - x = 0$ by inverse Hermite interpolation, starting with $x_0 = 0$, $x_1 = 1$.

**2** For the equation

$$f(x) := x^{1/3} - 2 = 0$$

inverse Hermite interpolation finds the exact solution $x = 8$ (apart from rounding errors) in one step, independently of the starting values.

**(a)** Verify, and explain.

**(b)** Describe classes of equations for which inverse Hermite interpolation converges in one step.

## §5.8 SPLINE INTERPOLATION

Here we return to the problem of approximating a given function by interpolation. We have seen that, if the function is smooth, and if the interpolating points may be chosen freely, then (after adapting the problem to the interval $[-1, 1]$) interpolation at the Chebyshev points often yields accurate and stable approximating polynomials. It is not always possible to freely choose the interpolating

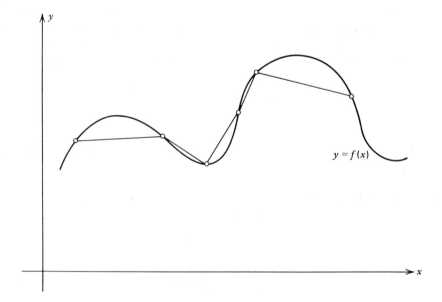

**Fig. 5.8a.** Piecewise linear interpolation.

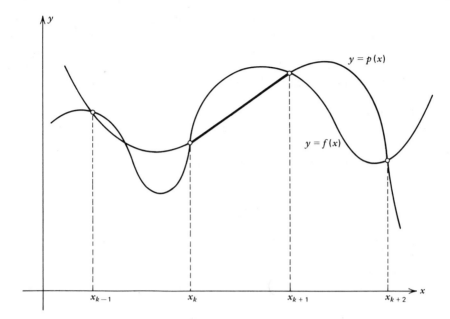

**Fig. 5.8b.**   Piecewise cubic interpolation.

points for the function to be approximated, however, particularly if the function is given in the form of a table or if it has been determined empirically, for instance, as a result of measurements. What should one do to interpolate such a function as smoothly as possible?

We already know that global interpolation (that is, interpolation by a single interpolating polynomial) should not be used without regard to the distribution of the interpolating points because of potential ill-conditioning and numerical instability.

Another option that offers itself (and that was almost universally used prior to 1950) is the piecewise use of interpolating polynomials of low degree, such as piecewise *linear* interpolation (see Figure 5.8a), or interpolation in the interval $[x_k, x_{k+1}]$ by a cubic interpolating polynomial using the interpolating points $x_{k-1}, x_k, x_{k+1}, x_{k+2}$. (See Figure 5.8b.)

As shown by the examples of §5.3, this kind of interpolation may be acceptable if the interpolation step is sufficiently small, and if the function $f$ to be interpolated is sufficiently smooth. Piecewise polynomial interpolation is questionable if nothing is known about the smoothness of $f$, which will be normal for empirical functions. One also has to accept the fact that, no matter how high the degree of the interpolating polynomials used for each subinterval, the *first* derivative of the global interpolant thus constructed will have discontinuities at the points $x_k$. Thus, the graph of the global interpolant will have corners—a situation which is totally unacceptable in certain industrial applications, such as the design of automobile bodies.

It is possible to obtain smooth interpolating functions by the technique of *spline interpolation*. It is true that to find a spline interpolant requires a some-what greater computational effort than to find the classical interpolating func-tion. Nevertheless, because of its numerical stability and its excellent condition, the technique of spline interpolation has become very widely used in recent years.

Our approach to spline interpolation will be *axiomatic*. We shall postulate a list of properties which the desired global interpolating function should have. We then shall show that there exists exactly one function with these properties.

Let $[a, b]$ be a closed finite interval, and let $x_0, x_1, \ldots, x_n$ be a set of interpolating points in that interval,

$$a = x_0 < x_1 < x_2 < \cdots < x_{n-1} < x_n = b.$$

We thus assume that the interpolating points are in their natural order, and that the endpoints of the interval are interpolating points. It is not assumed, however, that the interpolating points are equally spaced.

Let $f$ be a given function which is defined at least at all points $x_i$, $f(x_i) =: f_i$. We wish to approximate $f$ by a function $g$ having the following properties:

(i)  *$g$ interpolates $f$ at all interpolating points,*

$$g(x_i) = f_i, \qquad i = 0, 1, \ldots, n. \tag{1}$$

(ii)  *$g$ and $g'$ are continuous on the whole interval $[a, b]$.* This guarantees that the graph of $g$ has "no corners."

(iii)  *The remaining derivatives of $g$ are continuous on each subinterval $(x_i, x_{i+1})(i = 0, 1, \ldots, n-1)$ and have one-sided limits at the endpoints of each subinterval.* However, it is not required that the limits as $x$ approaches $x_i$ from the left and from the right are identical.

(iv)  *The functional*

$$J[g] := \int_a^b [g''(x)]^2 \, dx \tag{2}$$

*has the smallest possible value among all functions satisfying the conditions* (i), (ii), (iii). This condition expresses the desire that the interpolating function $g$ should be as smooth as possible. As is well known, the second derivative is a measure of the curvature of the graph. To make $J$ as small as possible is the same as to make the total curvature of the graph, measured in a certain way, as small as possible.

It might be asked why the integral $J$ has been chosen in the special form (2) and not in some other form, such as

$$\int_a^b |g''(x)| \, dx.$$

As in the method of least squares, one reason is mathematical convenience. Squares are much easier to handle than absolute values. Another reason is a mechanical analogy. Assume that the points $(x_i, f_i)$ are pivots that can turn freely, and assume that the graph of $g$ is an elastic rod (or *spline*, as such rods are called in naval architecture) which is passed through these pivots. According to the mechanical principle of least deformation energy, the spline will then assume precisely the form which minimizes $J$.

We specifically point out that it is not required that, at the points $x_k$, the slope of $g$ agrees with the slope of $f$. Indeed, the slope of $f$ need not even be defined.

**THEOREM 5.8:**  *There exists precisely one function g satisfying the postulates* (i), (ii), (iii), (iv) *stated above.*

The proof of Theorem 5.8 that will be given in the remainder of this section will not only establish the existence of $g$, but will also furnish an algorithm for constructing $g$. It will be seen that, after disposing of some preliminary work that can be done once and for all, to evaluate $g$ is no more difficult than to evaluate a third degree polynomial.

The logical argument by which the proof proceeds is this: We begin by *assuming* that a function $g$ satisfying the postulates (i) through (iv) exists. By a technique that in essence dates back to Euler and that is frequently used in the calculus of variations, we derive another set of properties that characterize $g$ uniquely. We then construct a function $g$ with these properties and show that it satisfies the postulates (i) through (iv).

Assume, then, that $g$ is a function which satisfies all four postulates. Then, if $g_1$ is any other function which satisfies (i), (ii), (iii),

$$J[g] \le J[g_1]. \tag{3}$$

Now assume $g_1$ in the special form

$$g_1 = g + \varepsilon h,$$

where $\varepsilon$ is a real parameter, and where $h$, if $g_1$ is to satisfy (i), (ii), (iii), also must satisfy (ii), (iii) and, in place of (i),

$$h(x_i) = 0, \qquad i = 0, 1, \dots, n. \tag{4}$$

Any $h$ satisfying this set of conditions will be called **admissible**. Now let

$$J(\varepsilon) := J[g_1] = \int_a^b [g''(x) + \varepsilon h''(x)]^2 \, dx. \tag{5}$$

The function $J(\varepsilon)$ is a quadratic polynomial in $\varepsilon$. By virtue of (3), this polynomial assumes its minimum for $\varepsilon = 0$. A necessary condition for this to be the case is

$$J'(0) = 0, \tag{6}$$

and this must hold for every admissible function $h$. Carrying out the square in (5) yields

$$J'(0) = 2 \int_a^b g''(x)h''(x) \, dx. \tag{7}$$

We transform this expression by integration by parts. By virtue of (iii), the integration has to be carried out separately for each subinterval $[x_i, x_{i+1}]$. The $i$-th subinterval contributes to the integral

$$\int_{x_i}^{x_{i+1}} g''(x)h''(x) \, dx = \left[ g''(x)h'(x) \right]_{x_i}^{x_{i+1}} - \int_{x_i}^{x_{i+1}} g^{(3)}(x)h'(x) \, dx.$$

(We use the notation $[f(x)]_a^b := f(b) - f(a)$.) Integrating by parts once more, this becomes

$$\left[ g''(x)h'(x) \right]_{x_i}^{x_{i+1}} - \left[ g^{(3)}(x)h(x) \right]_{x_i}^{x_{i+1}} + \int_{x_i}^{x_{i+1}} g^{(4)}(x)h(x) \, dx.$$

Here, the second term is zero because

$$h(x_i) = h(x_{i+1}) = 0$$

by virtue of (4). We denote by $g''(x_i +)$ and $g''(x_i -)$ the limits of $g''$ as $x$ approaches $x_i$ from the right and from the left, respectively. (Remember that these limits need not be the same.) Summing all contributions, we get

$$\begin{aligned} \tfrac{1}{2} J'(0) = & -g''(x_0 +)h'(x_0) + [g''(x_1 -) - g''(x_1 +)]h'(x_1) \\ & + [g''(x_2 -) - g''(x_2 +)]h'(x_2) + \cdots \\ & + g''(x_n -)h'(x_n) + \int_a^b g^{(4)}(x)h(x) \, dx. \end{aligned} \tag{8}$$

Two conclusions may be drawn from the fact that this is 0 for all admissible functions $h$.

(A) $\qquad\qquad g^{(4)}(x) = 0 \quad$ for all $\quad x \neq x_0, x_1, \ldots, x_n.$

If this were not so, say if $g^{(4)}(\xi) > 0$ at some $\xi$ interior to one of the subintervals, we could find an admissible $h$ satisfying $h(\xi) > 0$, $h(x) \geq 0$, and being $>0$ only in a small neighborhood of the point $\xi$. Then, in (8), only the integral would give a nonzero contribution, and this contribution would be positive. Thus, we would have $J'(0) > 0$ for a suitable admissible $h$, contradicting (6).

(B)
$$g''(x_0+) = g''(x_n-) = 0 \qquad (9)$$

and

$$g''(x_i-) = g''(x_i+), \qquad i = 1, 2, \ldots, n-1; \qquad (10)$$

that is, $g''$ is continuous on $[a, b]$ and zero at the endpoints. If this were not so, say if $g''(x_i -) - g''(x_i +) \neq 0$ for some $i$, we could find an $h$ so that $h' = 0$ at all nodes except at $x_i$, and it would turn out that $J'(0) \neq 0$ for that $h$, again contradicting (6).

In view of (A), we conclude that the function $g$ in each interval $[x_i, x_{i+1}]$ is represented by some cubic polynomial $g_i(x)$. From (i), (ii), and (B), we conclude that at the nodes $x_i$ these polynomials, as well as their first and second derivatives, fit together continuously,

$$g_i(x_i) = g_{i-1}(x_i), \qquad (11)$$
$$g_i'(x_i) = g_{i-1}'(x_i), \qquad (12)$$
$$g_i''(x_i) = g_{i-1}''(x_i), \qquad (13)$$

$i = 1, 2, \ldots, n-1$. Moreover from (9),

$$g_0''(x_0) = 0, \qquad g_{n-1}''(x_n) = 0. \qquad (14)$$

We now show that there exists one and only one set of cubic polynomials $(g_0, g_1, \ldots, g_{n-1})$ satisfying the foregoing conditions. Because a cubic polynomial has four coefficients, four conditions are required to fix each $g_i$. Two conditions are immediate in view of (i):

$$g_i(x_i) = f_i, \qquad g_i(x_{i+1}) = f_{i+1}. \qquad (15)$$

To formulate two additional conditions, we introduce the parameters

$$s_i := g'(x_i). \qquad (16)$$

These numbers are, at present, unknown; a method for determining them will emerge from our analysis. In terms of the $s_i$, the additional conditions are

$$g_i'(x_i) = s_i, \qquad g_i'(x_{i+1}) = s_{i+1}. \qquad (17)$$

It is convenient at this point to introduce the new variable

$$t := \frac{x - x_i}{h_i} \qquad (18)$$

where

$$h_i := x_{i+1} - x_i,$$

and to set

$$q_i(t) = g_i(x) = g_i(x_i + h_i t). \qquad (19)$$

The $q_i$ are again cubic polynomials. The conditions (15) and (17) in terms of the $q_i$ are

$$q_i(0) = f_i, \qquad q_i(1) = f_{i+1}, \qquad (20)$$
$$q_i'(0) = h_i s_i, \qquad q_i'(1) = h_i s_{i+1}. \qquad (21)$$

These are just the data required for two-point Hermite interpolation with data given at the points $t = 0$ and $t = 1$. By (9), §5.6, we have

$$q_i(t) = (1 - t)^2 f_i + t^2 f_{i+1}$$
$$+ t(1 - t)[(1 - t)(2f_i + h_i s_i) + t(2f_{i+1} - h_i s_{i+1})] \qquad (22)$$

Straightforward differentiation yields

$$q_i''(0) = 6(f_{i+1} - f_i) - 2h_i(2s_i + s_{i+1}),$$
$$q_i''(1) = -6(f_{i+1} - f_i) + 2h_i(s_i + 2s_{i+1}).$$

Using $g_i''(x) = q_i''(t)(dt/dx)^2 = q_i''(t)h_i^{-2}$, the conditions (13) and (14) asserting the continuity of $g''$, and its vanishing at the endpoints thus yield the relations

$$h_0^{-1}(2s_0 + s_1) = 3h_0^{-2}(f_1 - f_0);$$
$$h_{i-1}^{-1}(s_{i-1} + 2s_i) + h_i^{-1}(2s_i + s_{i+1})$$
$$= 3h_{i-1}^{-2}(f_i - f_{i-1}) + 3h_i^{-2}(f_{i+1} - f_i),$$
$$i = 1, 2, \ldots, n - 1;$$
$$h_{n-1}^{-1}(2s_{n-1} + s_n) = 3h_{n-1}^{-2}(f_n - f_{n-1}). \qquad (23)$$

The relations (23) represent a system of $n + 1$ equations for the $n + 1$ unknowns

$s_0$, $s_1$, ..., $s_n$, from which we may hope to determine these unknowns uniquely. In fact, the matrix of the system,

$$
\mathbf{A} := \begin{pmatrix}
\dfrac{2}{h_0} & & & & & & 0 \\[2ex]
\dfrac{1}{h_0} & \dfrac{2}{h_0}+\dfrac{2}{h_1} & \dfrac{1}{h_1} \\[2ex]
& \dfrac{1}{h_1} & \dfrac{2}{h_1}+\dfrac{2}{h_2} & \dfrac{1}{h_2} \\[1ex]
& \cdots\cdots\cdots\cdots\cdots\cdots\cdots\cdots\cdots\cdots\cdots \\[1ex]
& & & \dfrac{1}{h_{n-2}} & \dfrac{2}{h_{n-2}}+\dfrac{2}{h_{n-1}} & \dfrac{1}{h_{n-1}} \\[2ex]
0 & & & & \dfrac{1}{h_{n-1}} & \dfrac{2}{h_{n-1}}
\end{pmatrix}
$$

is non-singular, because it is easily seen that the quadratic form

$$Q(\mathbf{s}) := \mathbf{s}^T \mathbf{A} \mathbf{s}$$

is positive for every vector $\mathbf{s} \neq \mathbf{0}$, implying that $\mathbf{A}$ is positive definite. It follows that the system (23) always has a unique solution.

We conclude that there exists a function $g$ satisfying the postulates (i), (ii), (iii), (iv), and that there exists precisely one such function. In each interval $[x_i, x_{i+1}]$, this function is represented by a cubic polynomial $g_i(x)$. This polynomial is given by the formula

$$g_i(x) = q_i(t), \qquad t := \frac{x - x_i}{h_i},$$

where $q_i$ is given by (22), and where $s_0$, $s_1$, ..., $s_n$ denotes the solution of the system (23). The function $g$ thus constructed is called the **natural spline interpolant** for the data $(x_0, f_0)$, $(x_1, f_1)$, ..., $(x_n, f_n)$.

# PROBLEMS

1   *Periodic splines.* Repeat the analysis of this section for the case that a *periodic* function $f$ is to be approximated over a full period by a *periodic* spline function $g$. Assume equidistant interpolating points.

    **(a)**  Assuming the period to be 1, let $x_k := k/n$. The quantities

$s_i := g'(x_i)$ now satisfy the additional condition $s_0 = s_n$. Show that minimizing $J[g]$ now yields the system of equations

$$\mathbf{As} = \mathbf{b},$$

where

$$\mathbf{A} := \begin{pmatrix} 4 & 1 & & & & 1 \\ 1 & 4 & 1 & & & \\ & 1 & 4 & 1 & & \\ & & \cdots\cdots\cdots\cdots\cdots & & \\ & & & 1 & 4 & 1 \\ 1 & & & & 1 & 4 \end{pmatrix}, \quad \mathbf{s} := \begin{pmatrix} hs_0 \\ hs_1 \\ hs_2 \\ \vdots \\ hs_{n-1} \end{pmatrix}, \quad \mathbf{b} := 3\begin{pmatrix} f_1 - f_{n-1} \\ f_2 - f_0 \\ f_3 - f_1 \\ \vdots \\ f_0 - f_{n-2} \end{pmatrix}$$

**(b)** Construct the periodic spline interpolant for the function $f(x) := \sin 2\pi x$, using $n = 4$ subintervals.

---

## §5.9 CONSTRUCTION OF THE SPLINE INTERPOLANT

Here we discuss in some detail the solution of the linear system (23) of §5.8 in the case where the $x_i$ are equidistant, and therefore all

$$h_i = x_{i+1} - x_i =: h,$$

independently of $i$. In this case, only the quantities $hs_i$ (but not the $s_i$) are required for forming the polynomials $q_i$. It is thus reasonable to write the system (23) in a form which directly produces the $hs_i$. Multiplying (23) by $h^2$ we get

$$\mathbf{As} = \mathbf{b}, \tag{1}$$

where

$$\mathbf{A} := \begin{pmatrix} 2 & 1 & & & & 0 \\ 1 & 4 & 1 & & & \\ & 1 & 4 & 1 & & \\ & & \cdots\cdots\cdots\cdots\cdots & & \\ & & & 1 & 4 & 1 \\ 0 & & & & 1 & 2 \end{pmatrix}, \quad \mathbf{s} := \begin{pmatrix} hs_0 \\ hs_1 \\ hs_2 \\ \vdots \\ hs_{n-1} \\ hs_n \end{pmatrix},$$

$$\mathbf{b} := \begin{pmatrix} b_0 \\ b_1 \\ b_2 \\ \vdots \\ b_{n-1} \\ b_n \end{pmatrix} = \begin{pmatrix} 3f_1 - 3f_0 \\ 3f_2 - 3f_0 \\ 3f_3 - 3f_1 \\ \vdots \\ 3f_n - 3f_{n-2} \\ 3f_n - 3f_{n-1} \end{pmatrix}.$$

Because of the dominant diagonal terms, the system (1) is well conditioned, and Gaussian elimination may be used without interchanges. Proceeding directly to the $L$-$R$ decomposition, we therefore seek to find matrices

$$
\mathbf{L} := \begin{pmatrix}
l_0 & & & & & 0 \\
k_1 & l_1 & & & & \\
& k_2 & l_2 & & & \\
& & \cdots\cdots\cdots\cdots\cdots\cdots\cdots & & \\
& & & k_{n-1} & l_{n-1} & \\
0 & & & & k_n & l_n
\end{pmatrix},
$$

$$
\mathbf{R} := \begin{pmatrix}
1 & r_0 & & & & 0 \\
& 1 & r_1 & & & \\
& & 1 & r_2 & & \\
& & \cdots\cdots\cdots\cdots\cdots\cdots & & \\
& & & & 1 & r_{n-1} \\
0 & & & & & 1
\end{pmatrix},
$$

so that $\mathbf{LR} = \mathbf{A}$. Carrying out the multiplication and comparing elements yields

$$k_i = 1, \qquad i = 1, 2, \ldots, n;$$
$$l_0 = 2;$$
$$r_{i-1} k_i + l_i = 4, \qquad i = 1, 2, \ldots, n-1;$$
$$r_{n-1} k_n + l_n = 2;$$
$$l_i r_i = 1, \qquad i = 0, 1, \ldots, n-1.$$

There follows

$$r_0 = \tfrac{1}{2} \tag{2}$$

and from the fact that $r_i = l_i^{-1}$ and $l_i = 4 - r_{i-1}$,

$$r_i = \frac{1}{4 - r_{i-1}}, \qquad i = 1, 2, \ldots, n-1. \tag{3}$$

In view of the relations,

$$
\begin{aligned}
l_0 &= 2, \\
l_i &= 4 - r_{i-1}, \qquad i = 1, 2, \ldots, n-1, \\
l_n &= 2 - r_{n-1},
\end{aligned}
\tag{4}
$$

the elements of the matrix $\mathbf{L}$ may be expressed entirely in terms of the numbers $r_i$ defined by (2) and (3).

A time-saving finesse is incorporated into the algorithm by noting that the quantities $r_i$ arise by iterating the function

$$f(r) := \frac{1}{4 - r}$$

starting with $r_0 := 0.5$. It is easily seen that $f$ satisfies the hypotheses of Theorem 2.1 for the interval $[0, 0.5]$ with the value

$$L := f'(0.5) = (2/7)^2 = 0.081632653...$$

for the Lipschitz constant. Thus, the $r_i$ very rapidly tend towards the fixed point of $f$,

$$r_\infty = 2 - \sqrt{3} = 0.267949192.... \tag{5}$$

Because more than one decimal digit is gained at each iteration step, $r_m = r_\infty$ on a machine with an $m$-digit mantissa, and we may put $r_i := r_\infty$ for $i \geq m$ without loss of accuracy.

Along with **L** and **R**, we find the solution of the tridiagonal system

$$\mathbf{Ly} = \mathbf{b}. \tag{6}$$

In terms of the $r_i$ and the $b_i$, it is given by

$$\begin{aligned} y_0 &= r_0 b_0, \\ y_i &= r_i(b_i - y_{i-1}), \qquad i = 1, 2, \ldots, n-1, \\ y_n &= \frac{b_n - y_{n-1}}{2 - r_{n-1}}. \end{aligned} \tag{7}$$

The foregoing constitutes *Phase I* of the determination of the spline interpolant. The flow diagram of Phase I is shown in Figure 5.9a.

In *Phase II*, the solution vector **s** of (1) is found by back substitution; that is, as the solution of

$$\mathbf{Rs} = \mathbf{y}. \tag{8}$$

Written explicitly, this yields

$$\begin{aligned} hs_n &= y_n, \\ hs_i &= y_i - r_i(hs_{i+1}), \qquad i = n-1, n-2, \ldots, 0. \end{aligned} \tag{9}$$

Finally, in *Phase III* we evaluate the spline interpolant $g$ for an arbitrary $x \in [a, b]$ by performing the following steps:

(1) Calculate the integral part

$$i := \left[ \frac{x - x_0}{h} \right];$$

this determines the interval $[x_i, x_{i+1}]$ in which $x$ lies;

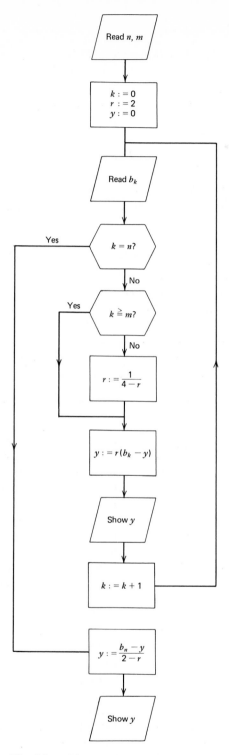

**Fig. 5.9a.** Phase I of spline algorithm.

(2)   Read $f_i, f_{i+1}, hs_i, hs_{i+1}$, and calculate

$$q(t) = (1 - t)^2 f_i + t^2 f_{i+1}$$
$$+ t(1 - t)\{(1 - t)(2f_i + hs_i) + t(2f_{i+1} - hs_{i+1})\},$$

where

$$t := \frac{x - x_i}{h}.$$

**DEMONSTRATION 5.9-1:**  This is designed to show the good condition of spline interpolation as opposed to Lagrangian interpolation. (See §5.5.) For the $n + 1 = 21$ interpolating points

$$x_k := -1 + 0.1k, \qquad k = 0, 1, 2, \ldots, 20,$$

we assume the data $f_{10} = 1$, all other $f_k = 0$. We recall that the resulting Lagrangian polynomial blows up near the endpoints of the interval $[-1, 1]$. For the spline interpolant, the results of the Phases I and II are as follows:

| $k$ | $f_k$ | $b_k$ | $y_k$ (Phase I) | $hs_k$ (Phase II) |
|---|---|---|---|---|
| 0  | 0 | 0  | 0.          | −0.00001145 |
| 1  | 0 | 0  | 0.          | 0.00002289 |
| 2  | 0 | 0  | 0.          | −0.00008013 |
| 3  | 0 | 0  | ↓ 0.        | 0.00029761 |
| 4  | 0 | 0  | 0.          | −0.00111032 |
| 5  | 0 | 0  | 0.          | 0.00414366 |
| 6  | 0 | 0  | 0.          | −0.01546433 |
| 7  | 0 | 0  | 0.          | 0.05771366 |
| 8  | 0 | 0  | 0.          | −0.21539031 |
| 9  | 0 | 3  | 0.80384758  | 0.80384758 |
| 10 | 1 | 0  | −0.21539031 | 0.00000000 |
| 11 | 0 | −3 | −0.74613392 | −0.80384758 |
| 12 | 0 | 0  | 0.19992598  | 0.21539031 |
| 13 | 0 | 0  | −0.05357001 | −0.05771366 |
| 14 | 0 | 0  | 0.01435404  | 0.01546433 |
| 15 | 0 | 0  | −0.00384615 | −0.00414366 |
| 16 | 0 | 0  | 0.00103057  | 0.00111031 |
| 17 | 0 | 0  | −0.00027614 | ↑ −0.00029761 |
| 18 | 0 | 0  | 0.00007399  | 0.00008013 |
| 19 | 0 | 0  | −0.00001983 | −0.00002290 |
| 20 | 0 | 0  | 0.00001145  | 0.00001145 |

In Phase III, the following values of the natural cubic spline interpolant are now generated at the midpoints of the intervals $[x_i, x_{i+1}]$:

| $\pm x$ | $g(x)$ | $\pm x$ | $g(x)$ |
|---------|---------|---------|---------|
| 0.05 | 0.60048095 | 0.55 | −0.00065675 |
| 0.15 | −0.12740474 | 0.65 | 0.00017599 |
| 0.25 | 0.03413800 | 0.75 | −0.00004722 |
| 0.35 | −0.00914725 | 0.85 | 0.00001288 |
| 0.45 | 0.00245100 | 0.95 | −0.00000429 |

Although the relative extrema of the spline interpolant need not occur exactly halfway between the interpolating points, it is clear that the absolute values of these extrema decrease rapidly towards the endpoints of the interval $[-1, 1]$.

# PROBLEMS

**1** Construct the spline interpolant $g$ for the function $f(x) := \sin x$, using the interpolating points

$$x_k := k \frac{\pi}{6}, \qquad k = 0, 1, \ldots, 6.$$

(a) Evaluate $g$ at the points halfway between the interpolating points, and record the errors $f(x) - g(x)$.

(b) Using the same data, also evaluate the polynomial interpolant (by means of the barycentric formula) at the same points. Compare the errors and comment.

**2** Perform the same comparison for the function

$$f(x) := \begin{cases} 1 - 3x, & 0 \le x \le \frac{1}{3}, \\ -\frac{1}{2} + \frac{3}{2}x, & \frac{1}{3} \le x \le 1, \end{cases}$$

using $3n + 1$ equidistant interpolating points in the interval $[0, 1]$. What happens as $n$ increases?

**3** Construct the spline interpolant for the data

| $x$ | 0.0 | 0.5 | 1.0 | 1.5 | 2.0 | 2.5 | 3.0 | 3.5 | 4.0 |
|-----|-----|-----|-----|-----|-----|-----|-----|-----|-----|
| $f$ | 0.0 | 0.4 | 1.0 | 0.4 | 0.0 | −0.4 | −1.0 | −0.4 | 0.0 |

and draw the graph of the resulting interpolant $g$. Also draw the graph
of the Lagrangian interpolating polynomial for the same data.


**Bibliographical Notes and Recommended Reading**

On approximation and interpolation in general, see Davis [1961] and Cheney
[1966]. The earliest reference to the barycentric formula appears to be Taylor
[1945]; see also Rutishauser [1976]. On spline approximation, see Sard and
Weintraub [1971].

The idea of minimizing a quadratic functional, which lies at the bottom of
spline interpolation, is also basic for the *finite element method* which has
developed into one of the most potent tools for solving problems in partial
differential equations. For a first orientation, see Strang and Fix [1973].

# CHAPTER 6

## Integration

Two topics will be discussed in Chapter 6; namely,
    **(i)**   the evaluation of definite integrals,

$$\int_a^b f(x)\, dx;$$

    **(ii)**   the solution of ordinary differential equations, such as

$$y' = f(x, y).$$

The evaluation of indefinite integrals, such as

$$\int_a^x f(x)\, dx$$

is a special case of (ii) and is treated as such numerically. Indeed, the indefinite integral of $f$ is a solution of the differential equation

$$y' = f(x).$$

## §6.1   DEFINITE INTEGRALS: THE PROBLEM POSED

Let $f$ be defined on the interval $[a, b]$. Here, we deal with the problem of evaluating the functional (Figure 6.1a)

$$I := \int_a^b f(x)\, dx. \tag{1}$$

    In the mathematical treatment of integration, several levels of sophistication may be discerned.

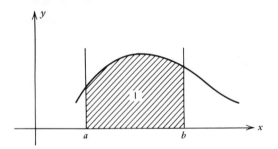

**Fig. 6.1a.** Definite integral.

(1)  The elementary theory of integration, as taught in calculus, is concerned with the evaluation of integrals involving certain types of elementary functions "in closed form." The emphasis on closed form integration sometimes generates the impression that almost all integrals can thus be evaluated, if only the right trick is found. In reality, the integrals that can be evaluated in closed form constitute a small minority among all conceivable integrals. Even very simple integrals, such as

$$\int_0^\pi \frac{\sin x}{x}\,dx \quad \text{or} \quad \int_0^1 e^{-x^2}\,dx$$

cannot be evaluated in closed form.

(2)  In the advanced theory of integration (measure and integration), the emphasis is *not*, as the uninitiated student might think, on the evaluation of more complicated integrals. In fact, no integrals are evaluated in this theory. The objective is the study of conditions, such as continuity or measurability, under which the integral of a function can be meaningfully defined. Generally, the theoretician will endeavor to define an integral and to prove theorems about it under the weakest possible conditions.

(3)  Here, we are concerned with the *numerical evaluation* of integrals. As elsewhere in numerical mathematics, the main concern is not the existence, but rather the efficient construction of the desired mathematical object. The problem of evaluating definite integrals occurs rather frequently in applications. For instance, in mechanical engineering one might wish to compute the moment of inertia of a body whose shape is defined in a complicated manner, or in electrical engineering the magnetic force exerted by a coil, or in operations research a probability that is defined geometrically. In such cases, the integral does not care whether it can be evaluated by a nice formula, and methods of numerical integration frequently are the only resort. Kepler was led to a formula for numerical integration by the problem of computing the volume of a barrel.

Which are the desirable properties that a good algorithm for numerical integration should have?

## (A)  Simple Applicability

There should be no restrictions on the integrand $f(x)$ other that it can be evaluated numerically at an arbitrary point $x$ of the interval of integration.

## (B)  Convergence for a Large Class of Integrands

The algorithm should yield the correct value of the integral (or more precisely, it should construct a sequence of numbers converging to the correct value) under as few hypotheses as possible on the function $f$ to be integrated.

## (C)  Good Convergence for Good Integrands

If the integrand is good-natured, that is, if it has many continuous derivatives, or if it is even analytic on the closed interval of integration, the algorithm should take advantage of this fact and furnish a sequence of approximations that converges rapidly to the value of the integral.

How can these wishes be satisfied?

It is easy to set up integration algorithms that satisfy (A) and (B). We divide the interval of integration into $n$ equal parts by means of the **sampling points**

$$x_k := a + kh, \qquad k = 0, 1, 2, \ldots, n,$$

where

$$h := \frac{1}{n}(b - a)$$

is the **integration step**. In each subinterval $[x_k, x_{k+1}]$, we now replace the function $f$ by its *linear* interpolant (Figure 6.1b) and approximate the integral over the subinterval by the integral of the interpolant:

$$\int_{x_k}^{x_{k+1}} f(x)\, dx \sim \frac{h}{2}[f(x_k) + f(x_{k+1})].$$

Adding the contributions of all subintervals then yields the approximation

$$T(h) := h[\tfrac{1}{2}f(x_0) + f(x_1) + f(x_2) + \cdots + f(x_{n-1}) + \tfrac{1}{2}f(x_n)] \qquad (2)$$

for the value of the entire integral. The number $T(h)$ is called the **trapezoidal value** of the integral $I$, calculated with the step $h$. The function $T(h)$ is defined for those values of $h$ for which $h^{-1}(b - a)$ is an integer.

It is clear that the trapezoidal value satisfies postulate (A) (simple computability). All that is required to obtain $T(h)$ is to evaluate $f$ at the sampling points.

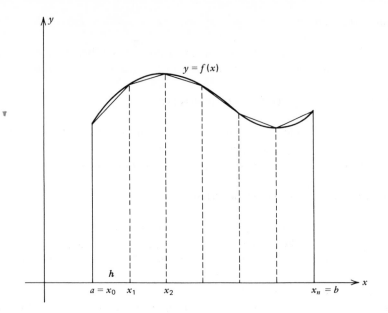

**Fig. 6.1 b.**   Trapezoidal value.

A satisfactory answer can also be given with regard to postulate (B). From the point of view of the mathematical theory of integration, $T(h)$ is a so-called *Riemann sum* for the integral $I$. Hence,

$$\lim_{h \to 0} T(h) = I$$

holds for every *Riemann-integrable* function $f$. The class of Riemann-integrable functions contains, in particular, all continuous or piecewise continuous functions.

Before discussing postulate (C), we mention another possibility of approximating the integral $I$. It consists in replacing $f$ in each subinterval by the value at the midpoint of the subinterval,

$$x_{k+1/2} := a + (k + \tfrac{1}{2})h,$$

$k = 0, 1, \ldots, n - 1$ (see Figure 6.1c). The value of the integral over the subinterval is then approximated by

$$\int_{x_k}^{x_{k+1}} f(x)\, dx \sim h f(x_{k+1/2}).$$

This is the so-called **midpoint rule** of numerical integration. Adding the contribu-

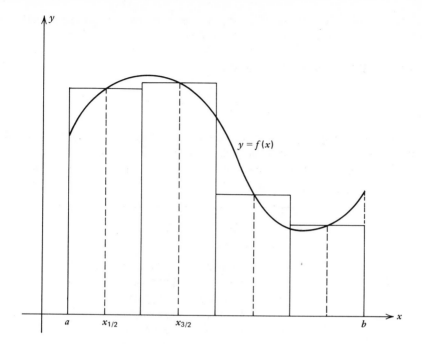

**Fig. 6.1c.**  Midpoint value.

tions of all subintervals yields the **midpoint value** of the integral $I$, calculated with the step $h$:

$$M(h) := h[f(x_{1/2}) + f(x_{3/2}) + \cdots + f(x_{n-1/2})]$$

$$= h \sum_{k=0}^{n-1} f(x_{k+1/2}). \tag{3}$$

Calculating $M(h)$ requires evaluating $f$ at the sampling points that are used for $T(h/2)$ but not for $T(h)$. There holds

$$T\left(\frac{h}{2}\right) = \tfrac{1}{2}T(h) + \tfrac{1}{2}M(h). \tag{4}$$

The midpoint value $M(h)$ also satisfies the postulates (A) and (B). Indeed, the relation

$$\lim_{h \to 0} M(h) = I$$

even holds for certain functions which at the endpoints $a$ and $b$ are merely improperly integrable, because the values of $f$ at the endpoints are not required to evaluate $M(h)$.

How about postulate (C) (rapid convergence for good-natured integrands)? Here the prognosis is less favorable.

**DEMONSTRATION 6.1-1:**  We evaluate

$$I := \int_0^\pi \frac{\sin x}{x} \, dx$$

by both the midpoint and the trapezoidal rule. The integrand (defined at 0 by its limit 1) is as good-natured as possible; in fact, considered as a function of the complex variable $x$, it has no singularities in the entire complex plane. The following values result:

| $n$ | $M(h)$ | $T(h)$ |
|-----|--------|--------|
| 1 | 2.00000000 | 1.57079633 |
| 2 | 1.88561808 | 1.78539816 |
| 4 | 1.86017649 | 1.83550812 |
| 8 | 1.85398597 | 1.84784231 |
| 16 | 1.85244860 | 1.85091414 |
| 32 | 1.85206490 | 1.85168137 |
| 64 | 1.85196901 | 1.85187314 |
| 128 | 1.85194504 | 1.85192107 |
| 256 | 1.85193905 | 1.85193306 |
| 512 | 1.85193755 | 1.85193605 |
| 1024 | 1.85193718 | 1.85193680 |
| 2048 | 1.85193708 | 1.85193699 |

The convergence of both $M(h)$ and $T(h)$ is seen to be slow: As many as 2048 evaluations of the integrand barely yield seven correct digits.

How could the speed of convergence be improved? The following possibilities come to mind.

(a)  Replacing $f$ by an interpolating polynomial of higher degree, using equidistant interpolating points. This is dangerous because, as we have seen, the sequence of these interpolating polynomials behaves in a highly unstable manner.

(b)  Replacing $f$ by an interpolating polynomial, using the Chebyshev points as interpolating points. This would result in a numerically stable process; however, the resulting integration formula would have complicated coefficients, thus violating postulate (A).

(c)  Approximating $f$ by a spline interpolant. This is discussed at some length by Forsythe, Malcolm, and Moler [1977], who recommend the method highly. However, the evaluation of each integral requires solving a tridiagonal system; moreover, it can be shown that especially for good-natured (for example,

analytic) integrands the approximation by the spline interpolant is not particularly accurate. (See also Problem 1, §5.9.)

Fortunately, however, there exists an integration method which satisfies
postulate (C) without having any of the disadvantages mentioned. This method
is based on a fundamental principle of numerical computation which we have
already encountered in connection with Aitken's $\Delta^2$ method: *Use an approximate
knowledge of the error to approximately eliminate the error.* "Approximate knowledge of the error" here as there means "knowledge of the qualitative behavior
of the error." The qualitative behavior of the error of the trapezoidal and midpoint values will be studied by a summation formula to be derived in the
following section. This formula is useful also in other contexts.

## PROBLEMS

1  Which integrals are approximated by the sums

(a) $$S_n := \sum_{k=1}^{n} \frac{1}{n+k},$$

(b) $$S_n := \sum_{k=1}^{n} \frac{n}{n^2 + k^2}\ ?$$

Therefore, which are the limits of these sums as $n \to \infty$? Try to determine the limits numerically, and experiment with convergence acceleration. [Consider the subsequences $t_k := s_n$ where $n = 2^k$.]

2  Compute the integrals

$$I_k := \int_0^{k\pi} \frac{\sin x}{x}\, dx, \qquad k = 1, 2, \ldots,$$

both by series expansion and by the trapezoidal rule.

(a)  An alternating series will be obtained by expanding the integrand in powers of $x^2$. How many digits will be lost due to smearing?
For which values of $k$ can you guarantee an error $< 10^{-8}$?

(b)  Comparing the computational effort ($=$running time on
calculator), which method is more economical for a given $k$ if the
maximum tolerable error is $10^{-8}$? $10^{-4}$?

(c)  Experiment with other semi-analytical approaches to the
integrals $I_k$. For instance, apply integration by parts to

$$I_{k+1} - I_k = \int_{k\pi}^{(k+1)\pi} \frac{\sin x}{x}\, dx,$$

which may yield accurate results for large values of $k$.

## §6.2   THE EULER-MACLAURIN SUM FORMULA

We begin by considering the seemingly very special problem of evaluating

$$I := \int_0^1 g(t)\, dt \tag{1}$$

by means of integration by parts. It is assumed that the integrand $g$ is good-natured in the sense that it has any desired number of continuous derivatives.

We write

$$I = \int_0^1 1 \cdot g(t)\, dt.$$

If $p_1$ is any polynomial such that

$$p_1'(t) = 1,$$

then we get

$$I = \left[ p_1(t)g(t) \right]_0^1 - \int_0^1 p_1(t)g'(t)\, dt.$$

We continue in the same manner, letting $p_2' = p_1$, $p_3' = p_2$, .... At the next step, this yields

$$I = \left[ p_1(t)g(t) \right]_0^1 - \left[ p_2(t)g'(t) \right]_0^1 + \int_0^1 p_2(t)g''(t)\, dt, \tag{2}$$

and two more integrations by parts furnish

$$
\begin{aligned}
I = {} & \left[ p_1(t)g(t) \right]_0^1 - \left[ p_2(t)g'(t) \right]_0^1 \\
& + \left[ p_3(t)g''(t) \right]_0^1 - \left[ p_4(t)g^{(3)}(t) \right]_0^1 \\
& + \int_0^1 p_4(t)g^{(4)}(t)\, dt.
\end{aligned}
\tag{3}
$$

Obviously, this process could be continued *ad infinitum.*

Up to this point, the polynomials $p_k$ only had to satisfy the condition $p_k' = p_{k-1}$. This condition does not fix the polynomials completely because an additive constant remains undetermined in $p_k$. We now impose on the $p_k$ the additional condition that

$$\int_0^1 p_k(t)\, dt = 0, \qquad k = 1, 2, \dots . \tag{4}$$

With this condition the $p_k$ are uniquely determined. Indeed, $p_0(t) = 1$ is given. If $p_k$ is fixed, then $p_{k+1}$ is fixed by $p'_{k+1} = p_k$ up to an additive constant, and there is only one value of that constant such that (4) holds. For instance, (4) implies that

$$p_1(t) = t - \tfrac{1}{2}.$$

The exact form of the polynomials $p_2, p_3, \dots$ is not important for the following; of importance, however, is the observation that the polynomial $p_1(t)$ evidently is an odd function of $t - \tfrac{1}{2}$. It follows that

$$p_2(t) = \int_{1/2}^t p_1(t)\, dt + \text{const}$$

is an even function of $t - \tfrac{1}{2}$, which implies that

$$p_2(0) = p_2(1).$$

The polynomial $p_3$ has the form

$$p_3(t) = \int_{1/2}^t p_2(t)\, dt + \text{const} = \text{odd function} + \text{const};$$

however, the constant is zero on account of (4) where $k = 3$, which permits us to conclude that $p_3$ is an odd function of $t - \tfrac{1}{2}$. This implies

$$p_3(1) = -p_3(0).$$

However, on the other hand, by (4) where $k = 2$

$$0 = \int_0^1 p_2(t)\, dt = p_3(1) - p_3(0)$$

or

$$p_3(1) = p_3(0),$$

and we have

$$p_3(0) = p_3(1) = 0.$$

In a similar way, there follows

$$p_4(0) = p_4(1)(\neq 0).$$

With this choice of the polynomials $p_k$, the formulas (2) and (3) take the simpler form

$$I = \int_0^1 g(t)\, dt = \tfrac{1}{2}[g(0) + g(1)] + A[g'(0) - g'(1)] + \int_0^1 p_2(t)g''(t)\, dt \qquad (5)$$

and

$$I = \int_0^1 g(t)\, dt = \tfrac{1}{2}[g(0) + g(1)] + A[g'(0) - g'(1)]$$

$$+ B[g^{(3)}(0) - g^{(3)}(1)] + \int_0^1 p_4(t)g^{(4)}\, dt. \qquad (6)$$

The precise values of the numbers $A := p_2(1)$ and $B := p_4(1)$ will not be needed.

From (5) and (6) we easily deduce the desired sum formulas. For instance, from (5) we get, by shifting the variable in the integral to the interval $[k, k+1]$, where $k$ is any integer,

$$\tfrac{1}{2}[g(k) + g(k+1)] = \int_k^{k+1} g(t)\, dt - A[g'(k) - g'(k+1)]$$

$$- \int_0^1 p_2(t)g''(k+t)\, dt.$$

This we sum from $k = 0$ to $k = n - 1$. In the expression involving the constant $A$, all intermediate terms cancel, and, in view of

$$\int_0^1 + \int_1^2 + \cdots + \int_{n-1}^n = \int_0^n,$$

we obtain

$$\tfrac{1}{2}g(0) + g(1) + g(2) + \cdots + g(n-1) + \tfrac{1}{2}g(n)$$

$$= \int_0^n g(t)\, dt + A[g'(n) - g'(0)]$$

$$- \int_0^1 p_2(t)\{g''(t) + g''(t+1) + \cdots + g''(t+n-1)\}\, dt. \qquad (7)$$

In a similar way, we get from (6)

$$\tfrac{1}{2}g(0) + g(1) + g(2) + \cdots + g(n-1) + \tfrac{1}{2}g(n)$$

$$= \int_0^n g(t)\, dt + A[g'(n) - g'(0)] + B[g^{(3)}(n) - g^{(3)}(0)]$$

$$- \int_0^1 p_4(t)\{g^{(4)}(t) + g^{(4)}(t+1) + \cdots + g^{(4)}(t+n-1)\}\, dt. \tag{8}$$

The formulas (7) and (8) are simple versions of the **Euler-Maclaurin sum formula** which has many applications in classical analysis.

   The connection with the trapezoidal rule is established by the observation that the sum on the left of (7) and (8) is the trapezoidal value, computed with the step $h = 1$, of the integral

$$I := \int_0^n g(t)\, dt$$

that occurs on the right. Thus, the additional terms on the right express the error that is committed when the integral is approximated by the trapezoidal value with step $h = 1$. This will be adapted in §6.3 to the approximation of arbitrary integrals by trapezoidal values $T(h)$, where $h$ is arbitrary.

   Here we pause to illustrate the use of (7) by a proof of Stirling's formula,

$$n! \sim \sqrt{2\pi n}\left(\frac{n}{e}\right)^n \tag{9}$$

which has been used repeatedly in preceding chapters. The mathematical meaning of the sign $\sim$ is that

$$\lim_{n \to \infty} \frac{n!}{\sqrt{2\pi n}\left(\dfrac{n}{e}\right)^n} = 1; \tag{10}$$

thus, the *relative* error committed when $n!$ is approximated by the expression on the right of (9) tends to 0 as $n \to \infty$.

   To try to discover Stirling's formula, we should look for a simple way of expressing the product

$$n! = 1 \cdot 2 \cdot 3 \cdots n.$$

We do not know about products, but we have just learned how to express sums.

A product is converted into a sum by taking logarithms; thus, we are led to consider

$$\text{Log } n! = \text{Log } 1 + \text{Log } 2 + \cdots + \text{Log } n.$$

The sum on the right has the form of the sum on the left of (7) where $g(t) :=$ Log $t$, except for the limits of summation and for the factors $\frac{1}{2}$. Adjusting both, we get

$$\text{Log } 1 + \text{Log } 2 + \cdots + \text{Log } n = \tfrac{1}{2} \text{Log } n + \int_1^n \text{Log } t \, dt$$

$$+ A\left[\frac{1}{n} - 1\right] + \int_0^1 p_2(t) \sum_{k=1}^{n-1} \frac{1}{(k+t)^2} \, dt.$$

Using

$$\int_1^n \text{Log } t \, dt = n \text{ Log } n - n + 1,$$

there follows

$$\text{Log}(n!) = (n + \tfrac{1}{2})\text{Log } n - n + 1 + A\left(\frac{1}{n} - 1\right) + \int_0^1 p_2(t) \sum_{k=1}^{n-1} (k+t)^{-2} \, dt.$$

Because

$$(n + \tfrac{1}{2})\text{Log } n - n = \text{Log}\left\{\sqrt{n}\left(\frac{n}{e}\right)^n\right\},$$

this yields

$$\text{Log}\frac{n!}{\sqrt{n}\left(\dfrac{n}{e}\right)^n} = 1 + A\left(\frac{1}{n} - 1\right) + \int_0^1 p_2(t) \sum_{k=1}^{n-1} (k+t)^{-2} \, dt.$$

Here we let $n \to \infty$. The limit of the expression on the right obviously exists, because the series

$$\sum_{k=1}^{\infty} (k+t)^{-2}$$

converges uniformly for $0 \le t \le 1$ by comparison with the convergent series*
$\sum k^{-2}$. It thus follows that

$$\lim_{n \to \infty} \text{Log} \frac{n!}{\sqrt{n} \left(\dfrac{n}{e}\right)^n} = \text{Log } C, \tag{11}$$

where

$$\text{Log } C := 1 - A + \int_0^1 p_2(t) \sum_{k=1}^{\infty} (k + t)^{-2} \, dt.$$

This implies that

$$n! \sim C\sqrt{n} \left(\frac{n}{e}\right)^n \tag{12}$$

for some positive constant $C$. The numerical value of $C$ cannot be computed by the above method. It can be deduced, however, from the following result of calculus which is due to J. Wallis:

$$\frac{\pi}{2} = \lim_{n \to \infty} w_n, \quad \text{where} \quad w_n := \frac{2 \cdot 2 \cdot 4 \cdot 4 \cdot 6 \cdot 6 \cdots 2n \cdot 2n}{1 \cdot 3 \cdot 3 \cdot 5 \cdot 5 \cdot 7 \cdots (2n - 1)(2n + 1)}.$$

Using the obvious relations

$$2 \cdot 4 \cdot 6 \cdots \cdot 2n = 2^n n!,$$

$$1 \cdot 3 \cdot 5 \cdots \cdot (2n - 1) = \frac{(2n)!}{2^n n!},$$

we have

$$w_n = \frac{2^{4n}(n!)^4}{[(2n)!]^2(2n + 1)},$$

or on using the approximation (12),

$$w_n \sim \frac{2^{4n} C^4 n^2 \left(\dfrac{n}{e}\right)^{4n}}{C^2 2n \left(\dfrac{2n}{e}\right)^{4n} (2n + 1)} \sim \tfrac{1}{4} C^2. \tag{13}$$

---

*See Salas and Hille [1978], p. 528; Anton [1980], p. 597.

The fact that $w_n$ tends to $\pi/2$ as $n \to \infty$ is compatible with (13) only if $C^2/4 = \pi/2$. Because $C$ is positive, there follows

$$C = \sqrt{2\pi}.$$

**DEMONSTRATION 6.2-1:** We test the accuracy of Stirling's formula by computing for various values of $n$, the exact value of $n!$, the **Stirling approximation** (9), and the ratio of the two.

| $n$ | $n!$ | $\sqrt{2\pi n}\left(\dfrac{n}{e}\right)^n$ | ratio |
|---|---|---|---|
| 1 | 1 | 0.922137 | 1.084437 |
| 2 | 2 | 1.919004 | 1.042207 |
| 3 | 6 | 5.836209 | 1.028064 |
| 4 | 24 | 23.506175 | 1.021008 |
| 5 | 120 | 118.019168 | 1.016783 |
| 10 | $3.628800 * 10^6$ | $3.598695 * 10^6$ | 1.008365 |
| 20 | $2.432902 * 10^{18}$ | $2.422786 * 10^{18}$ | 1.004175 |
| 30 | $2.652528 * 10^{32}$ | $2.645170 * 10^{32}$ | 1.002781 |

## PROBLEMS

1  Compute the polynomials $p_2, p_3, p_4$, and deduce the values

$$A = \tfrac{1}{12}, \qquad B = -\tfrac{1}{720}.$$

2  By applying the Euler-Maclaurin sum formula to the functions $g(t) := t$, $t^2, t^3, t^4$, obtain closed expressions for the sums

$$\sum_{k=1}^{n} k, \qquad \sum_{k=1}^{n} k^2, \qquad \sum_{k=1}^{n} k^3, \qquad \sum_{k=1}^{n} k^4.$$

3  Using Stirling's formula, find an approximation for the central binomial coefficient

$$\binom{2n}{n} = \frac{(2n)!}{(n!)^2}.$$

Compute the error of the approximation obtained for various values of $n$.

**4** Wallis' infinite product

$$\frac{\pi}{2} = \frac{2}{1}\frac{2}{3}\frac{4}{3}\frac{4}{5}\frac{6}{5}\frac{6}{7}\frac{8}{7}\cdots$$

was used above to determine the constant factor in Stirling's formula. Establish **Pippenger's product**

$$\frac{e}{2} = \left(\frac{2}{1}\right)^{1/2}\left(\frac{2}{3}\frac{4}{3}\right)^{1/4}\left(\frac{4}{5}\frac{6}{5}\frac{6}{7}\frac{8}{7}\right)^{1/8}\cdots$$

(Amer. Math. Monthly 87, p. 391). Study the convergence properties of Pippenger's product, and try extrapolation to the limit.

---

## §6.3 THE ROMBERG ALGORITHM

The connection of equation (8), §6.2, with the integral

$$I := \int_a^b f(x)\, dx$$

is established by the substitution

$$t = \frac{x - a}{h}, \qquad x = a + ht, \qquad \text{where} \quad h = \frac{b - a}{n}.$$

As $x$ moves from $a$ to $b$, $t$ moves from 0 to $n$, and to integer values $t = k$ there correspond the points $x_k = a + kh$. Setting $f(x) = g(t)$, the trapezoidal value

$$T(h) = h\{\tfrac{1}{2}f(x_0) + f(x_1) + f(x_2) + \cdots + f(x_{n-1}) + \tfrac{1}{2}f(x_n)\}$$

thus equals

$$h\{\tfrac{1}{2}g(0) + g(1) + g(2) + \cdots + g(n-1) + \tfrac{1}{2}g(n)\},$$

and by (8), §6.2, this may be expressed as

$$h\int_0^n g(t)\, dt + hA[g'(n) - g'(0)] + hB[g^{(3)}(n) - g^{(3)}(0)]$$

$$-\int_0^1 p_4(t) \sum_{k=0}^{n-1} g^{(4)}(t + k)\, dt.$$

Expressing this in terms of $f$, and taking into account that

$$dx = h\, dt, \qquad g^{(k)}(t) = h^k f^{(k)}(x),$$

we obtain

$$T(h) = \int_a^b f(x)\, dx + Ah^2[f'(b) - f'(a)] + Bh^4[f^{(3)}(b) - f^{(3)}(a)] + R_4(h),$$

where the term

$$R_4(h) := h^5 \int_0^1 p_4(t) \sum_{k=0}^{n-1} f^{(4)}(a + kht)\, dt$$

is bounded by a constant times $h^4$. Thus, we see that the trapezoidal value of the integral is related to the exact value $I$ by a relation of the form

$$T(h) = I + C_1 h^2 + O(h^4). \tag{1}$$

Here $C_1$ is given by

$$C_1 := A[f'(b) - f'(a)].$$

In line with postulate ($A$) of §6.1, we do not wish the derivatives of $f$ to be known or computable. The value of $C_1$ must therefore be regarded as unknown; however, this constant is independent of $h$.

The mere fact that (1) holds will enable us to speed up the convergence of the trapezoidal values for good integrands. Let the trapezoidal values be computed for

$$h := 2^{-n}(b - a), \qquad n = 0, 1, 2, \dots .$$

We put

$$T_n := T(2^{-n}(b - a)). \tag{2}$$

By virtue of (1),

$$T_n = I + D_1 2^{-2n} + O(2^{-4n}) \tag{3}$$

as $n \to \infty$, where $D_1 := C_1(b - a)^2$ does not depend on $n$. The idea now is to form a linear combination

$$T_n^* := \alpha T_n + \beta T_{n-1},$$

where the constants $\alpha$ and $\beta$ are to be determined so that the sequence $\{T_n^*\}$ converges to $I$ more rapidly than the sequence $\{T_n\}$. Using (3),

$$T_n^* = (\alpha + \beta)I + D_1(\alpha + 4\beta)2^{-2n} + O(2^{-4n}),$$

where we have used the relations

$$O(2^{-4n+4}) = O(16 \cdot 2^{-4n}) = O(2^{-4n}),$$
$$O(2^{-4n}) + O(2^{-4n}) = O(2^{-4n}),$$

satisfied by the $O$-symbol. Because the sequence $\{T_n^*\}$ still should converge to $I$, we want

$$\alpha + \beta = 1.$$

Because it should converge faster and therefore the term involving $2^{-2n}$ should vanish, we want

$$\alpha + 4\beta = 0.$$

The solution of this system of two equations for the two unknowns $\alpha$, $\beta$ is

$$\alpha = \tfrac{4}{3}, \qquad \beta = -\tfrac{1}{3}.$$

Thus, it follows that the sequence of values

$$T_n^* := \frac{4T_n - T_{n-1}}{3}, \qquad n = 1, 2, \ldots, \tag{4}$$

in general (that is, unless $D_1 = 0$ accidentally) converges to $I$ faster than the sequence $\{T_n\}$. For purposes of computation it is often convenient to write (4) as

$$T_n^* = T_n - \tfrac{1}{3}(T_{n-1} - T_n). \tag{4'}$$

**DEMONSTRATION 6.3-1:** We consider the example of Demonstration 6.1-1 and list both the original and the accelerated values.

| $n$ | $T_n$ | $T_n^*$ |
|---|---|---|
| 0 | 1.57079633 | |
| 1 | 1.78539816 | 1.85693211 |
| 2 | 1.83550812 | 1.85221144 |
| 3 | 1.84784231 | 1.85195370 |
| 4 | 1.85091414 | 1.85193809 |
| 5 | 1.85168137 | 1.85193712 |
| 6 | 1.85187313 | 1.85193705 |
| 7 | 1.85192107 | 1.85193705 |

Using a mere $2^6 + 1 = 65$ values of the function $f$, we now obtain a value of the integral which is more accurate than the unaccelerated value formerly obtained with 2048 evaluations of $f$, that is, with 32 times as much work.

Obviously the method can be extended further. Using the same analytical technique, it may be shown that for sufficiently differentiable functions $f$, the asymptotic formula (1) may be generalized as follows:

$$T(h) = I + C_1 h^2 + C_2 h^4 + \cdots + C_k h^{2k} + O(h^{2k+2}) \qquad (5)$$

as $h \to 0$. Here $k$ is any positive integer, and $C_1, C_2, \ldots$ are constants that involve derivatives of $f$, and hence are unknown, but that do not depend on $h$. By forming linear combinations of consecutive $T$-values, it is possible to eliminate not only $C_1$ but also any desired number of further $C_k$ from the error formula. For instance, the quantities

$$T_n^{**} := \frac{16 T_n^* - T_{n-1}^*}{16 - 1} = T_n^* - \tfrac{1}{15}(T_{n-1}^* - T_n^*)$$

already satisfy

$$T_n^{**} = I + O(2^{-6n}).$$

To formulate a general law, we write

$$T_{n,0}, \qquad T_{n,1}, \qquad T_{n,2}, \ldots$$

in place of $T_n$, $T_n^*$, $T_n^{**}$, .... . The general recurrence relations then are

$$T_{n,0} := T[2^{-n}(b-a)], \qquad n = 0, 1, 2, \ldots,$$

$$T_{n,m} := \frac{4^m T_{n,m-1} - T_{n-1,m-1}}{4^m - 1} \qquad (6)$$

$$= T_{n,m-1} - \frac{1}{4^m - 1}(T_{n-1,m-1} - T_{n,m-1}),$$

$$m = 0, 1, 2, \ldots, n.$$

The numbers $T_{n,m}$ may be thought as being arranged in a triangular array, called the **Romberg scheme**:

$$
\begin{array}{ccccc}
T_{0,0} & & & & \\
T_{1,0} & T_{1,1} & & & \\
T_{2,0} & T_{2,1} & T_{2,2} & & \\
T_{3,0} & T_{3,1} & T_{3,2} & T_{3,3} & \\
T_{4,0} & T_{4,1} & T_{4,2} & T_{4,3} & T_{4,4} \; . \\
\downarrow & \downarrow & \downarrow & \downarrow & \downarrow \\
I & I & I & I & I
\end{array}
$$

By induction one easily proves,

**THEOREM 6.3:** *For each fixed $m = 0, 1, 2, \ldots$ there holds*

$$T_{n,m} = I + O(2^{-2mn}), \qquad n \to \infty. \tag{7}$$

*If in the expansion (5) all $C_k \neq 0$, then each column of the Romberg scheme converges to I faster than the preceding column.*

To appreciate the efficacy of this **Romberg algorithm**, one should realize that the main work involved in constructing the Romberg scheme is done in the first column $(m = 0)$. Here the trapezoidal values $T(h)$ are computed for $h := 2^{-n}(b - a)$, which for each $n$ means that the values of $f$ must be calculated at $2^{n-1}$ new sampling points. Thus, the work necessary for obtaining the $T_{n,0}$ roughly doubles at each step. The remaining columns of the scheme are obtained from the very simple relation (6) involving $2\alpha + 1\delta$, and thus are practically free.

A completely analogous theory holds for the midpoint values $M(h)$. From (4), §6.1, there follows

$$M(h) = 2T\left(\frac{h}{2}\right) - T(h). \tag{8}$$

Substituting for $T(h)$ the expansion (5), it follows that an analogous expansion holds for $M(h)$,

$$M(h) = I + C_1' h^2 + C_2' h^4 + \cdots + C_k' h^{2k} + O(h^{2k+2}), \tag{9}$$

with certain unknown constants $C_k'$. Thus, the theoretical basis for applying the Romberg algorithm exists, and the scheme defined by (6) can be set up also with the first column

$$M_{n,0} := M[2^{-n}(b - a)],$$

with similar effects.

In applications of either form of the Romberg algorithm, it is rarely necessary to go beyond $m = 2$ or 3, because within the limits of machine accuracy the convergence acceleration becomes already fully effective in the third or fourth column.

**DEMONSTRATION 6.3-2:** This shows the full Romberg scheme for the midpoint values of the integral considered in the Demonstrations 6.1-1 and 6.3-1.

| | | | | |
|---|---|---|---|---|
| 2.00000000 | | | | |
| 1.88561808 | 1.84749078 | | | |
| 1.86017649 | 1.85169596 | 1.85197630 | | |
| 1.85398598 | 1.85192247 | 1.85193757 | 1.85193695 | |
| 1.85244861 | 1.85193615 | 1.85193706 | 1.85193705 | 1.85193705 |
| 1.85206489 | 1.85193699 | 1.85193705 | 1.85193706 | 1.85193706 |

A result that is correct with all digits given has been reached in the column $m = 2$ already with 15 evaluations of the function $f$. The convergence acceleration that theoretically takes place in the columns further to the right is not actually visible due to the limited word length.

The following Demonstrations show some situations where the Romberg algorithm fails to be effective.

**DEMONSTRATION 6.3-3:**   For the integral

$$I := -\int_0^1 \frac{\text{Log}(1 - x)}{x} \, dx = \tfrac{1}{6}\pi^2$$

the midpoint version of the Romberg algorithm produces the following results:

| | | |
|---|---|---|
| 1.38629436 | | |
| 1.49956039 | 1.53731573 | |
| 1.56685639 | 1.58928840 | 1.59275324 |
| 1.60417207 | 1.61661063 | 1.61843211 |
| 1.62402131 | 1.63063772 | 1.63157286 |
| 1.63431824 | 1.63775056 | 1.63822475 |
| 1.63957947 | 1.64133322 | 1.64157206 |

$$\tfrac{1}{6}\pi^2 = 1.64493407$$

No appreciable convergence acceleration takes place here, because the integrand fails to have a continuous first derivative (let alone higher derivatives) on the closed interval of integration.

**DEMONSTRATION 6.3-4:**   Here we compute the elliptic integral

$$I := \int_0^{\pi/2} \frac{1}{\sqrt{1 - k^2 \sin^2 x}} \, dx$$

for given values of $k^2$. Using midpoint values, we obtain for $k^2 = 0.5$ the Romberg scheme

| | | |
|---|---|---|
| 1.85319133 | | |
| 1.85407413 | 1.85436839 | |
| 1.85407468 | 1.85407486 | 1.85405529 |
| 1.85407468 | 1.85407468 | 1.85407467 |

Again, no convergence acceleration can be discerned. Here this is due to the fact that the integrand is a periodic analytic function which is (effectively) integrated over a full period. Because the constants $C_k$ in (5) all are zero, already the convergence of the unaccelerated values (first column) is excellent, and no further gain is possible by the Romberg method. This is confirmed by the values for $k^2 = 0.99$:

| | | |
|---|---|---|
| 2.21041688 | | |
| 2.84440876 | 3.05573938 | |
| 3.36904152 | 3.54391910 | 3.57646442 |
| 3.64259476 | 3.73377917 | 3.74643651 |
| 3.69405727 | 3.71121144 | 3.70970693 |
| 3.69563554 | 3.69616162 | 3.69515830 |
| 3.69563734 | 3.69563795 | 3.69560304 |
| 3.69563742 | 3.69563744 | 3.69563741 |

Convergence is slower here due to the near presence of a singularity on the interval of integration; nevertheless, the unaccelerated values are as good as the accelerated values.

If, however, we evaluate the incomplete elliptic integral

$$I := \int_0^1 \frac{1}{\sqrt{1 - k^2 \sin^2 x}} \, dx,$$

where the interval of integration is not commensurable with a period of the integrand, we obtain for $k^2 = 0.5$:

| | | |
|---|---|---|
| 1.06294262 | | |
| 1.07849294 | 1.08367638 | |
| 1.08206607 | 1.08325711 | 1.08322916 |
| 1.08293105 | 1.08321938 | 1.08321687 |
| 1.08314547 | 1.08321694 | 1.08321677 |
| 1.08319895 | 1.08321678 | 1.08321677 |

Here, the positive effect of the algorithm on the speed of convergence is clearly discernible.

We finally mention that the applications of the Romberg algorithm are not limited to sequences of numbers that arise in numerical integration. By virtue of its derivation, the algorithm is applicable to any sequence of numbers $\{y_k\}$ that converges to its limit $s$ so that an error law of the form

$$y_k = s + C_1 4^{-k} + C_2 4^{-2k} + \cdots + C_m 4^{-mk} + O(4^{-(m+1)k}) \tag{10}$$

holds with constants $C_i$ that, however unknown, are known to be independent of $k$.

**DEMONSTRATION 6.3-5:**   A law of the form (10) holds for the numbers

$$y_k := 2^k \sin(2^{-k}\pi)$$

studied in Demonstration 1.3-4. The resulting Romberg scheme,

| | | | |
|---|---|---|---|
| 2.00000000 | | | |
| 2.82842712 | 3.10456950 | | |
| 3.06146746 | 3.13914757 | 3.14145277 | |
| 3.12144515 | 3.14143771 | 3.14159039 | 3.14159257 |
| 3.13654849 | 3.14158293 | 3.14159261 | 3.14159265 |

$(\pi = 3.14159265 \ldots)$ clearly shows the effectiveness of the algorithm.

**Multiple Integrals**

Such integrals frequently occur in physical applications. The only effective method that is usually available to deal with such integrals numerically is the technique by which such integrals are evaluated analytically (if this evaluation is possible), namely, the reduction to lower dimensional integrals. (See Figure 6.3a.) Because of the many integrals that are to be evaluated, it is particularly important in such cases to be able to deal with one-dimensional integrals efficiently.

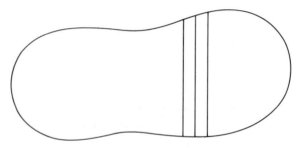

**Fig. 6.3a.**   Multiple integral.

# PROBLEMS

1  Construct the full Romberg scheme for the integral

$$\int_0^1 x^9 \, dx.$$

Predict theoretically, and confirm experimentally, which columns of the scheme will have *all* their entries exact.

2  The Romberg algorithm behaves abnormally for the integrals

**(a)** $\displaystyle\int_0^{\pi/2} (\cos x)^2 \, dx,$

**(b)** $\displaystyle\int_0^1 x \, \mathrm{Log}(1 - x) \, dx,$

**(c)** $\displaystyle\int_0^{\pi/2} \frac{1}{2 - (\cos x)^2} \, dx.$

Explain, and suggest a cure where necessary.

3  Will Romberg acceleration be effective for the integral

$$\int_{-5}^5 e^{-x^2} \, dx?$$

4  The functions

$$f_\alpha(x) := \int_0^x t^\alpha e^{-t} \, dt,$$

which were considered in Problem 2, §1.4, for integer values of $\alpha$, are of interest in statistics also for $\alpha = n - \frac{1}{2}$ ($n$ an integer $\geq 0$), where they cannot be evaluated in closed form.

   **(a)**  Show that the Romberg algorithm does not always work for such values of $\alpha$.

   **(b)**  Show that by the substitution $t = s^2$ the convergence already of the trapezoidal values is improved, and that the Romberg algorithm brings a further improvement if $x$ is not too large.

   **(c)**  Which of the methods of Problem 1, §1.5, can you make to work in this case?

5  Suppose that, due to singularities in the integrand, the trapezoidal formula yields for $\int_a^b f(x) \, dx$ values $T(h)$ satisfying

$$T(h) = I + C_1 h^{1/2} + C_2 h + C_3 h^{3/2} + \cdots .$$

(a)   How would the Romberg algorithm have to be modified to speed up the convergence?

(b)   Experiment with the integral of Problem 4 *before* transforming variables.

(c)   Which algorithm would you use if

$$T(h) = I + C_1 h^\alpha + C_2 h^{2\alpha} + C_3 h^{3\alpha} + \cdots ?$$

6   The Romberg algorithm also can be used for purposes of numerical differentiation. Suppose $f$ is analytic at 0, and we wish to approximate $f'(0)$ by the symmetric difference quotient

$$D(h) := \frac{f(h) - f(-h)}{2h}.$$

(a)   Show that $D(h)$ has an expansion similar to (5), and that by applying the Romberg algorithm to the numbers

$$D_k := D(2^{-k})$$

$f'(0)$ may be computed more rapidly and with less danger of numerical instability than by using $D(h)$.

(b)   Experiment with $f(x) := a^x$ for various positive values of $a$.

(c)   Find a recurrence relation for the numbers $D_k$ in the case just mentioned, and establish the connection with Problem 4, §1.3.

7   Devise an algorithm for the rapid evaluation of integrals of the form

$$I := \int_0^\infty \frac{g(y)}{e^{2\pi y} - 1} dy,$$

where the function $g$ is differentiable at 0, $g(0) = 0$. Such integrals occur in connection with the Plana summation formula (see Henrici [1974], p. 274). For certain sufficiently nice functions $f$ there holds

$$\sum_{n=0}^\infty f(n) = \tfrac{1}{2}f(0) + \int_0^\infty f(x)\, dx + \int_0^\infty \frac{g(y)}{e^{2\pi y} - 1} dy,$$

where

$$g(y) := i\{f(iy) - f(-iy)\}.$$

Use this to sum

(a)                    $\displaystyle\sum_{n=1}^\infty \frac{1}{n^2}$        $\left( f(x) := \frac{1}{(1+x)^2} \right),$

**(b)**
$$\sum_{n=1}^{\infty} \frac{1}{n^3}.$$

Exact result for (a): $\frac{1}{6}\pi^2$.

**8** Show that the first application of the Romberg algorithm yields the values

$$S(h) := \frac{h}{6}\{f_0 + 4f_1 + 2f_2 + 4f_3 + 2f_4 + \cdots + 2f_{2n-2} + 4f_{2n-1} + f_{2n}\},$$

where

$$f_k := f\left(x_0 + k\frac{b-a}{2n}\right).$$

[The values $S(h)$ are known as the **Simpson values** of the integral $I$; they could be obtained by approximating the basic integral

$$\int_{-1}^{1} g(t)\, dt$$

by $\frac{1}{3}g(-1) + \frac{4}{3}g(0) + \frac{1}{3}g(1)$. Conclude that Simpson integration is a mere special case of Romberg integration.]

**9** The Romberg scheme may be constructed either by using the formula

$$T_{n,m} = \frac{4^m T_{n,m-1} - T_{n-1,m-1}}{4^m - 1}, \tag{11}$$

or by using the mathematically equivalent formula

$$T_{n,m} = T_{n,m-1} - \frac{1}{4^m - 1}(T_{n-1,m-1} - T_{n,m-1}). \tag{12}$$

Show that if $n$ is large, the coefficients of relative error propagation (see §1.6) are approximately

$$\rho = 3 + \frac{1}{4^m - 1}$$

for (11), and

$$\rho = 1 + \frac{1}{4^m - 1}$$

for (12), and conclude that (12) is the more stable formula.

## §6.4  INITIAL VALUE PROBLEMS FOR ORDINARY DIFFERENTIAL EQUATIONS: ONE-STEP METHODS

We begin by considering the *initial value problem* for a single ordinary differential equation (ODE) of order one:

$$y' = f(x, y),$$
$$y(a) = b. \tag{1}$$

We refer to textbooks on differential equations (Boyce and DiPrima [1977]) for questions of existence and uniqueness of a solution of (1). Both existence and uniqueness are guaranteed if $f$, in addition to being merely continuous, satisfies a Lipschitz condition with respect to $y$, i.e., if there exists a constant $L$ (which here need not satisfy $L < 1$) such that for all $y_1$, $y_2$ and for all $x$ under consideration

$$|f(x, y_1) - f(x, y_2)| \le L|y_1 - y_2|. \tag{2}$$

It will always be assumed that this condition is satisfied and, in addition, that $f$ possesses as many continuous derivatives with respect to both variables as are required by the context.

Remarks similar to those made at the beginning of §6.1 also apply to the theory of differential equations. The differential equations that can be solved in closed form constitute a small minority in the family of all differential equations that are of technical interest. For instance, the differential equations of celestial mechanics, important as they are for space navigation, cannot be solved in closed form.

In order to discover a simple method for integrating a given ODE numerically, we recall the geometric interpretation of an ODE as a *direction field*. (See Figure 6.4a.) At each point $(x, y)$ of its domain of definition, the function $f$ prescribes a slope of value $f(x, y)$. The problem of solving the initial value problem (1) is the same as the problem of finding a function $y = y(x)$ which

(a) at each point $(x, y(x))$ of its graph has the slope which is defined by the direction field at that point, and

(b) which at the point $x = a$ assumes the value $b$ or, in other words, whose graph passes through the point $(a, b)$.

The simplest way to approximately satisfy these requirements is given by the **method of Euler**. (This method is mentioned here for didactic reasons only; its use in actual applications is not recommended, because it is too inaccurate.) We choose an integration step $h > 0$ and put

$$x_n := a + nh, \qquad n = 0, 1, 2, \ldots.$$

We now construct values $y_n$ intended to approximate the values $y(x_n)$ of the exact solution, as follows. We have $x_0 = a$, and hence put $y_0 = b$. Starting at the

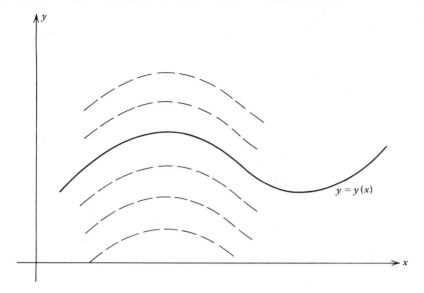

**Fig. 6.4a.** Direction field.

point $(x_0, y_0)$, we now step in the direction of the tangent of the exact solution at the point $(x_0, y_0)$. Let $h$ be the length of the projection of the step onto the x-axis. We then arrive at a point $(x_1, y_1)$ where

$$x_1 = x_0 + h,$$
$$y_1 := y_0 + hf(x_0, y_0).$$

At the point $(x_1, y_1)$, we take a step of the same projected length in the direction of the exact solution at the point $(x_1, y_1)$. This yields a point $(x_2, y_2)$ where

$$x_2 = x_1 + h,$$
$$y_2 := y_1 + hf(x_1, y_1).$$

Continuing in this manner (Figure 6.4b) the exact solution is approximated by a polygon with vertices $(x_n, y_n)$ determined by the recurrence relation

$$x_{n+1} = x_n + h,$$
$$y_{n+1} := y_n + hf(x_n, y_n), \tag{3}$$

$n = 0, 1, 2, \ldots$ .

What can be said about the *convergence* of Euler's method? What, indeed, is to be meant by convergence? We evidently want to be able to approximate the true solution as closely as we like by choosing the step $h$ sufficiently small.

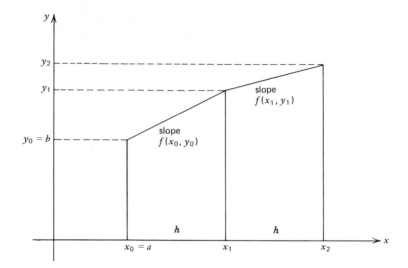

**Fig. 6.4b.** Euler's method.

Expressed in symbols: at each point $x$ where the exact solution $y(x)$ exists, there
should hold

$$\lim_{\substack{h \to 0 \\ n \to \infty}} y_n = y(x). \tag{4}$$

so that $nh = x - a$

(The limit here is somewhat delicate; $h$ tends to zero, and simultaneously $n$ tends
to $\infty$ so that $y_n$ always represents the approximation to the exact solution at the
point $x_n = x$.)

It can be shown that Euler's method is convergent in the sense just mentioned (Figure 6.4c); for a proof see Henrici [1964], §14.3. (Using Euler's method,
it is even possible to prove the *existence* of a solution of the initial value problem
(1).) Instead of proving convergence, we consider, as is frequently done in numerical mathematics, a *model problem* where the performance of the numerical
method is easily studied. For initial value problems, the problem

$$y' = qy,$$
$$y(0) = 1 \tag{5}$$

where $q$ is a real or complex constant, is frequently used as a model problem.
The exact solution of the model problem of course is

$$y(x) = e^{qx}.$$

We note in passing that the model problem is not chosen to show numerical
methods in a particularly favorable light. Since the values of the exact solution
are transcendental numbers for rational values of $qx$, they cannot be computed

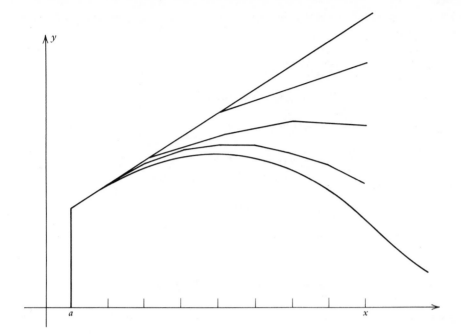

**Fig. 6.4c.** Convergence of Euler's method.

in a finite number of steps. The problem of approximating the solution of the model problem is genuine.

How does Euler's method perform in the model problem? The recurrence relations (3) here are

$$y_0 = 1, \qquad y_{n+1} = y_n + hqy_n = (1 + hq)y_n,$$

and induction yields

$$y_n = (1 + hq)^n.$$

We now have to perform the limit (4). That is, for a fixed $x > 0$, we want to let $h \to 0$ and $n \to \infty$ so that $nh = x$ remains fixed. Letting $h = (1/n)x$, we obtain

$$y_n = \left(1 + \frac{qx}{n}\right)^n$$

and using a classical limit of the differential calculus we indeed obtain

$$\lim_{\substack{h \to 0 \\ n \to \infty \\ nh = x}} y_n = e^{qx}.$$

Thus, we obtain the correct value; *Euler's method is convergent*, at least if applied to the model problem.

The model problem also permits a statement about the *speed of convergence* of Euler's method. Writing

$$y_n = (1 + hq)^{x/h}$$

and expanding in powers of $h$, we obtain

$$(1 + hq)^{x/h} = e^{(x/h)\text{Log}(1 + hq)}$$

$$= e^{(x/h)\{hq - (h^2q^2/2) + O(h^3)\}}$$

$$= e^{qx\{1 - (hq/2) + O(h^2)\}}$$

$$= e^{qx}e^{-(q^2xh/2) + O(h^2)},$$

and thus finally

$$(1 + hq)^{x/h} = e^{qx}\{1 - (q^2xh/2) + O(h^2)\}.$$

Thus, we see that the error at a fixed $x > 0$ is given by

$$\underset{\substack{\text{approximate} \\ \text{value at } x = x_n}}{y_n} - \underset{\substack{\text{value of} \\ \text{exact solution} \\ \text{at } x = x_n}}{e^{qx}} = -\tfrac{1}{2}q^2xe^{qx}h + O(h^2). \tag{6}$$

Thus in particular,

$$y_n - y(x_n) = O(h);$$

the error tends to zero only linearly with $h$. This result can be shown to hold quite generally for Euler's method. The convergence of the method thus is slow. In order to compute $e = 2.71828\ldots$ with an error $< 10^{-3}$ by solving the model problem where $q = 1$, more than $10^3$ integration steps would be necessary.

In order to find better methods of integration, we consider the Euler method from a more general point of view. Let $z(x)$ be the exact solution of the given ODE passing through the point $(x_n, y_n)$. (This solution in general does not coincide with the exact solution $y(x)$ which passes through $(x_0, y_0)$.) We then have

$$z(x_n) = y_n, \qquad z'(x_n) = f(x_n, y_n). \tag{7}$$

The Taylor series of $z(x)$ at the point $x = x_n$ is, writing $h := x - x_n$,

$$z(x_n + h) = z(x_n) + hz'(x_n) + \frac{h^2}{2}z''(x_n) + \cdots,$$

or by virtue of (7),

$$z(x_n + h) = y_n + hf(x_n, y_n) + \frac{h^2}{2} z''(x_n) + \cdots .$$

Euler's method results by truncating the Taylor series after the linear term. (The *local* error of $O(h^2)$ produces a *global* error of $O(h)$.) Is it possible to push Taylor's expansion farther?

It is a basic fact in the theory of differential equations that in an ODE of the form (1) not only the first derivative, but also all higher derivatives of the solution, are uniquely determined by, and can be expressed in terms of, the function $f$. By differentiating the identity

$$y'(x) = f(x, y(x)),$$

we obtain, using the chain rule,

$$y''(x) = f_x(x, y(x)) + f_y(x, y(x))y'(x).$$

By virtue of $y'(x) = f(x, y(x))$, the expression on the right equals

$$f_x(x, y(x)) + f_y(x, y(x))f(x, y(x)).$$

Thus, the second derivative of the solution passing through the point $(x, y)$ at this point has the value

$$f_2(x, y) := f_x(x, y) + f_y(x, y)f(x, y), \tag{8}$$

and thus is expressed in terms of $f$. In a similar manner expressions $f_3, f_4, \ldots,$ for the higher derivatives are obtained. If the functions $f_2, \ldots, f_m$ have been computed, it is possible to set up an integration formula which agrees with the Taylor polynomial of order $m$:

$$y_{n+1} = y_n + hf(x_n, y_n) + \frac{h^2}{2} f_2(x_n, y_n) + \cdots + \frac{h^m}{m!} f_m(x_n, y_n).$$

Studying the performance of this method for the model problem as above, it is found that

$$y_n - y(x_n) = O(h^m). \tag{9}$$

Again, this result holds for general differential equations. The convergence behavior of this method is for $m > 1$ substantially better than that of Euler's method. By choosing $m$ sufficiently large, it is possible to obtain convergence of any desired order as $h \to 0$.

Unfortunately, the above method suffers from the disadvantage that the actual computation of the higher derivatives $f_2$, $f_3$, ... from the function $f$ may be tedious. Even if, as is possible today, algebraic manipulators are used, a large number of different functions may have to be evaluated. If systems of differential equations are integrated, the number of partial derivatives that has to be computed in order to obtain $f_m$ even grows exponentially with $m$. For these reasons, the methods that are based on the direct use of the Taylor polynomial, excepting special cases where the evaluation of the derivatives is easy, have not found widespread acceptance. Fortunately, however, it is possible to simulate the effect of Taylor's expansion in a simple manner that requires only evaluations of the function $f$ itself. This device, which originally is due to Runge and Kutta, consists in using iterated values of the function $f$.

We begin by considering the general **one-step method** of the form

$$y_{n+1} = y_n + hF(x_n, y_n; h).$$

Here $F$ denotes the **increment function**; for Euler's method,

$$F(x, y; h) = f(x, y),$$

for the Taylor method of order 2,

$$F(x, y; h) := f(x, y) + \frac{h}{2} f_2(x, y)$$

$$= f(x, y) + \frac{h}{2} \{ f_x(x, y) + f_y(x, y) f(x, y) \}. \tag{10}$$

In order to explain the idea of Runge and Kutta in its simplest manifestation, we consider an increment function of the form

$$F(x, y; h) = af(x, y) + bf(x + \alpha h, y + \beta hf(x, y)), \tag{11}$$

with constants $a$, $b$, $\alpha$, $\beta$ to be determined. Evidently, the evaluation of $F$ merely requires evaluations of $f$ (two such evaluations to be precise). We now endeavor to determine the parameters $a$, $b$, $\alpha$, $\beta$ so that $F$ agrees as well as possible with the increment function of the Taylor method. Here agreement as well as possible is taken to mean: agreement to the highest possible power of $h$ if both functions are expanded in powers of $h$. Using Taylor's formula for functions of two variables, we have

$$f(x + \alpha h, y + \beta hf(x, y)) = f(x, y) + f_x(x, y)\alpha h + f_y(x, y)\beta hf(x, y) + O(h^2).$$

Collecting terms involving like powers of $h$, we obtain

$$F(x, y; h) = (a + b)f(x, y) + \{ b\alpha f_x(x, y) + b\beta f_y(x, y)f(x, y) \}h + O(h^2).$$

If this is to agree with (10), at least through the linear term in $h$, we must have

$$a + b = 1, \qquad b\alpha = \tfrac{1}{2}, \qquad b\beta = \tfrac{1}{2}.$$

This system of three equations for the four unknowns has a solution which depends on one free parameter. Selecting $b$ for the role of the parameter, the solution is

$$\alpha = \beta = \frac{1}{2b}, \qquad a = 1 - b.$$

Substituting into (11) yields a one-parameter family of increment functions

$$F(x, y; h) = (1 - b)f(x, y) + bf\left(x + \frac{h}{2b}, \; y + \frac{h}{2b} f(x, y)\right).$$

For each choice of $b \neq 0$, this agrees with the increment function for Taylor's second order method up to terms $O(h^2)$, although the evaluation of $F$ does not require computing any derivatives. The values of $b$ selected most frequently are $b = 1$ and $b = \tfrac{1}{2}$. This yields the methods

$$y_{n+1} = y_n + hf\left(x_n + \frac{h}{2}, \; y_n + \frac{h}{2} f(x_n, y_n)\right)$$

**(improved Euler method)** and

$$y_{n+1} = y_n + \frac{h}{2} \{f(x_n, y_n) + f(x_n + h, \; y_n + hf(x_n, y_n))\}$$

**(simplified Runge-Kutta method)**. The order of the global error for both methods is 2; that is, (9) holds with $m = 2$.

Using the same principle, Runge-Kutta methods of order higher than two may be constructed. The algebraic manipulations required to *derive* these methods become quite formidable as $m$ increases; however, the resulting methods are very easy to apply. The most popular of these methods is the **classical Runge-Kutta method**, whose increment function agrees with the Taylor polynomial up to and including the term involving $h^4$. This method may be written in the form

$$y_{n+1} = y_n + \tfrac{1}{3}(k_1 + 2k_2 + 2k_3 + k_4) \tag{12}$$

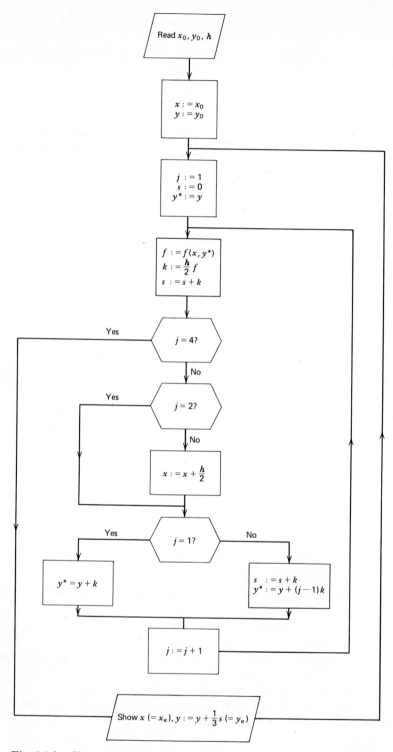

**Fig. 6.4d.** Classical Runge–Kutta method.

where

$$k_1 := \frac{h}{2} f(x_n, y_n),$$

$$k_2 := \frac{h}{2} f\left(x_n + \frac{h}{2}, y_n + k_1\right),$$

$$k_3 := \frac{h}{2} f\left(x_n + \frac{h}{2}, y_n + k_2\right),$$

$$k_4 := \frac{h}{2} f(x_n + h, y_n + 2k_3).$$

An implementation of the method which is suitable for calculators with minimal storage facilities is shown in the flow diagram in Figure 6.4d.

**DEMONSTRATION 6.4-1:** We use the (classical) Runge-Kutta method to solve the initial value problem

$$y' = x^2 - y^2, \qquad y(0) = 1.$$

Before attempting to solve a differential equation, it is always a good idea to discuss the qualitative behavior of the solution. Here, the direction field prescribes negative slopes for $|y| > |x|$, positive slopes for $|y| < |x|$, and slope zero for $y = \pm x$. We thus expect our solution first to decrease, then to cross the line $y = x$ with a horizontal tangent, and then to rise again. The following values result by using various steps $h$:

| $x$ | $y$ | | |
|-----|----------|----------|----------|
|  | $h = 0.20$ | $h = 0.10$ | $h = 0.05$ |
| 0.0 | 1.00000000 | 1.00000000 | 1.00000000 |
| 0.2 | 0.83579258 | 0.83578553 | 0.83578505 |
| 0.4 | 0.73273656 | 0.73272762 | 0.73272703 |
| 0.6 | 0.68424027 | 0.68423005 | 0.68422940 |
| 0.8 | 0.68976156 | 0.68974863 | 0.68974786 |
| 1.0 | 0.75003570 | 0.75001684 | 0.75001577 |
| 1.2 | 0.86368082 | 0.86365109 | 0.86364947 |
| 1.4 | 1.02495311 | 1.02490782 | 1.02490542 |
| 1.6 | 1.22340074 | 1.22334016 | 1.22333710 |
| 1.8 | 1.44581302 | 1.44574745 | 1.44574433 |
| 2.0 | 1.67951278 | 1.67946132 | 1.67945911 |

Impressive as these numbers may be, it is possible to squeeze even more accuracy out of them by a variant of the Romberg algorithm. If $f$ is sufficiently differentiable, then it may be shown that there exist functions $e_4(x)$, $e_5(x)$, ... so that the Runge-Kutta values $y_n$, in addition to relation (9) where $m = 4$, satisfy an asymptotic expansion of the form

$$y_n \approx y(x_n) + e_4(x_n)h^4 + e_5(x_n)h^5 + \cdots \tag{13}$$

as $h \to 0$. Thus, if we denote by $y(h)$ the numerical value obtained at a fixed point $x$, using the step $h$, the convergence of $y(h)$ to the exact value of the solution may be speeded up by first forming

$$y^*(h) := y(h) - \tfrac{1}{15}\{y(2h) - y(h)\}$$

and then

$$y^{**}(h) := y^*(h) - \tfrac{1}{31}\{y^*(2h) - y^*(h)\}.$$

In this way, the following values result, for instance, at $x = 2$:

| $h$ | $y(h)$ | $y^*(h)$ | $y^{**}(h)$ |
|------|-----------|-----------|-----------|
| 0.20 | 1.67951278 | | |
| 0.10 | 1.67946132 | 1.67945789 | |
| 0.05 | 1.67945911 | 1.67945896 | 1.67945900 |

Although rounding errors come into play, we may feel reasonably confident that 1.6794590 is an accurate value of the solution at $x = 2$.

# PROBLEMS

1  Use the method applied to analyze Euler's method to determine the asymptotic behavior of the error ($x$ fixed, $h \to 0$) when the model problem

$$y' = qy, \qquad y(0) = 1$$

is numerically integrated by the Taylor expansion method of order 2. *How does this generalize to the Taylor method of arbitrary order $p$?

**2** Show that in the integration of the model problem the results of the Runge-Kutta methods are (analytically) *identical* with the results obtained by the Taylor method of corresponding order.

**3** Write a program to integrate the general initial value problem

$$y' = f(x, y), \qquad y(x_0) = y_0$$

by the classical Runge-Kutta method. Apply the program to the problem

$$y' = -\frac{1}{1 + x^2} y, \qquad y(0) = 1,$$

using the steps $h = 0.2, 0.1, 0.05$, and compare the solution obtained with the exact solution

$$y(x) = e^{-\arctan x}.$$

Letting the calculator run, how accurately do you obtain the limit of the solution as $x \to \infty$?

**4** Sketch the direction field defined by the differential equation

$$y' = \cos\left(\frac{\pi x y}{2}\right).$$

Find the curves where $y'$ has the constant values $1, 0, -1$. On the basis of your sketch, determine the qualitative behavior of the solutions satisfying (a) $y(0) = 0$, (b) $y(0) = 4$. Then determine the solutions accurately by the Runge-Kutta method.

---

# §6.5 SYSTEMS OF ORDINARY DIFFERENTIAL EQUATIONS

In applications, systems of differential equations occur more frequently than single equations. The characteristic feature of systems is the occurrence of *several* unknown functions of the independent variable which are linked by differential equations. Generally, in order to fix the solution, as many differential equations must be given as there are unknown functions, plus the required initial conditions. A system of differential equations of the first order for $m$ unknown functions $y_1, y_2, \ldots, y_m$ thus has the form

$$y'_1 = f_1(x,\ y_1,\ \ldots,\ y_m)$$

$$y'_2 = f_2(x,\ y_2,\ \ldots,\ y_m) \tag{1}$$

$$\cdots\cdots\cdots\cdots\cdots\cdots$$

$$y'_m = f_m(x,\ y_1,\ \ldots,\ y_m)$$

where $f_1, f_2, \ldots, f_m$ are given functions. Normally, the solution functions are uniquely determined by the $m$ initial conditions

$$y_i(a) = b_i, \qquad i = 1, 2, \ldots, m. \tag{2}$$

The theoretical *and* the numerical treatment of such systems is considerably simplified by vector notation. Let

$$\mathbf{y} := \begin{pmatrix} y_1 \\ y_2 \\ \vdots \\ y_m \end{pmatrix}, \qquad \mathbf{b} := \begin{pmatrix} b_1 \\ b_2 \\ \vdots \\ b_m \end{pmatrix}$$

and let $\mathbf{f}$ be the vector-valued function

$$\mathbf{f}\colon (x,\ \mathbf{y}) \rightarrow \mathbf{f}(x,\ \mathbf{y}) := \begin{pmatrix} f_1(x,\ \mathbf{y}) \\ f_2(x,\ \mathbf{y}) \\ \vdots \\ f_m(x,\ \mathbf{y}) \end{pmatrix}.$$

Then, the differential equations (1) and the initial conditions (2) may be written in the compact form

$$\mathbf{y}' = \mathbf{f}(x,\ \mathbf{y}), \tag{3}$$

$$\mathbf{y}(a) = \mathbf{b}. \tag{4}$$

The methods discussed in §6.4 for the integration of a single ODE are equally applicable to the integration of systems. Denoting by $\mathbf{y}_n$ the vector furnished by the numerical method that approximates the solution vector $\mathbf{y}(x_n)$ at the point $x_n = a + nh$, we thus obtain
the *Euler method*,

$$\mathbf{y}_{n+1} = \mathbf{y}_n + h\mathbf{f}(x_n,\ \mathbf{y}_n); \tag{5}$$

the *improved Euler method*,

$$\mathbf{y}_{n+1} = \mathbf{y}_n + h\mathbf{f}\left(x_n + \frac{h}{2},\ \mathbf{y}_n + \frac{h}{2}\,\mathbf{f}(x_n,\ \mathbf{y}_n)\right); \tag{6}$$

the *simplified Runge-Kutta method,*

$$\mathbf{y}_{n+1} = \mathbf{y}_n + \frac{h}{2}\{\mathbf{f}(x_n, \mathbf{y}_n) + \mathbf{f}(x_n + h, \mathbf{y}_n + h\mathbf{f}(x_n, \mathbf{y}_n))\};\tag{7}$$

and the *classical Runge-Kutta method,*

$$\mathbf{y}_{n+1} = \mathbf{y}_n + \tfrac{1}{3}(\mathbf{k}_1 + 2\mathbf{k}_2 + 2\mathbf{k}_3 + \mathbf{k}_4),\tag{8}$$

where

$$\mathbf{k}_1 := \frac{h}{2}\mathbf{f}(x_n, \mathbf{y}_n),$$

$$\mathbf{k}_2 := \frac{h}{2}\mathbf{f}\left(x_n + \frac{h}{2}, \mathbf{y}_n + \mathbf{k}_1\right),$$

$$\mathbf{k}_3 := \frac{h}{2}\mathbf{f}\left(x_n + \frac{h}{2}, \mathbf{y}_n + \mathbf{k}_2\right),$$

$$\mathbf{k}_4 := \frac{h}{2}\mathbf{f}(x_n + h, \mathbf{y}_n + 2\mathbf{k}_3).$$

We illustrate these formulas in the case $m = 2$ by using a notation that avoids subscripts and vectors. Denoting the two unknown functions by $y$ and $z$ and the given right hand sides by $f$ and $g$, the system then may be written

$$y' = f(x, y, z),$$
$$z' = g(x, y, z).\tag{9}$$

In this notation, the simplified Runge-Kutta method, for instance, takes the form

$$y_{n+1} = y_n + \frac{h}{2}\{f(x_n, y_n, z_n) + f(x_n + h, y_n + hf(x_n, y_n, z_n), z_n + hg(x_n, y_n, z_n))\}$$

$$z_{n+1} = z_n + \frac{h}{2}\{g(x_n, y_n, z_n) + g(x_n + h, y_n + hf(x_n, y_n, z_n), z_n + hg(x_n, y_n, z_n))\}.$$

Many models of mathematical physics and of engineering involve differential equations of order higher than one. For instance, dynamical problems almost always involve the second derivative with respect to time. Such differential equations may be reduced to systems of first-order equations by introducing derivatives as new unknown functions. For instance, the initial value problem

$$y'' = g(x, y, y'), \qquad y(a) = b, \qquad y'(a) = c,\tag{10}$$

involving a differential equation of the second order is equivalent to the first order system

$$y' = z$$
$$z' = g(x, y, z),$$

(11)

subject to the initial conditions

$$y(a) = b, \qquad z(a) = c.$$

Thus, the function $f$ here takes the trivial form $f(x, y, z) = z$.

**DEMONSTRATION 6.5-1:** We consider a *mathematical pendulum* of length $L$ and mass $m$. If $x$ is the time and $y$ is the deviation of the pendulum from the vertical position (Figure 6.5a), then Newton's law implies that the differential equation

$$mLy'' = -mg \sin y$$

holds, where $g$ is the gravitational acceleration, $g = 9.81$ meter/sec$^2$ in metric units. Thus,

$$y'' = -\frac{g}{L} \sin y,$$

(12)

which is a differential equation of the form (10). In the elementary theory, one assumes $y$ to be small and replaces $\sin y$ by $y$. The solutions of (12) then are harmonic oscillations with the period $T = 2\pi(L/g)^{1/2}$, independently of the amplitude of the oscillation. Here we wish to determine the period of the solutions of the exact equation (12), without this linearization.

If the pendulum is released at time $x = 0$ from the initial position $y_0$ with initial velocity zero, then for reasons of symmetry the period of the resulting motion

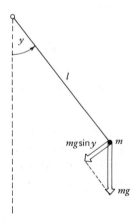

**Fig. 6.5a.** Mathematical pendulum.

is four times the time required for the pendulum to pass for the first time through the position $y = 0$. Thus, we wish to solve the equation

$$y(x) = 0, \tag{13}$$

where $y(x)$ is the solution of (12) satisfying the initial conditions $y(0) = y_0$, $y'(0) = 0$.

In solving equation (13), we make use of the fact that nothing keeps us from changing the step $h$ in the numerical integration of a differential equation by a one-step method. Even negative steps are permissible. We thus integrate (12) with a constant step $h$ until the first time when a value $y < 0$ appears. Calling $x_0$ the corresponding value of $x$, we obtain further approximations to the solution of (13) by Newton's method,

$$x_{n+1} = x_n - \frac{y(x_n)}{y'(x_n)},$$

which, if the second-order equation is written as a first-order system of the form (11), takes the particularly simple form

$$x_{n+1} = x_n - \frac{y(x_n)}{z(x_n)} \sim x_n - \frac{y_n}{z_n}.$$

Carrying out this process for the values $L = 1$ meter, $y_0 = \frac{1}{3}\pi = 60°$, using the simplified Runge-Kutta method with an initial time step $h = 0.0125$, yields the following values:

| $x_n$ | $y_n$ |
|---|---|
| 0.550000 | −0.037194 |
| 0.538116 | 0.000026 |
| 0.538125 | 0.000000 |

For the full period this yields the value $T = 2.1525$, a substantial deviation from the value $T = 2.0066$ predicted by the linear theory. It may be mentioned that in the problem under consideration the period may also be calculated by means of elliptic integrals. This yields the exact value $T = 2.1529$, which is in satisfactory agreement with the value obtained by means of numerical integration.

# PROBLEMS

1  **Prey-predator model in population dynamics** (see Batschelet [1975], p. 369). Consider a situation in population dynamics where one species, called predator, feeds on another species, called prey. It is assumed that the prey population find ample food at all times, while the food of the

predator population consists entirely of the prey population. It is then clear that when the prey population becomes more numerous, the predatory species obtain a larger food base. The growing demand for food will then reduce the prey population, and after some time the food becomes scarce for the predator, so that its propagation is inhibited. The size of the predator population then declines, and the prey population will experience a boom. It is thus apparent that a cyclic behavior in the size of both populations may be expected.

A mathematical model of the prey-predator system was introduced by Volterra and (independently) by Lotka around 1925. Denoting the prey and the predator populations by $y$ and $z$, respectively, they assumed the birth rate of the prey to be proportional to $y$, and that of the predator proportional to $yz$. The death rate of the prey is assumed proportional to $yz$, while the death rate of the predator is proportional to $z$ alone. The following system of differential equations thus results:

$$\frac{dy}{dt} = ay - byz,$$

$$\frac{dz}{dt} = cyz - dz.$$

Here $a, b, c, d$ are positive constants.

(a) Solve the system for the constants $a = 1$, $b = 0.002$, $c = 0.00001$, $d = 0.08$, and the initial conditions

$$y(0) = 20{,}000, \qquad z(0) = 500.$$

Using the simplified Runge-Kutta method, begin the integration with the step $h = 1$.

(b) Find the approximate period of the process.

---

## §6.6   TRAPEZOIDAL RULE; STABILITY

The methods discussed in the previous section use or simulate the *Taylor series* of the solution sought. We now consider a method for the solution of ODEs which uses *numerical integration.* The exact solution of a single D.E. for a single unknown function by definition satisfies

$$y'(x) = f(x, y(x)).$$

Integrating between $x_n$ and $x_{n+1}$, we get

$$y(x_{n+1}) - y(x_n) = \int_{x_n}^{x_{n+1}} f(x, y(x))\, dx.$$

Approximating the integral by means of the trapezoidal rule yields

$$\frac{h}{2}\{f(x_n, y(x_n)) + f(x_{n+1}, y(x_{n+1}))\}.$$

We now assume that $y_n = y(x_n)$. Then it is possible to determine an approximate value $y_{n+1}$ for $y(x_{n+1})$ from the equation

$$y_{n+1} - y_n = \frac{h}{2}\{f(x_n, y_n) + f(x_{n+1}, y_{n+1})\}. \tag{1}$$

Equation (1) represents an implicit equation for the new value $y_{n+1}$. This equation has the general form

$$y = F(y); \tag{2}$$

in the present case,

$$F(y) = y_n + \frac{h}{2} f(x_n, y_n) + \frac{h}{2} f(x_{n+1}, y) = \text{const} + \frac{h}{2} f(x_{n+1}, y),$$

and we see that the desired value $y_{n+1}$ is a *fixed point* of $F$. Disregarding the possibility that (2) can be solved analytically—this in general will be possible only if $f$ is linear or quadratic in $y$—the fixed point has to be determined numerically, and iteration suggests itself for this purpose. What can be said about the convergence of this method? According to the theory presented in §2.1, convergence will take place if

$$\left| \frac{dF}{dy} \right| < 1;$$

the smaller the absolute value of this derivative, the more rapid convergence will be. In the present case,

$$\frac{dF}{dy} = \frac{h}{2} f_y(x_{n+1}, y).$$

Because $f$ by hypothesis satisfies a Lipschitz condition and thus has a bounded derivative with respect to $y$, convergence will surely take place if the integration step is chosen sufficiently small.

In order to carry through the iteration procedure, it is necessary to secure a starting value $y_{n+1}^{(0)}$. A very crude starting value would be $y_{n+1}^{(0)} = y_n$. A better starting value is provided by Euler's method:

$$y_{n+1}^{(0)} = y_n + hf(x_n, y_n).$$

Better yet, one may use the value obtained from the midpoint formula,

$$y_{n+1}^{(0)} = y_{n-1} + 2hf(x_n, y_n),$$

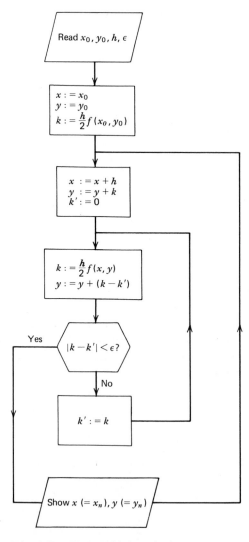

**Fig. 6.6a.**  Trapezoidal method.

which requires keeping track of $y_{n-1}$ and thus leads to a slightly more complicated program. Moreover, the midpoint value cannot be used for $n = 0$.

No matter how started, the iteration proceeds according to the formula

$$y_{n+1}^{(m+1)} = y_n + \frac{h}{2}\{f(x_n, y_n) + f(x_{n+1}, y_{n+1}^{(m)})\}, \qquad m = 0, 1, \dots . \qquad (3)$$

Unless special properties of $f$ permit a machine-independent convergence criterion, the iteration will be terminated artificially by a criterion of the form

$$\left| y_{n+1}^{(m+1)} - y_n^{(m)} \right| < \varepsilon, \qquad (4)$$

where $\varepsilon$ is some preassigned error tolerance. If the criterion is met, then

$$y_{n+1} := y_{n+1}^{(m+1)}$$

is accepted as the definite value of $y_{n+1}$, and the whole process is continued with $n$ increased by 1.

In the flow diagram shown in Figure 6.6a, the starting value $y_{n+1}^{(0)} = y_n + (h/2)f(x_n, y_n)$ is used in order to obtain a simple program.

**DEMONSTRATION 6.6-1:**  We consider the problem of Demonstration 6.4-1,

$$y' = x^2 - y^2, \qquad y(0) = 1.$$

The following values were obtained for various steps $h$, using the termination criterion (4) with $\varepsilon = 10^{-9}$:

| $x$ | $y$ | | |
| --- | --- | --- | --- |
| | $h = 0.20$ | $h = 0.10$ | $h = 0.05$ |
| 0.0 | 1.000000 | 1.000000 | 1.000000 |
| 0.2 | 0.834381 | 0.835443 | 0.835700 |
| 0.4 | 0.731284 | 0.732373 | 0.732639 |
| 0.6 | 0.683139 | 0.683961 | 0.684163 |
| 0.8 | 0.688999 | 0.689562 | 0.689702 |
| 1.0 | 0.749371 | 0.749854 | 0.749975 |
| 1.2 | 0.862777 | 0.863429 | 0.863594 |
| 1.4 | 1.023569 | 1.024569 | 1.024821 |
| 1.6 | 1.221575 | 1.222896 | 1.223227 |
| 1.8 | 1.443874 | 1.445279 | 1.445628 |
| 2.0 | 1.677871 | 1.679067 | 1.679361 |

Because the iteration (3) converges faster when $h$ is smaller, the work is less than doubled when $h$ is halved. Because the accumulated error of the trapezoidal method is $O(h^2)$, the above values are less accurate than those obtained by the Runge-Kutta method. We may try extrapolation, however. It may be shown that the values obtained by means of the trapezoidal method satisfy an asymptotic relation of the form

$$y_n \approx y(x_n) + e_2(x_n)h^2 + e_4(x_n)h^4 + \cdots \tag{5}$$

as $h \to 0$, with functions $e_k(x)$ that do not depend on $h$. Because only even powers of $h$ appear in the foregoing expansion, the Romberg algorithm is applicable in its original form when the values of $h$ are successively halved. For instance, at $x_n = 2$, this yields the Romberg scheme, constructed with the full nine-digit values to minimize the effect of rounding,

$$
\begin{array}{lll}
1.677871265 & & \\
1.679066690 & 1.679465165 & \\
1.679361196 & 1.679459365 & 1.679458979
\end{array}
$$

The final value agrees well with the extrapolated Runge-Kutta value obtained earlier.

The trapezoidal method is but one example of an ODE solver based on numerical integration. Many other such methods will be found in more elaborate treatments of the subject. Contrary to most of these methods, however, the trapezoidal method enjoys the property of being *absolutely stable*.

The notion of numerical stability has been treated in a general way in §1.5. In the context of differential equations, there exist several more specific definitions of numerical stability. (See, for example, Henrici [1962].) The notion of **absolute stability** is defined by means of our model problem,

$$y' = qy, \qquad y(0) = 1,$$

with the exact solution $y(x) = e^{qx}$. If the trapezoidal method is applied to the model problem, the recurrence relation for the $y_n$ reads

$$y_{n+1} = y_n + \frac{h}{2}(qy_n + qy_{n+1}),$$

or, solving for $y_{n+1}$,

$$y_{n+1} = \frac{1 + (hq/2)}{1 - (hq/2)} y_n,$$

which in view of $y_0 = 1$ yields

$$y_n = \left[\frac{1 + (hq/2)}{1 - (hq/2)}\right]^n,$$

$n = 0, 1, 2, \ldots$ . Let now $q < 0$. The exact solution then decreases exponentially. Because for $q < 0$

$$\left|\frac{1 + (hq/2)}{1 - (hq/2)}\right| < 1$$

for all positive $h$, the numerical solution also tends to zero for $n \to \infty$.

An integration method with the property that the numerical values $y_n$ generated for the model problem where $q < 0$ tend to zero as $n \to \infty$ for *all* values of $h > 0$ is called **absolutely stable**.

Not all methods are absolutely stable. For instance, the Euler method for the model problem yields the approximation

$$y_n = (1 + qh)^n.$$

If $q < 0$ and $h > -2/q$, then

$$|1 + hq| > 1,$$

and $|y_n| \to \infty$. If too large a step has been chosen, this may be recognized from strong oscillations of the numerical solution. (See Figure 6.6b.)

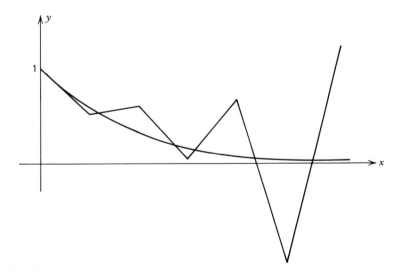

**Fig. 6.6b.**   Unstable solution.

It may be shown that the Runge-Kutta methods also fail to be absolutely stable, as do all methods based on Taylor's expansion.

Lack of absolute stability is no serious problem if only a single (scalar) differential equation is to be integrated, because in this case the step must be chosen small for reasons of accuracy as well as stability. Absolute stability is a very desirable property in the integration of systems, however, particularly if some components of the solution vector carry out rapid oscillations which are unimportant for the system as a whole, or if some components are subject to strong damping. Such systems are called **stiff**. In such cases one would like to choose an integration step which is appropriate for the components that change slowly. No great precision is required for the rapidly changing components; however, no disproportionate errors should be introduced by integrating the fast components sloppily. Only an absolutely stable method satisfies this requirement.

If the trapezoidal method is applied to the integration of a stiff system, ordinary integration may fail to solve the implicit equation (2) because of a large Lipschitz constant. It is then necessary to use some other method such as Newton's method.

# PROBLEMS

1  An electric coil with a certain non-linear magnetic characteristic is excited by an alternating voltage. (See Figure 6.6c.) Measured in

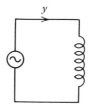

**Fig. 6.6c.**   Electric coil.

appropriate units the current $y$ in the coil as a function of the time $x$ then satisfies the differential equation

$$y' = -\frac{e^y - 1}{e^y + 1} + \sin 2x.$$

(a)  Use the trapezoidal method to determine the current as a function of time over a full period of the exciting voltage, assuming there is no current at time 0. An error $< 10^{-4}$ is required.

(b)  Also solve the problem by the Runge-Kutta method and compare the amount of work ($=$ computing time).

**(c)** Compare the solutions obtained with the analytic solution of the linearized equation

$$y' = -\tfrac{1}{2}y + \sin 2x.$$

**2** The system of two linear differential equations

$$y' = -77y + 114z$$
$$z' = -38y + 56z$$

under the initial conditions

$$y(0) = z(0) = 1$$

has the exact solution

$$y(x) = 3e^{-x} - 2e^{-20x},$$
$$z(x) = 2e^{-x} - e^{-20x}.$$

Thus, except for very small positive $x$, the exact solution behaves like

$$y(x) \sim 3e^{-x}, \qquad z(x) \sim 2e^{-x}.$$

Solve the initial value problem numerically using the simplified Runge-Kutta method. Show that with the step $h = 0.10$ the exact solution is reproduced with fair accuracy, whereas for $h > 0.15$ the numerical solution bears no resemblance to the exact solution.

**3**⊕ Write a program for solving the linear system

$$y' = ay + bz,$$
$$z' = cy + dz$$

by means of the trapezoidal method, solving the implicit equation for $y_{n+1}$, $z_{n+1}$ by Gaussian elimination. Apply the program to the system of Problem 2 and show that the numerical solution is stable for all $h > 0$.

**4** If the trapezoidal method is applied to the integration of very stiff systems, the numerical solution will contain components of the approximate form $y_n = \text{const} \cdot (-1)^n$. Explain, and show that the unwanted oscillations can be ironed out by forming the smoothed values

$$\tilde{y}_n := \tfrac{1}{4}(y_{n+1} + 2y_n + y_{n-1}).$$

Try smoothing on the numerical solution obtained in Problem 3.

5   **Simpson's method.**   The idea of solving a differential equation by
numerical integration can also be realized in the following way: In the
relation

$$y(x_{n+1}) - y(x_{n-1}) = \int_{x_{n-1}}^{x_{n+1}} f(x, y(x)) \, dx,$$

we approximate the integral by Simpson's rule (see Problem 8, §6.3),
using the three values $f_{n-1}, f_n, f_{n+1}$, where $f_k := f(x_k, y_k)$. The formula

$$y_{n+1} - y_{n-1} = \frac{h}{3} (f_{n-1} + 4f_n + f_{n+1})$$

results, which again represents an implicit equation for $y_{n+1}$. Two starting values $y_0$ and $y_1$ are now required, but otherwise the flow diagram
of the method is similar to that of the trapezoidal rule.

(a)   Implement the Simpson method for an arbitrary function $f$.

(b)   Apply the method to the differential equation of Problem 1,
and compare the effort required for comparable accuracy.

(c)   Determine the order of Simpson's method experimentally by
solving the same D.E. with several values of $h$. (Here the second starting value $y_1$ should be sufficiently accurate in the sense that its error is
at most $O(h^5)$.)

(d)   Use the method to solve the model problem

$$y' = -y, \qquad y(0) = 1$$

and show that for this problem the method suffers from a mild form of
numerical instability.

(e)   Explain the numerical instability by explicitly solving the
difference equation for the $y_n$ resulting from the model problem.

6   **Special second order equations.**   Differential equations of the form

$$y'' = f(x, y) \tag{6}$$

(first derivative missing in the righthand term) are called **special** second
order equations. A straightforward method for integrating special
second order equations is obtained by approximating $y''(x_n)$ by

$$\frac{1}{h^2} (y_{n+1} - 2y_n + y_{n-1}),$$

which results in the explicit recurrence relation

$$y_{n+1} = 2y_n - y_{n-1} + h^2 f_n, \tag{7}$$

where $f_n := f(x_n, y_n)$. (See also Demonstration 1.7-15.) Two starting values $y_0$ and $y_1$ are required, which is consistent with the fact that two initial conditions $y(x_0) = a$, $y'(x_0) = b$ are required for the exact solution.

(a)  Apply the method (7) to the model problem

$$y'' = -y, \qquad y(0) = 1, \qquad y'(0) = 0.$$

Solve the resulting difference equation analytically, and by letting $h \to 0$ while $nh = x$ is fixed, determine the order of the method.

(b)  Also solve the equation numerically for various values of $h$, and experiment with extrapolation to the limit.

7  **Numerov's method.**  A more sophisticated method to integrate special second order equations is obtained by replacing (7) by an expression of the form

$$y_{n+1} = 2y_n - y_{n-1} + h^2(\alpha f_{n-1} + \beta f_n + \alpha f_{n+1}), \tag{8}$$

where $f_k := f(x_k, y_k)$.

(a)  Determine the best possible values of $\alpha$ and $\beta$ in (8) by the following requirement: If an exact solution of the differential equation is substituted into (8), so that $f(x_k, y_k) = y''(x_k)$, and if the resulting expressions are expanded at $x = x_n$ in powers of $h$, then the disagreement between the expressions on the right and on the left should have the highest possible order in $h$. Show that this yields the unique values

$$\alpha = \tfrac{1}{12}, \qquad \beta = \tfrac{10}{12}.$$

(b)  Implement the resulting method for an arbitrary function $f(x, y)$, assuming that two starting values $y_0$ and $y_1$ are given. Provide for an inner iteration to solve equation (8) for the new value $y_{n+1}$.

8  The lateral vibrations of a small body suspended between two springs

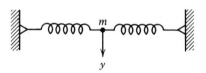

**Fig.  6.6d.**  Lateral  vibration  of spring.

(Figure 6.6d) in suitable units as a function of the time $x$ for small amplitudes obey the special second order equation

$$y'' = -y^3. \tag{9}$$

Assuming that the body at time $x = 0$ is moved from its rest position $y = 0$ by giving it a push of velocity $v_0$, equation (9) is to be solved under the initial condition

$$y(0) = 0, \qquad y'(0) = v_0.$$

**(a)** Obtain the value $y_1$ by assuming the solution in the form of a power series,

$$y(h) = v_0 h(1 + a_1 h^4 + a_2 h^8 + \cdots)$$

and determining at least the coefficients $a_1$ and $a_2$.

**(b)** Using the value $y_1$ thus obtained, solve the differential equation numerically for the value $v_0 = 0.025$. Experiment both with the method of Problem 6 and with Numerov's method, and compare the accuracy obtained.

**(c)** The motion of the spring may be shown to be periodic. Determine, for various values of $v_0$, the period $T$ of the motion ($T/2$ = first positive zero of $y(x)$).

---

## §6.7 SECURING ACCURACY

Once more, we consider the numerical solution of the initial value problem,

$$y' = f(x, y), \qquad y(a) = b,$$

the exact solution of which we denote by $y(x)$. We assume that a one-step method is being used, which furnishes approximate values $y_n$ for the solution at $x = x_n = a + nh$ by a formula

$$y_{n+1} = y_n + hF(x_n, y_n; h),$$

$n = 0, 1, 2, \ldots$, where $F$ denotes the increment function. Almost always in problems of numerical analysis, the basic question to be answered is this: *How accurate* are the values $y_n$?

As always, two kinds of errors must be distinguished:

(1) the *rounding error*

$$r_n := \tilde{y}_n - y_n;$$

(2) the *discretization error*

$$e_n := y_n - y(x_n).$$

Here we denote by

$\tilde{y}_n$   the values generated on the machine, working with a finite number of $b$-ary places;

$y_n$   the ideal numerical values, which would be obtained on a machine with infinitely many decimal places;

$y(x_n)$   the mathematically defined values of the exact solution at the point $x_n$.

It is of interest to compare the role of these errors in numerical linear algebra and in the numerical solution of differential equations. Roughly, the following may be said:

|  | Linear algebra | Differential equations |
|---|---|---|
| rounding errors | important | not important |
| discretization errors | do not exist | important |

In the solution of differential equations, the *rounding error* can always be made negligible by working with double precision. Almost the same effect can be reached with much less effort by working with **partial double precision**, however. Here only the values $y_n$ are accumulated in double precision. All other computations are done in simple precision, according to the following scheme:

$$\underbrace{y_{n+1}}_{\substack{\text{double}\\\text{precision}}} = \underbrace{y_n}_{\substack{\text{double}\\\text{precision}}} + \underbrace{h * F(x_n, \underbrace{y_n,}_{\substack{\text{simple}\\\text{precision}}} h)}_{}$$

$$\underset{\text{simple precision}}{\underbrace{\qquad\qquad}}$$

$$\underset{\text{simple precision}}{\underbrace{\qquad\qquad}}$$

$$\underset{\text{double precision}}{\underbrace{\qquad\qquad}}$$

Another device to reduce rounding error consists in not forming the additions $y_n + hF(x_n, y_n; h)$ at each step, but accumulating the sum $\sum F(x_k, y_k; h)$ separately and generating $y_n = y_0 + h \sum F(x_k, y_k; h)$ each time when required.

The main problem, as mentioned, is *discretization error*. In addition to the global error $e_n := y_n - y(x_n)$, we must consider the **local discretization error**

$$d_n := y_{n+1} - z(x_{n+1}),$$

where $z(x)$ denotes the solution passing through the point $(x_n, y_n)$. In formulas,

$$d_n = hF(x_n, y_n; h) - \{z(x_{n+1}) - y_n\}.$$

In principle, it is always possible to determine the local discretization error of a method by expanding the expression on the right in powers of $h$. The result can always be written

$$d_n = h^{p+1}g(x_n, y_n) + O(h^{p+2}),$$

where the integer $p$ is determined uniquely by the condition that the function $g$ does not vanish identically for all sufficiently smooth functions $f$. The integer thus determined is called the **order** of the method. The higher $p$, the more accurate (in principle) the method. For the methods discussed earlier the following table results:

| Method | Order |
|---|---|
| Euler | 1 |
| Simplified Runge-Kutta | 2 |
| Classical Runge-Kutta | 4 |
| Trapezoidal | 2 |

The function $g$ is called the **local error function**. This function is of crucial importance for the theoretical study of the convergence of one-step methods. However, $g$ is at least as hard to compute as the higher derivatives of the function $f$, and it therefore cannot be used directly for the purpose of estimating local discretization errors. More easily, these errors are estimated as follows: Let

$y_{n+1}$   denote the value computed from $y_n$, using a *single* step of length $h$; and let

$y^*_{n+1}$   denote the value computed from $y_n$, using *two* steps of length $h/2$ (Figure 6.7a).

Then, up to terms of higher order in $h$,

$$d_n \sim \frac{1}{1 - 2^{-p}}(y_{n+1} - y^*_{n+1}).$$

If the local errors are known, it is possible to estimate the global discretization error by means of a rather elaborate theory (see Henrici [1962]). This theory

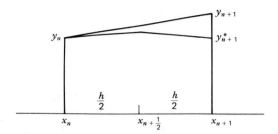

**Fig. 6.7a.** Estimating the local discretization
error.

is probably too cumbersome to apply, however, except in situations where a very great value is placed on the reliability of the numerical result. For this reason it has become customary to study error propagation under the following simplifying assumption:

$$\text{Global error} = \text{sum of all local errors}$$

that is,

$$e_n = \sum_{k=0}^{n-1} d_k. \tag{1}$$

It is easy to see that this assumption is not mathematically correct, not even in an asymptotic sense. For instance, the assumption would imply that the global error always grows if the local errors have constant sign. However, we have seen that in the integration of the model problem $y' = -y$ by Euler's method, the global error is approximately given by

$$e_n = -\tfrac{1}{2}xe^{-x}h + O(h^2),$$

where $x = x_n$. As a discussion of the function $e(x) := xe^{-x}$ shows, the global error grows in absolute value up to the point $x = 1$, and then decreases exponentially, in spite of the fact that the local errors $d_n \sim -\tfrac{1}{2}y''(x)h^2$ here have constant sign. As a practical working hypothesis, our assumption nevertheless frequently furnishes useful results.

A practical demand that is often made is that the global error after integration over an interval of length $L$ should not exceed a prescribed tolerance $\eta$. Under our hypothesis (1), this demand can be met by trying to achieve a uniform distribution of the errors, which means that the error per unit length should not exceed $\eta/L$. This means that the local error (committed by integrating over a step of length $h$) should satisfy

$$|d_n| < h\eta/L. \tag{2}$$

One thus will endeavor to choose the step $h$ so that (2) is always satisfied. Practically, this is accomplished by estimating the local error from time to time by the method described above and checking whether (2) holds. If the local error is too big, the step must be made smaller; if the local error is smaller than the bound given by (2), a larger step can be chosen.

The above method of error control, although somewhat rough-hewn and of doubtful theoretical validity, is very popular and has been incorporated in many differential equation solvers. Some of the more sophisticated programs even contain a viscosity element which prevents the machine from spending all its computing time on the search for the correct step size. For problems on a smaller scale, where the solution can be expected not to experience any violent changes of its degree of smoothness, one should not disregard the time-honored device of solving the equation several times with various constant steps and of observing the convergence of the numerical values. (See Demonstrations 6.4-1 and 6.6-1.) It also should be noted that Romberg type extrapolation to the limit is possible only if the integration step is *not* adjusted automatically.

## PROBLEMS

1   Apply the method of error control described here to the differential equation of Problem 1, §6.6, if the error at the point $x = 3$ should not exceed $10^{-5}$. Experiment with both the Runge-Kutta method and with the trapezoidal rule.

$2^{\oplus}$  Make an experimental study of the propagation of rounding error, simulating single precision by artificial four-digit precision as in Chapter 4. For instance, consider the initial value problem

$$y' = -2xy, \qquad y(x_0) = y_0,$$

with the exact solution $y(x) = y_0 e^{-(x^2 - x_0^2)}$.

(a)   For some chosen value of $x_0$ and for $y_0 = 1$, obtain the numerical solution at $x_n = b > x_0$ by means of a one-step method, such as the improved Euler method, using a step $h$ commensurable with $b - x_0$ and working with full machine precision.

(b)   For the same $x_0$ and $h$ and using the same method, compute the numerical values $\tilde{y}_{n, k}$ at $x_n = b$ corresponding to the starting values $y_{0, k} := y_{0, 0} + k\Delta$, $k = 1, 2, \ldots, K$, where $y_{0, 0}$ and $\Delta$ are nonzero constants, working with simulated low ($=$artificial four-digit) precision. (A subroutine that rounds a number in the $X$ register correctly to $p$ digits is required.) For each $k$, record the relative rounding error

$$\varepsilon_k := \frac{\tilde{y}_{n, k} - y_{n, k}}{y_{n, k}}$$

where $y_{n,k}$ is the numerical value obtained with full precision. Note that the values $y_{n,k}$ need not be computed for each $k$, because by virtue of the linearity of the problem

$$y_{n,k} = y_{0,k} \cdot y_n,$$

where $y_n$ is the value obtained in part (a). Calculate the mean and the standard deviation of the $\varepsilon_k$.

(c) Repeat part (b) using partial double precision, simulating double precision by full precision.

(d) The data of this problem may be varied in many ways. Do so, and discover laws of statistical error propagation. As far as the average growth of the relative rounding errors is concerned, does it make a difference whether the exact solution increases $(x < 0)$ or decreases $(x > 0)$?

$3^{\oplus}$ Repeat the foregoing problem using the midpoint rule,

$$y_{n+1} = y_{n-1} + 2hf(x_n, y_n),$$

using exact starting values $y_0$ and

$$y_1 = y_0 e^{-h(2x_0 + h)}.$$

What is now the answer to Problem 2, part (d)?

---

### Bibliographical Notes and Recommended Reading

On the evaluation of definite integrals, see Davis and Rabinowitz [1967]. They treat, in particular, the theory of Gaussian quadrature, which in spite of its importance is not dealt with in this volume because the method is not suitable for small calculators. For some further applications of the Euler-Maclaurin formula, see Henrici [1977].

Numerous texts deal with theory and practice of the solution of ordinary differential equations, see Henrici [1962], Lambert [1973], Shampine and Gordon [1975] and, in particular regard to stiff problems, Lapidus and Seinfield [1971]. Numerical instability in solution methods for ordinary differential equations was discovered by Rutishauser [1952]. The concept of absolute stability was introduced by Dahlquist [1963].

For obvious reasons we do not deal here with the numerical solution of partial differentiation equations, for which Forsythe and Wasow [1960] and Richtmyer and Morton [1967] are two classical references. For a more elementary treatment see Smith [1978].

# CHAPTER 7

## Periodicity

Science requires mathematical models for many phenomena that are exactly or approximately periodic. Some simple examples are alternating current in electrical engineering, the current measured in the brain and in the heart by means of electroencephalograms and electrocardiagrams, geophysical phenomena, such as daily temperatures and tides, and celestial phenomena, such as the motion of a planet around the sun or of a moon around a planet.

Obviously, periodic phenomena are best modeled by functions that are themselves periodic. If the period is $L$, the smoothest functions with that period—apart from the constant function—are the trigonometric functions

$$f(\tau) = \cos\frac{2\pi\tau}{L}, \qquad f(\tau) = \sin\frac{2\pi\tau}{L}$$

or, more generally,

$$f(\tau) = \cos\frac{2\pi n\tau}{L}, \qquad f(\tau) = \sin\frac{2\pi n\tau}{L},$$

where $n$ is any positive integer.

In this chapter we study the problem of how to approximately represent periodic functions. Two approaches will be considered: (i) *Interpolation* by means of linear combinations of trigonometric periodic functions, and (ii) *expansion* in terms of these functions. Fourier analysis, which forms the mathematical basis for the second approach, compels us to be concerned with the *spectrum* of a given periodic function. In periodic and quasiperiodic phenomena of science, the spectrum often has a direct physical meaning, which is explored, for instance, in the analysis of *time series*. The numerical computation of spectra is much facilitated by an algorithm called the *Fast Fourier Transform*.

## §7.1 TRIGONOMETRIC INTERPOLATION

For ease of mathematical presentation, we shall usually assume that the period of the phenomena to be studied is 1. The case of an arbitrary period $L$ can be reduced to this special case by introducing the new variable $\tau := x/L$.

Let $m$ be a positive integer. A function of the form

$$t(\tau) = a_0 + \sum_{k=1}^{m} (a_k \cos 2\pi k\tau + b_k \sin 2\pi k\tau) \tag{1}$$

where $a_0, a_1, b_1, \ldots, a_m, b_m$ are real or complex constants, is called a **trigonometric polynomial** of degree **m** (and period 1). The trigonometric polynomial is said to be **balanced** if $b_m = 0$, that is, if the sine term of period $1/m$ is missing.

Now let $f$ be a real or complex valued function of period 1. Let $n$ be a positive integer, and let, for any integer $k$,

$$\tau_k := \frac{k}{n}.$$

The points $\tau_k$ are called the **sampling points**. We wish to construct a trigonometric polynomial $t$ which interpolates $f$ at the $n$ sampling points $\tau_0, \tau_1, \ldots, \tau_{n-1}$, that is, which satisfies

$$t(\tau_k) = f_k := f(\tau_k), \qquad k = 0, 1, \ldots, n-1. \tag{2}$$

Because both $t$ and $f$ are periodic with period 1, a trigonometric polynomial satisfying (2) will automatically satisfy $t(\tau_k) = f_k$ at *all* sampling points, not just at those belonging to one period interval.

What is the minimum degree of a trigonometric polynomial $t(\tau)$ that can be expected to satisfy (2)? A trigonometric polynomial of degree $m$ contains the $2m + 1$ constants $a_0, a_1, b_1, \ldots, a_m, b_m$. Thus, by choosing these constants appropriately, we may expect to be able to satisfy $n = 2m + 1$ conditions. Thus, if $n$ is odd, we expect the degree of a $t$ satisfying (2) to be $m := (n-1)/2$. If $n$ is even, the trigonometric polynomial of degree $m = n/2$ has one parameter too many. If $m = n/2$, however, the last sine term in $t$ satisfies

$$\sin 2\pi m\tau_k = \sin \pi n\tau_k = \sin \pi k = 0.$$

Thus, if a trigonometric polynomial of degree $m$ satisfies (2), it will do so for arbitrary values of $b_m$. In the interest of economy, it is reasonable to ask that for $n = 2m$, the conditions (2) be satisfied with a *balanced* trigonometric polynomial of degree $m$.

In the following we usually assume that $n$, the number of sampling points per period interval, is *even*. The corresponding results and formulas for an odd number of sampling points are dealt with in the problem section.

Our goal is to show that for every even $n = 2m$ and for arbitrary complex data $f_0, f_1, \ldots, f_{n-1}$, there exists precisely one balanced trigonometric polynomial $t$ of degree $m$ which satisfies the $n$ relations

$$t(\tau_k) = f_k, \qquad k = 0, 1, \ldots, n - 1. \tag{3}$$

It will simplify later formulas if we assume $t$ in the form

$$t(\tau) = \tfrac{1}{2}a_0 + \sum_{k=1}^{m-1} (a_k \cos 2\pi k\tau + b_k \sin 2\pi k\tau) + \tfrac{1}{2}a_m \cos 2\pi m\tau. \tag{4}$$

To carry out the necessary computations as economically as possible, it is convenient to express $t(\tau)$ in terms of complex exponentials. Substituting Euler's relations

$$\cos 2\pi k\tau = \tfrac{1}{2}(e^{2\pi i k\tau} + e^{-2\pi i k\tau}),$$

$$\sin 2\pi k\tau = \frac{1}{2i}(e^{2\pi i k\tau} - e^{-2\pi i k\tau})$$

in (4) and rearranging, there results the expression, called a **complex trigonometric polynomial** of degree $m$,

$$t(\tau) = \tfrac{1}{2}c_{-m}e^{-2\pi i m\tau} + \sum_{k=-m+1}^{m-1} c_k e^{2\pi i k\tau} + \tfrac{1}{2}c_m e^{2\pi i m\tau}, \tag{5}$$

where

$$c_k := \tfrac{1}{2}(a_k - ib_k),$$
$$c_{-k} := \tfrac{1}{2}(a_k + ib_k), \tag{6}$$

$k = 0, 1, \ldots, m$. Because $t$ is balanced, $c_{-m} = c_m$. Conversely, if $t$ is any function represented in the form (5) with arbitrary constants $c_{-m}, c_{-m+1}, \ldots, c_{m-1}$ and $c_m = c_{-m}$, there results on using

$$e^{2\pi i k\tau} = \cos 2\pi k\tau + i \sin 2\pi k\tau$$

and rearranging a balanced trigonometric polynomial (4) where

$$a_k = c_k + c_{-k}, \qquad b_k = i(c_k - c_{-k}), \tag{7}$$

$k = 0, 1, \ldots, m$. It thus suffices to show the unique existence of a complex trigonometric polynomial (5) satisfying the given interpolating conditions (3).
     To simplify the notation, we write

$$z := e^{2\pi i\tau}$$

and, if $j$ is any integer,

$$z_j := e^{2\pi i \tau_j} = \exp\frac{2\pi i j}{n}.$$

(The fact that $z_j$ also depends on $n$ is not stressed in this notation.) We note that for any integer $j$,

$$z_j = z_1^j, \tag{8}$$

and therefore

$$z_j^n = z_1^{jn} = (z_1^n)^j = 1; \tag{9}$$

furthermore, if $n = 2m$,

$$z_m = z_{-m} = z_1^m = z_1^{-m} = -1. \tag{10}$$

In this new notation, $t$ takes the form

$$t(\tau) = \tfrac{1}{2}c_{-m}z^{-m} + \sum_{k=-m+1}^{m-1} c_k z^k + \tfrac{1}{2}c_m z^m,$$

and the interpolating conditions (3) become

$$\tfrac{1}{2}c_{-m}z_j^{-m} + \sum_{k=-m+1}^{m-1} c_k z_j^k + \tfrac{1}{2}c_m z_j^m = f_j, \qquad j = 0, 1, \ldots, n-1.$$

In view of (8) and (10), using $c_{-m} = c_m$, this may be written

$$\sum_{k=-m+1}^{m} c_k z_1^{jk} = f_j, \qquad j = 0, 1, \ldots, n-1. \tag{11}$$

The $n$ equations (11) represent a system of $n$ linear equations for the $n$ unknowns $c_{-m+1}, \ldots, c_m$. Assume that a solution exists, and let $l$ denote one of the integers $-m+1, \ldots, m$. Multiplying the $j$-th equation by $z_l^{-j} = z_1^{-lj}$ and adding all equations, we get on the right

$$\sum_{j=0}^{n-1} f_j z_1^{-lj}.$$

On the left we get as a factor of $c_l$

$$z_1^{0(l-l)} + z_1^{1(l-l)} + \cdots + z_1^{(n-1)(l-l)} = n,$$

and as a factor of $c_k$ where $k \neq l$,

$$z_1^{0(k-l)} + z_1^{1(k-l)} + \cdots + z_1^{(n-1)(k-l)} = \frac{1 - z_1^{n(k-l)}}{1 - z_1^{k-l}}$$

by the well-known formula for a finite geometric sum. The last expression equals zero in view of (9), and there remains a formula for $c_l$,

$$c_l = \frac{1}{n} \sum_{j=0}^{n-1} f_j z_1^{-lj}, \tag{12}$$

where $z_1 := \exp(2\pi i/n)$ as always. This formula holds for $l = -m + 1, -m + 2,$ $\ldots, m$ and also for $l = -m$ in view of (10). Equation (12) was derived under the assumption that a solution of the system (11) exists. If the $c_l$ are *defined* by (12), however, an easy computation shows that they do satisfy the system (11). Hence, the trigonometric polynomial (5), or the polynomial (4) where the coefficients $a_k$ and $b_k$ are given by (7), does satisfy the interpolating conditions (3).

Altogether we have proved:

**THEOREM 7.1:** *Let $n = 2m$ be an even positive integer, and let $f_0, f_1, \ldots,$ $f_{n-1}$ be arbitrary complex numbers. There exists precisely one balanced trigonometric polynomial $t(\tau)$ of degree $m$ and period 1 satisfying the $n$ interpolating conditions*

$$t\left(\frac{j}{n}\right) = f_j, \qquad j = 0, 1, \ldots, n - 1.$$

*The polynomial is represented in complex form by (5), where the coefficients $c_k$ are given by (12), or in real form by (4), where the $a_k$ and $b_k$ are given by (7).*

## Real Data

In many applications, the data $f_0, f_1, \ldots, f_{n-1}$ will be real numbers. In view of $z_1^{-1} = \overline{z_1}$ (the bar denoting the complex conjugate number) it then follows from (12) that

$$c_{-k} = \frac{1}{n} \sum_{j=0}^{n-1} f_j z_1^{kj} = \frac{1}{n} \sum_{j=0}^{n-1} f_j \overline{z_1^{-kj}},$$

and hence that

$$c_{-k} = \overline{c_k}, \qquad k = 0, 1, \ldots, m. \tag{13}$$

It follows that for real data only the coefficients $c_0$, $c_1$, ..., $c_m$ need to be computed; the remaining coefficients are obtained by conjugation. Moreover, from $c_0 = \overline{c_0}$ we see that $c_0$ is real. In view of $c_{-m} = c_m = \overline{c_m}$, the coefficient $c_m$ also is real.

### Evaluation of $t(\tau)$

A possible method to evaluate $t(\tau)$ for a given value of $\tau$ would consist in computing the required values of $\cos 2\pi k\tau$ and $\sin 2\pi k\tau$ $(k = 1, 2, ..., m)$, for instance, by means of the recurrence relations studied in Problem 8, §3.1, and then computing the sum (4). However, a more elegant way consists in writing

$$t(\tau) = t^+(\tau) + t^-(\tau),$$

where

$$t^+(\tau) := \tfrac{1}{2}c_0 + \sum_{k=1}^{m-1} c_k z^k + \tfrac{1}{2}c_m z^m,$$

$$t^-(\tau) := \tfrac{1}{2}c_0 + \sum_{k=1}^{m-1} c_{-k} z^{-k} + \tfrac{1}{2}c_{-m} z^{-m},$$

and using Horner's rule to evaluate

$$
\begin{aligned}
t^+(\tau) &= \tfrac{1}{2}c_0 + z(c_1 + z(c_2 + \cdots + z(c_{m-1} + \tfrac{1}{2}zc_m)\cdots)), \\
t^-(\tau) &= \tfrac{1}{2}c_0 + z^{-1}(c_{-1} + z^{-1}(c_{-2} + \cdots + z^{-1}(c_{-m+1} + \tfrac{1}{2}z^{-1}c_{-m})\cdots)).
\end{aligned}
\tag{14}
$$

If the data are real, then

$$t^-(\tau) = \overline{t^+(\tau)},$$

and we have

$$t(\tau) = c_0 + 2\,\mathrm{Re}\{z(c_1 + z(c_2 + \cdots + z(c_{m-1} + \tfrac{1}{2}zc_m)\cdots))\}. \tag{15}$$

### Simultaneous Computation of Two Sets of $c_k$'s for Real Data

Suppose two sets of real data are given,*

$$f_0', f_1', ..., f_{n-1}' \quad \text{and} \quad f_0'', f_1'', ..., f_{n-1}'',$$

---

*Primes do not denote derivatives here.

and let $c_k'$ and $c_k''$ denote the corresponding coefficients defined by (12),

$$c_k' := \frac{1}{n} \sum_{j=0}^{n-1} f_j' z_1^{-kj},$$

$$c_k'' := \frac{1}{n} \sum_{j=0}^{n-1} f_j'' z_1^{-kj}.$$

Because the data are real, we have by (13)

$$c_k' = \overline{c_{-k}'}, \qquad c_k'' = \overline{c_{-k}''}, \qquad k = -m, -m+1, \ldots, m. \qquad (16)$$

Let now, on the other hand,

$$f_j := f_j' + i f_j'',$$

$$c_k := \frac{1}{n} \sum_{j=0}^{n-1} f_j z_1^{-kj}, \qquad (17)$$

$k = -m, -m+1, \ldots, m$. Clearly,

$$c_k = c_k' + i c_k''. \qquad (18)$$

Using (16),

$$\overline{c_{-k}} = \overline{c_{-k}'} - i \, \overline{c_{-k}''} = c_k' - i c_k''.$$

There follows for $k = -m, -m+1, \ldots, m$

$$c_k' = \tfrac{1}{2}(c_k + \overline{c_{-k}}),$$

$$c_k'' = \frac{1}{2i}(c_k - \overline{c_{-k}}). \qquad (19)$$

In other words, the coefficients for two sets of real data $f_j'$ and $f_j''$ can be obtained from the coefficients for one set of complex data $f_j := f_j' + i f_j''$ at almost no additional cost.

**DEMONSTRATION 7.1-1:** Suppose the following temperatures (in celsius) have been recorded at three-hour intervals in Johnson City, New Mexico:

| Time | Mid-night | 3 A.M. | 6 A.M. | 9 A.M. | Noon | 3 P.M. | 6 P.M. | 9 P.M. |
|------|-----------|--------|--------|--------|------|--------|--------|--------|
| $f$  | 11        | 10     | 17     | 24     | 32   | 26     | 23     | 19     |

We wish to interpolate these temperatures by a balanced trigonometric polynomial $t(\tau)$ of degree 4 with period one day. Equation (12) yields the following values for the coefficients $c_k$ $(k = 0, 1, \ldots, 4)$:

$$
\begin{aligned}
c_0 &= \phantom{-}20.25000000 + i0.00000000 \\
c_1 &= -4.48115530 + i1.72227184 \\
c_2 &= \phantom{-}0.37500000 + i0.87500000 \\
c_3 &= -0.76884470 + i0.22227183 \\
c_4 &= \phantom{-}0.50000000 + i0.00000001
\end{aligned}
$$

As predicted, $c_0$ and $c_4$ are real. (The error of one unit in the last digit of Im $c_4$ is due to rounding.) With these $c_k$, the interpolating polynomial is given by

$$t(\tau) = c_0 + 2 \operatorname{Re}\{z(c_1 + z(c_2 + z(c_3 + \tfrac{1}{2}zc_4)))\},$$

where

$$z = \exp(2\pi i\tau),$$

$\tau$ being measured in units of one day. Which temperature results at 1:45 PM? This corresponds to a fraction of

$$\tau = \frac{1}{2} + \frac{1}{24} + \frac{45}{2 \cdot 12 \cdot 60} = 0.57296667$$

of one day. The above formula yields

$$t(\tau) = 29.49.$$

As a check, we may also evaluate

$$
\begin{aligned}
t(0) &= 9.00, \\
t(\tfrac{1}{8}) &= 10.00, \\
t(\tfrac{2}{8}) &= 17.00, \qquad \text{and so on.}
\end{aligned}
$$

# PROBLEMS

1    Show how to evaluate the complex coefficients $c_l$ defined by (12) by Horner's rule,

$$c_l = \frac{1}{n}(f_0 + z_1^{-l}(f_1 + z_1^{-l}(f_2 + \cdots + z_1^{-l}f_{n-1}))),$$

and write a program which, given real data $f_0, f_1, \ldots, f_{n-1}$, evaluates the real and imaginary parts of $c_0, \ldots, c_m$. [On pocket calculators, efficient use may be made of the $\rightarrow R$ and $\rightarrow P$ instructions for doing complex arithmetic.]

2   On a certain day the tide at Oceanville, Oregon, behaves according to the following schedule:

|  |  |
|---|---|
| 6:30 AM | 3 ft 10 in (low tide) |
| 8:30 | 5 ft 2 in |
| 10:30 | 7 ft 6 in |
| 12:30 PM | 10 ft 4 in |
| 2:30 | 8 ft 11 in |
| 4:30 | 6 ft 3 in |
| 6:30 | 3 ft 10 in |

Assuming a 12-hour period, interpolate this table by a balanced trigonometric polynomial of degree $m = 3$, and compute the tide at 9:30 AM and at 3:30 PM.

3   Prove the analog of Theorem 7.1 for odd values of $n$, $n = 2m + 1$. Show that there exists precisely one (unbalanced) trigonometric polynomial $t(\tau)$ of degree $m$, satisfying the given interpolating conditions, and show that the coefficients of $t$ are again given by (12).

4   What happens if the function $f(\tau) := \cos(10\pi\tau)$ is interpolated by a balanced trigonometric interpolating polynomial of degree $n = 4$?

## §7.2   A BARYCENTRIC FORMULA

The representation (5), §7.1, of the trigonometric interpolating polynomial is the analog of representing the ordinary interpolating polynomial $p(x)$ in the form $a_0 + a_1 x + \cdots + a_n x^n$. This representation, while not recommended for ordinary interpolation, is feasible for trigonometric interpolation because the linear system (11), §7.1, has the simple explicit solution (12).

The Lagrangian approach works in the trigonometric case as well and has its own advantages. Let again $n$ be even, $n = 2m$, and let $z$, $\tau_k$, and $z_k$ have the same meaning as in §7.1. We seek to represent the balanced trigonometric polynomial of degree $m$ interpolating the data $f_0, f_1, \ldots, f_{n-1}$ in the form

$$t(\tau) = \sum_{k=0}^{n-1} l_k(z) f_k, \qquad z := e^{2\pi i \tau}, \tag{1}$$

where the functions $l_0(z), \ldots, l_{n-1}(z)$ are also balanced trigonometric polynomials of degree $m$, satisfying the conditions

$$l_k(z_j) = \begin{cases} 1, & j = k, \\ 0, & j \neq k. \end{cases} \tag{2}$$

We begin by determining $l_0(z)$. It is easy enough to find a *polynomial* which vanishes at $z_1, \ldots, z_{n-1}$ and has the value 1 at $z_0 = 1$. Because the points $z_0, z_1, \ldots, z_{n-1}$ are just the $n$ $n$-th roots of unity, the polynomial

$$z^n - 1$$

vanishes at all these points. The polynomial

$$\frac{z^n - 1}{z - 1} = 1 + z + z^2 + \cdots + z^{n-1}$$

vanishes at $z_1, \ldots, z_{n-1}$. At $z = 1$, it has the value $n$. Thus,

$$\frac{1}{n} \frac{z^n - 1}{z - 1}$$

is the polynomial that equals 1 at $z = 1$ and vanishes at $z_1, \ldots, z_{n-1}$. On setting $z = e^{2\pi i t}$, this becomes a trigonometric polynomial, but the degree is $n - 1$ in place of $m$. We obtain a trigonometric polynomial of degree $m$ by forming

$$\hat{l}_0(z) := z^{-m+1} \frac{1}{n} \frac{z^n - 1}{z - 1}; \tag{3}$$

multiplication by $z^{-m+1}$ does not change the value at $z = 1$ nor the vanishing at $z_1, \ldots, z_{n-1}$. However, $\hat{l}_0$ is not yet balanced. To balance $\hat{l}_0$, we note that if $\hat{l}_0(z)$ is a trigonometric polynomial of degree $m$, satisfying the interpolating conditions (2) where $k = 0$, then so is $\hat{l}_0(z^{-1})$. We obtain a balanced polynomial by forming the arithmetic mean. Thus, finally,

$$l_0(z) := \tfrac{1}{2}\{\hat{l}_0(z) + \hat{l}_0(z^{-1})\} \tag{4}$$

is the (unique) balanced trigonometric polynomial of degree $m$, satisfying (2) for $k = 0$.

Having found $l_0(z)$, it is easy to construct the remaining $l_k(z)$. We assert that

$$l_k(z) := l_0(z_k^{-1} z) \tag{5}$$

has the required properties. Indeed, this clearly is a trigonometric polynomial of degree $m$. The polynomial is balanced because the coefficient of $z^m$ in $l_0$ has been multiplied by $z_k^{-m}$ and that of $z^{-m}$ by $z_k^m$, which is the same; finally,

$$l_k(z_k) = l_0(1) = 1$$

whereas for $j \neq k$

$$l_k(z_j) = l_0(z_k^{-1}z_j) = l_0(z_{k-j}) = 0,$$

by (2).

It remains to find a manageable expression for the $l_k(z)$. From (3) we have, using $n = 2m$,

$$l_0(z) = \frac{1}{n} z \frac{z^m - z^{-m}}{z - 1}.$$

Equation (4) now yields

$$l_0(z) = \frac{1}{2n} (z^m - z^{-m}) \left| \frac{z}{z - 1} - \frac{1}{1 - z} \right|$$

$$= \frac{1}{2n} (z^m - z^{-m}) \frac{z + 1}{z - 1},$$

and from (5) there follows in view of $z_k^{-m} = z_k^m = (-1)^k$

$$l_k(z) = \frac{(-1)^k}{2n} (z^m - z^{-m}) \frac{z + z_k}{z - z_k}.$$

(See Equation (10) of §7.1.) To express these functions in trigonometric form, we note that the relations

$$\tfrac{1}{2}(z^m - z^{-m}) = \tfrac{1}{2}(e^{2\pi im\tau} - e^{-2\pi im\tau}) = i \sin 2\pi m\tau$$

and

$$\frac{z + z_k}{z - z_k} = \frac{e^{2\pi i(\tau - \tau_k)} + 1}{e^{2\pi i(\tau - \tau_k)} - 1} = \frac{1}{i} \cot[\pi(\tau - \tau_k)]$$

imply for $k = 0, 1, \ldots, n - 1$

$$l_k(e^{2\pi i\tau}) = \frac{\sin \pi n\tau}{n} (-1)^k \cot[\pi(\tau - \tau_k)]. \tag{6}$$

In view of (1) the following representation for $t(\tau)$ now results:

$$t(\tau) = \frac{\sin \pi n \tau}{n} \sum_{k=0}^{n-1} (-1)^k \cot[\pi(\tau - \tau_k)] f_k. \tag{7}$$

In spite of its elegance, the representation (7) is not yet well suited for numerical computation, because for values of $\tau$ near an interpolating point $\tau_k$, even small errors in the numerical value of the factor $\sin \pi n \tau$ may be blown beyond proportion by the corresponding term $\cot[\pi(\tau - \tau_k)]$.

**DEMONSTRATION 7.2-1:** We interpolate the function $f(\tau) = 1$ at $n = 12$ points by the balanced interpolating polynomial $t(\tau)$ of degree 6. Mathematically, the values $t(\tau) = 1$ should result, because a constant function is represented exactly by every trigonometric interpolating polynomial. Using (7), however, the following numerical values result:

| $\tau$ | $t(\tau)$ |
|---|---|
| 0.9 | 0.999999984 |
| 0.99 | 1.000000011 |
| 0.999 | 1.000000033 |
| 0.9999 | 0.999997850 |
| 0.99999 | 0.999978677 |
| 0.999999 | 1.000105258 |
| 0.9999999 | 0.998187972 |
| 0.99999999 | 0.963099612 |
| 0.999999999 | 0.877474253 |
| 0.9999999999 | $-1.835587010$ |

The instability shown in Demonstration 7.2-1 is easily corrected by using a *barycentric version* of (7). By the mathematical fact mentioned, the identity

$$1 = \frac{\sin \pi n \tau}{n} \sum_{k=0}^{n-1} (-1)^k \cot[\pi(\tau - \tau_k)]$$

holds for $\tau \neq \tau_k$. Dividing (7) by the identity, there follows

$$t(\tau) = \frac{\sum_{k=0}^{n-1} (-1)^k \cot[\pi(\tau - \tau_k)] f_k}{\sum_{k=0}^{n-1} (-1)^k \cot[\pi(\tau - \tau_k)]}. \tag{8}$$

**THEOREM 7.2a:** *If $n$ is even, $n = 2m$, the balanced trigonometric polynomial of degree $m$ that interpolates given data $f_k$ at the points $\tau_k := k/n$ for $k = 0, 1, \ldots, n - 1$ is for $\tau \neq \tau_k$ represented by (8).*

For $n$ odd, $n = 2m + 1$, the interpolating trigonometric polynomial of degree $m$ is given by

$$t(\tau) = \frac{\sum_{k=0}^{n-1} (-1)^k \sec[\pi(\tau - \tau_k)] f_k}{\sum_{k=0}^{n-1} (-1)^k \sec[\pi(\tau - \tau_k)]}. \tag{9}$$

(See Problem 1.)

For the example considered in Demonstration 7.2-1, the foregoing barycentric form of the trigonometric interpolating polynomial is stable by construction. For arbitrary data, while it still may be the case that the coefficients $\cot[\pi(\tau - \tau_k)]$ are evaluated with large relative errors if $\tau - \tau_k$ is small, these errors are harmless if the same numerical values of these coefficients are used in the numerator and in the denominator.

**DEMONSTRATION 7.2-2:** Here we represent the function $f(\tau) := \cos \pi n \tau$ by a balanced trigonometric interpolating polynomial of degree $m = n/2$. Mathematically, because $f(\tau)$ itself is such a polynomial, the representation should be exact. Numerically, the following values result for $n = 12$ if the form (8) is used:

| $\tau$ | $t(\tau)$ | $\cos 12\pi\tau$ |
|---|---|---|
| 0.9 | −0.809016995 | −0.809016995 |
| 0.99 | 0.929776486 | 0.929776488 |
| 0.999 | 0.999289473 | 0.999289473 |
| 0.9999 | 0.999992894 | 0.999992894 |
| 0.99999 | 0.999999930 | 0.999999929 |
| 0.999999 | 0.999999999 | 0.999999999 |
| 0.9999999 | 1.000000000 | 1.000000000 |
| 0.99999999 | 1.000000000 | 1.000000000 |
| 0.999999999 | 1.000000000 | 1.000000000 |

In programming the barycentric form of $t(\tau)$, one must provide, of course, for the possibility that $\tau - \tau_k = 0$ exactly for some $k$ in order to avoid dividing by zero. This can be done either by returning $f_k$, which is the mathematically correct value, or by replacing $\tau - \tau_k$ by a small machine number such as $10^{-30}$, which (as long as $|f_k|$ is not exceedingly small but $\neq 0$) will again result in the value $f_k$ by virtue of the fact that all other terms in numerator and denominator of (8) will then be negligible in comparison with the term involving $f_k$.

### Recursive Generation of $t(\tau)$

Let $n$ be a power of 2, $n = 2^l$, where $l = 0, 1, 2, \ldots$ . We denote by $t_l(\tau)$ the balanced trigonometric polynomial interpolating at the $2^l$ points $\tau_k := 2^{-l}k$,

$k = 0, 1, \ldots, 2^l - 1$. We split the sum in the numerator of (8) into the sums of the terms where $k$ is even and of those where $k$ is odd:

$$a_l(\tau) := \sum_{\substack{k=0 \\ k\ \text{even}}}^{n-2} f\left(\frac{k}{n}\right)\cot\left[\pi\left(\tau - \frac{k}{n}\right)\right],$$

$$r_l(\tau) := \sum_{\substack{k=1 \\ k\ \text{odd}}}^{n-1} f\left(\frac{k}{n}\right)\cot\left[\pi\left(\tau - \frac{k}{n}\right)\right], \qquad n := 2^l,$$

and similarly for the denominator:

$$b_l(\tau) := \sum_{\substack{k=0 \\ k\ \text{even}}}^{n-2} \cot\left[\pi\left(\tau - \frac{k}{n}\right)\right],$$

$$s_l(\tau) := \sum_{\substack{k=1 \\ k\ \text{odd}}}^{n-1} \cot\left[\pi\left(\tau - \frac{k}{n}\right)\right], \qquad n := 2^l.$$

The numerator then equals $a_l(\tau) - r_l(\tau)$, and the denominator equals $b_l(\tau) - s_l(\tau)$, and it is clear that

$$t_l(\tau) = \frac{a_l(\tau) - r_l(\tau)}{b_l(\tau) - s_l(\tau)}. \tag{10}$$

If we now proceed to the polynomial $t_{l+1}(\tau)$, using twice as many points, then clearly

$$a_{l+1}(\tau) = a_l(\tau) + r_l(\tau),$$
$$b_{l+1}(\tau) = b_l(\tau) + s_l(\tau), \tag{11}$$

and the sums that need to be computed are those with the odd-numbered terms,

$$r_{l+1}(\tau) := \sum_{\substack{k=1 \\ k\ \text{odd}}}^{2^{l+1}-1} f(2^{-l-1}k)\cot[\pi(\tau - 2^{-l-1}k)],$$

$$s_{l+1}(\tau) := \sum_{\substack{k=1 \\ k\ \text{odd}}}^{2^{l+1}-1} \cot[\pi(\tau - 2^{-l-1}k)]. \tag{12}$$

The equations (10) and (11) are correct for $l = 0$ if we put

$$a_0(\tau) := f(0)\cot \pi\tau, \qquad b_0(\tau) := \cot \pi\tau,$$
$$r_0(\tau) := 0, \qquad\qquad s_0(\tau) := 0. \tag{13}$$

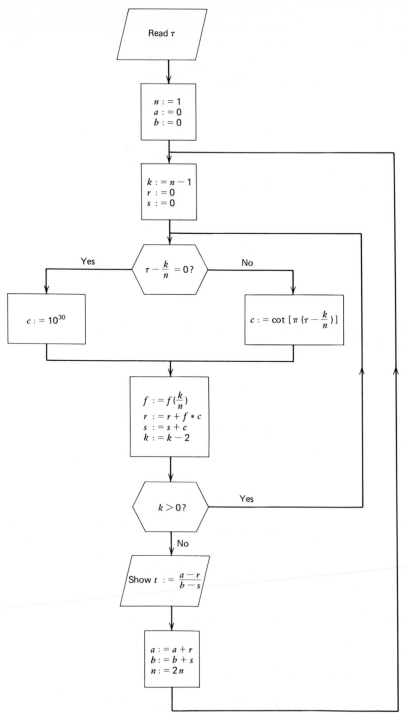

**Fig. 7.2a.** Algorithm of Theorem 7.2b.

Thus, we have

**THEOREM 7.2b:**  *The balanced trigonometric polynomial $t_l(\tau)$ of period 1 interpolating $f$ at the $n = 2^l$ points $\tau_k = 2^{-l}k$, $k = 0, 1, \ldots, 2^l - 1$, is for $l = 0$, 1, 2, ... given by* (10), *where the quantities $a_l$, $b_l$, $r_l$, $s_l$ are initialized by* (13) *and computed recursively by* (11) *and* (12).*

A flow diagram showing the algorithm described in Theorem 7.2b is shown in Figure 7.2a. This algorithm is convenient, especially if one wishes to study the convergence behavior of the sequence $\{t_l(\tau)\}$ for a fixed value of $\tau$.

**DEMONSTRATION 7.2-3:**   For the function $f$ of period 1 shown in Figure 7.2b, the values of the polynomials $t_l(\tau)$ at $\tau = 0.2$ are as follows:

| $n = 2^l$ | $t_l(0.2)$ |
|---|---|
| 1 | 1.000000000 |
| 2 | 0.654508497 |
| 3 | 0.654508497 |
| 8 | 0.572642354 |
| 16 | 0.595367437 |
| . . . . . . . . . . . . . . . . . . . . . . |
| 2048 | 0.599999524 |
| 4096 | 0.599999925 |
| 8192 | 0.600000028 |
| 16384 | 0.600000003 |

The convergence to the exact value $f(0.2) = 0.6$ is clearly evident, although the convergence is slow due to the fact that $f$ has a discontinuous derivative (see §7.4). The demonstration also shows the excellent stability properties of the algorithm.

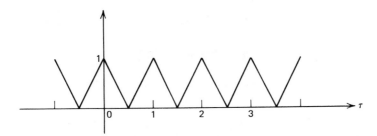

**Fig. 7.2b.**   Sawtooth current.

**Computational Effort**

For $n$ data points, the use of the algorithm of §7.1 requires $n^2$ complex $\mu$ merely to evaluate the coefficients $c_k$ by means of the formulas (12) of §7.1. An additional $n$ complex $\mu$ (or, if the data are real, $n\mu/2$) are then required to evaluate

$t(\tau)$ for a given $\tau$. In contrast to this, the barycentric formulas described here require only $O(n)$ operations to evaluate $t(\tau)$ from scratch, including $n$ evaluations of trigonometric functions. (The latter could be reduced to the evaluation of the single value $\cot \pi\tau$ by using rational recurrence relations for the numbers $\cot[\pi(\tau - k/n)]$.) It thus appears to be clear that the barycentric formula is to be preferred if $t(\tau)$ is to be evaluated only for a few values of $\tau$. This comparison does not take into account the existence of an algorithm, to be explained in §7.5 and §7.6, however, for the evaluation of the $c_k$ which requires a total of only $O(n \operatorname{Log} n)$ operations.

## PROBLEMS

**1** Prove the barycentric representation (9) of the interpolating trigonometric polynomial of degree $m$ for an odd number of $n = 2m + 1$ data points. [Begin by computing the Lagrangian factors $l_k(z)$ satisfying (2).]

**2** Do Problem 2, §7.1, by means of the barycentric formula.

**3** Devise an analog of the algorithm described in Theorem 7.2b for constructing the sequence of trigonometric polynomials, interpolating at $n = 3 \cdot 2^l$, $l = 0, 1, \ldots$, data points.

**4** In Lego City, Western Ontario, the sun in a non-leap year rises according to the following schedule:

| | | |
|---|---|---|
| January | 1 | 8 : 14 AM |
| January | 29 | 7 : 49 |
| February | 26 | 7 : 02 |
| March | 26 | 6 : 06 |
| April | 23 | 5 : 13 |
| May | 21 | 4 : 36 |
| June | 18 | 4 : 29 |
| July | 16 | 4 : 51 |
| August | 13 | 5 : 27 |
| September | 10 | 6 : 03 |
| October | 8 | 6 : 42 |
| November | 5 | 7 : 24 |
| December | 3 | 8 : 02 |

(a) Assuming a year of 364 days (omitting December 31), when does the sun rise on June 21? On December 21? Check various other dates, using your own local data if desired.

(b) How well are the foregoing data interpolated by a trigonometric polynomial of the form

$$t(\tau) = a_0 + a_1 \cos 2\pi\tau,$$

using the data of June 21 and of December 21?

## §7.3  FOURIER SERIES

The reader is presumed to be vaguely familiar with the fact that a function $f$, which is periodic with period 1, can under liberal conditions be represented as a **Fourier series**,

$$f(\tau) = \tfrac{1}{2}a_0 + \sum_{k=1}^{\infty} (a_k \cos 2\pi k\tau + b_k \sin 2\pi k\tau), \tag{1}$$

where the $a_k$ and the $b_k$ are the **Fourier coefficients** of $f$, given by the formulas

$$a_k = 2 \int_0^1 f(\tau)\cos 2\pi k\tau \; d\tau, \tag{2a}$$

$$b_k = 2 \int_0^1 f(\tau)\sin 2\pi k\tau \; d\tau, \tag{2b}$$

$k = 0, 1, 2, \ldots$ . By using Euler's formulas the series (1) may also be written in complex form,

$$f(\tau) = \sum_{k=-\infty}^{\infty} c_k e^{2\pi i k\tau}, \tag{3}$$

where the coefficients $c_k$ may either be calculated from the $a_k$ and the $b_k$ by means of the formulas

$$c_k = \tfrac{1}{2}(a_k - ib_k),$$
$$c_{-k} = \tfrac{1}{2}(a_k + ib_k), \tag{4}$$

$k = 0, 1, 2, \ldots$, or directly from the formula

$$c_k = \int_0^1 f(\tau)e^{-2\pi i k\tau} \; d\tau \tag{5}$$

valid for any integer $k$.

We begin our treatment of the numerical aspects of Fourier series by establishing a link with the method of least squares. Suppose $f$ is a real function with period 1, and $t$ is a (not necessarily balanced) trigonometric polynomial of degree $m$, which we choose to write in the form

$$t(\tau) = \tfrac{1}{2}a_0 + \sum_{k=1}^{m} (a_k \cos 2\pi k\tau + b_k \sin 2\pi k\tau). \tag{6}$$

We intend to approximate $f$ by $t$. For reasons similar to those advanced when discussing the approximation by splines, we measure the goodness of this approximation by the integral over one period of the *square* of the error:

$$\varepsilon := \int_0^1 [f(\tau) - t(\tau)]^2 \, d\tau. \tag{7}$$

Clearly, $\varepsilon$ is a function of the parameters $a_0, \ldots, a_m$ and $b_1, \ldots, b_m$ on which $t$ depends. We now ask: *How do we have to choose these parameters in order to obtain the smallest possible value of $\varepsilon$?*

On expanding the square in (7) it is seen that $\varepsilon$ is a quadratic (and hence everywhere differentiable) function of these parameters. Therefore, for the best possible choice of the parameters, the relations

$$\frac{\partial \varepsilon}{\partial a_j} = 0, \qquad j = 0, 1, \ldots, m;$$

$$\frac{\partial \varepsilon}{\partial b_j} = 0, \qquad j = 1, 2, \ldots, m \tag{8}$$

must hold. Differentiating under the integral sign we have, for instance,

$$\frac{\partial \varepsilon}{\partial a_j} = 2 \int_0^1 [f(\tau) - t(\tau)] \frac{\partial t(\tau)}{\partial a_j} \, d\tau.$$

Using the explicit representation (5) of $t(\tau)$ and observing that for every fixed $\tau$,

$$\frac{\partial t(\tau)}{\partial a_j} = \begin{cases} \frac{1}{2}, & j = 0 \\ \cos 2\pi j\tau, & j = 1, 2, \ldots, m, \end{cases}$$

there follows for $j = 0$

$$\frac{\partial \varepsilon}{\partial a_0} = 2 \int_0^1 \left[ f(\tau) - \tfrac{1}{2}a_0 - \sum_{k=1}^m (a_k \cos 2\pi k\tau + b_k \sin 2\pi k\tau) \right] \tfrac{1}{2} \, d\tau,$$

and for $j = 1, 2, \ldots, m$,

$$\frac{\partial \varepsilon}{\partial a_j} = 2 \int_0^1 \left[ f(\tau) - \tfrac{1}{2}a_0 - \sum_{k=1}^m (a_k \cos 2\pi k\tau + b_k \sin 2\pi k\tau) \right] \cos 2\pi j\tau \, d\tau.$$

The foregoing expressions are greatly simplified by means of the well-known orthogonality relations for the trigonometric functions,

$$\int_0^1 \cos 2\pi k\tau \cos 2\pi j\tau \, d\tau = \begin{cases} 1, & k = j = 0; \\ \frac{1}{2}, & k = j \neq 0; \\ 0, & k \neq j; \end{cases}$$

$$\int_0^1 \cos 2\pi k\tau \sin 2\pi j\tau \, d\tau = 0;$$

$$\int_0^1 \sin 2\pi k\tau \sin 2\pi j\tau \, d\tau = \begin{cases} 0, & k = j = 0; \\ \frac{1}{2}, & k = j \neq 0; \\ 0, & k \neq j. \end{cases}$$

We thus obtain $\dfrac{\partial \varepsilon}{\partial a_0} = \int_0^1 f(\tau) \, d\tau - \frac{1}{2}a_0$,

$$\frac{\partial \varepsilon}{\partial a_j} = 2 \int_0^1 f(\tau)\cos 2\pi j\tau \, d\tau - a_j,$$

$j = 1, 2, \ldots, m$, and, proceeding in a similar fashion,

$$\frac{\partial \varepsilon}{\partial b_j} = 2 \int_0^1 f(\tau)\sin 2\pi j\tau \, d\tau - b_j.$$

On setting these partial derivatives equal to zero, we obtain

$$a_j = 2 \int_0^1 f(\tau)\cos 2\pi j\tau \, d\tau,$$

$$b_j = 2 \int_0^1 f(\tau)\sin 2\pi j\tau \, d\tau,$$

that is, the coefficients $a_j$ and $b_j$ are just the *Fourier coefficients* of $f$. We thus have proved:

**THEOREM 7.3a:** *For any integer $m \geq 0$, the coefficients of the trigonometric polynomial $t(\tau)$ which best approximates a given function $f$ of period 1 in the sense that the mean square $\varepsilon$ of the error is minimized, are just the Fourier coefficients of $f$.*

It is remarkable in this result that the values of the best coefficients $a_k$ and $b_k$ of a trigonometric polynomial of degree $m \geq k$ do not depend on $m$. Thus, whether we want to approximate $f$ by a trigonometric polynomial of degree 5 or 27, the value of $a_2$ will be the same. Expressed differently, for each $m$ the $m$-th partial sum of the Fourier series (1) is identical with the best least squares approximation of $f$ by means of a trigonometric polynomial of degree $m$.

Denoting by $t_m$ the best approximating trigonometric polynomial of degree $m$ and by $\varepsilon_m$ the corresponding error measure (7), the question now arises under what conditions $\varepsilon_m \to 0$ as $m \to \infty$, or even $t_m(\tau) \to f(\tau)$ at every $\tau$. The discussion of these questions we must leave to more purely mathematical treatments of Fourier series. It may suffice here to mention that $\varepsilon_m \to 0$ for every square integrable function $f$ while $t_m(\tau) \to f(\tau)$, for instance, for all functions $f$ of bounded variation. This even includes functions with simple discontinuities provided that the value at the discontinuity is defined as the arithmetic mean of the limits from the left and from the right.

We shall address, however, the question of how to compute the Fourier coefficients of a periodic function that is given empirically. Suppose $n$ is a positive integer, and $f$ is a function of period 1 whose values are known only at the sampling points $\tau_k := k/n$ for integer values of $k$. We write $f_k := f(\tau_k)$. For ease of presentation, we consider the determination of the complex Fourier coefficients $c_k$ from which the real coefficients may be obtained by inverting the relations (4). How, then, should one approximate the integrals

$$c_k := \int_0^1 f(\tau) e^{-2\pi i k \tau} \, d\tau ? \tag{9}$$

The first thing that comes to mind is to evaluate the integral (9) by means of the trapezoidal rule and then to apply Romberg acceleration. The trapezoidal value of (9) is in view of $h = 1/n$

$$\hat{c}_k := \frac{1}{n} \{ \tfrac{1}{2} f(\tau_0) + f(\tau_1) e^{-2\pi i k \tau_1} + \cdots$$

$$+ f(\tau_{n-1}) e^{-2\pi i k \tau_{n-1}} + \tfrac{1}{2} f(\tau_n) e^{2\pi i k \tau_n} \}.$$

In view of $\tau_j = j/n$, we have

$$e^{-2\pi i k \tau_j} = z_1^{-kj}$$

where $z_1 := \exp(2\pi i/n)$ as in §7.1. Evidently $z_1^{-kn} = 1$ in view of $z_1^n = 1$, and since $f(\tau_n) = f(1) = f(0)$ because of the periodicity of $f$, the foregoing reduces to

$$\hat{c}_k = \frac{1}{n} \sum_{j=0}^{n-1} f_j z_1^{-jk}. \tag{10}$$

We recognize that *the trapezoidal values of the complex Fourier coefficients are identical with the coefficients of the complex form of the trigonometric polynomial (5) of §7.1 interpolating $f$ at the sampling points $\tau_k$.*

What will the Romberg acceleration do to the trapezoidal values $\hat{c}_k$? Romberg acceleration is a device to eliminate the constants $C_1$, $C_2$, ... from the expansion

$$T(h) = I + C_1 h^2 + C_2 h^4 + \cdots \tag{11}$$

which holds for the trapezoidal values $T(h)$ for a definite integral

$$I := \int_a^b g(\tau) \, d\tau,$$

provided that the integrand is sufficiently differentiable. We recall that $C_k$ equals a numerical constant times

$$g^{(2k-1)}(b) - g^{(2k-1)}(a). \tag{12}$$

In the present case, $a = 0$, $b = 1$, and the integrand

$$g(\tau) := f(\tau)e^{-2\pi i k \tau}$$

has period one, implying that all the differences (12)—if the derivatives involved exist at all—are zero. Thus Romberg acceleration would only eliminate constants that are already zero, and *the values $\hat{c}_k$ obtained by means of a simple application of the trapezoidal rule are already as good as those obtained in an ordinary integral after an arbitrary number of Romberg accelerations.*

In spite of this excellent result, we cannot generally expect that $\hat{c}_k = c_k$ because the $\hat{c}_k$ depend only on a finite number of sampling values, whereas the exact Fourier coefficients $c_k$ depend on all values of the function $f$. In fact, since

$$z_1^{-j(k+n)} = z_1^{-jk} z_1^{-jn} = z_1^{-jk},$$

it follows from (10) that *the sequence of trapezoidal values $\hat{c}_k$ is periodic with period n,* that is,

$$\hat{c}_{k+n} = \hat{c}_k \tag{13}$$

for all integers $k$, while the exact Fourier coefficients of any integrable function are known to tend to 0 for $k \to \pm\infty$. It follows that the $\hat{c}_k$ can be expected to bear a reasonable resemblance to the $c_k$ only if $|k| < n/2$.

To estimate the error $\hat{c}_k - c_k$ more closely, we assume that $f$ is represented by a complex Fourier series

$$f(\tau) = \sum_{m=-\infty}^{\infty} c_m e^{2\pi i m \tau} \tag{14}$$

which converges absolutely,

$$\sum_{m=-\infty}^{\infty} |c_m| < \infty. \tag{15}$$

By the theory of Fourier series, this will be the case, for instance, if there exist constants $\mu$ and $\alpha > 0$ so that

$$|f(\sigma) - f(\tau)| \le \mu |\sigma - \tau|^\alpha$$

for all real $\sigma$ and $\tau$. Let us now express the sampling values $f_j = f(\tau_j)$ in terms of the Fourier series (14). The exponential in the $m$-th term equals

$$e^{2\pi i m \tau_j} = e^{2\pi i j m/n} = z_1^{jm},$$

hence,

$$f_j = \sum_{m=-\infty}^{\infty} c_m z_1^{jm}$$

for arbitrary integers $j$. Substituting this in (10), we get

$$\hat{c}_k = \frac{1}{n} \sum_{j=0}^{n-1} \left( \sum_{m=-\infty}^{\infty} c_m z_1^{jm} \right) z^{-jk},$$

or on reversing the order of the summations,

$$\hat{c}_k = \frac{1}{n} \sum_{m=-\infty}^{\infty} c_m \sum_{j=0}^{n-1} z_1^{(m-k)j}.$$

Because $z_1^{m-k}$ is an $m$-th root of unity, the inner sum is zero except when $m - k$ is an integral multiple of $n$ in which case the sum equals $n$. We thus find

$$\hat{c}_k = \sum_{l=-\infty}^{\infty} c_{k+ln} \tag{16}$$

or, in a more suggestive notation,

$$\hat{c}_k = \cdots + c_{k-2n} + c_{k-n} + c_k + c_{k+n} + c_{k+2n} + \cdots.$$

From this we readily obtain,

**THEOREM 7.3b:** *The n-point trapezoidal values $\hat{c}_k$ of the Fourier coefficients of a function f represented by an absolutely convergent Fourier series (14) satisfy*

$$\hat{c}_k - c_k = \sum_{l=1}^{\infty} (c_{k+ln} + c_{k-ln}). \tag{17}$$

Equation (17) represents the error of the approximate Fourier coefficients $\hat{c}_k$ in terms of the exact Fourier coefficients $c_k$. Although the exact values of the $c_k$ are not known in general, much is known about their qualitative behavior as $|k| \to \infty$. For instance, it is known (and implicit in our assumption (15)) that $c_k \to 0$ as $|k| \to \infty$ for any integrable function $f$. Making the somewhat stronger assumption that $|c_{-k}| = |c_k|$ and that $|c_k| \to 0$ monotonically as $k \to \infty$, we see that the largest terms on the right of (17) are $c_{k-n}$ if $k > 0$ and $c_{k+n}$ if $k < 0$. If $|k| > n/2$, the error $\hat{c}_k - c_k$ thus may be larger than $|c_k|$, confirming our earlier observation that the approximations $\hat{c}_k$ are useful only if $|k| < n/2$.

If $n$ is even and $k = n/2$, relation (16) permits us to conclude that

$$\hat{c}_{n/2} - (c_{n/2} + c_{-n/2}) = \sum_{l=1}^{\infty} (c_{n/2+ln} + c_{-n/2-ln}),$$

showing that $\hat{c}_{n/2}$ is a good approximation for $c_{n/2} + c_{-n/2}$. If $f$ is real, then $c_{-k} = \bar{c}_k$, and if $f$ is an even function, then $c_{-k} = c_k$. Thus, if $f$ is both real and even, then

$$\tfrac{1}{2}\hat{c}_{n/2} - c_{n/2} = \sum_{l=1}^{\infty} c_{n/2+ln}; \tag{18}$$

that is, $c_{n/2}$ is approximated by $\tfrac{1}{2}\hat{c}_{n/2}$.

The theory of Fourier series also offers information about the *speed* with which the Fourier coefficients tend to 0 for $k \to \pm\infty$. Using (17), this information can be used to assess the speed with which the $\hat{c}_k$ approach $c_k$ if $k$ is fixed and $n \to \infty$. We consider three typical situations.

(I) Assume that $f$ is such that, for some integer $p > 0$, the derivative $f^{(p)}$ is continuous with the exception of finitely many points in each period interval where it has simple jump discontinuities. Then it may be shown that there exists a constant $K > 0$ such that

$$|c_m| \le \frac{K}{|m|^{p+1}} \qquad \text{for } |m| = 1, 2, \dots . \tag{19}$$

Using this estimate on the right of (17) and assuming $-n < k < n$, there follows

$$|\hat{c}_k - c_k| \le K \sum_{l=1}^{\infty} \left| \frac{1}{(ln+k)^{p+1}} + \frac{1}{(ln-k)^{p+1}} \right|$$

$$= \frac{K}{n^{p+1}} \sum_{l=1}^{\infty} \left\{ \frac{1}{\left(l+\dfrac{k}{n}\right)^{p+1}} + \frac{1}{\left(l-\dfrac{k}{n}\right)^{p+1}} \right\}.$$

By drawing the appropriate graph, the reader will convince himself that if $\alpha > -\frac{1}{2}$,

$$\sum_{l=1}^{\infty} \frac{1}{(l+\alpha)^{p+1}} < \int_{1/2}^{\infty} \frac{1}{(x+\alpha)^{p+1}} \, dx = \frac{1}{p} \frac{1}{(\frac{1}{2}+\alpha)^p}.$$

Using this with $\alpha = k/n$ we get, if $-n/2 < k < n/2$,

$$|\hat{c}_k - c_k| \le \frac{K}{pn^{p+1}} \left\{ \frac{1}{\left(\frac{1}{2}+\frac{k}{n}\right)^p} + \frac{1}{\left(\frac{1}{2}-\frac{k}{n}\right)^p} \right\}. \tag{20}$$

For $n \to \infty$, the expression in braces tends to a constant, and the error bound thus tends to zero like $n^{-p-1}$. If the bound correctly reflects the asymptotic behavior of the error, then, on doubling $n$, the error will be approximately reduced by $2^{-p-1}$, and Romberg acceleration of the $\hat{c}_k$ may be feasible.

**DEMONSTRATION 7.3-1:** The function $f$ represented in $[0, 1]$ by $f(\tau) := |2\tau - 1|$ satisfies the foregoing hypotheses for $p = 1$. (See Figure 7.2b.) Indeed, its exact Fourier series is

$$f(\tau) = \frac{1}{2} + \sum_{\substack{m=1 \\ m \text{ odd}}}^{\infty} \frac{4}{\pi^2 m^2} \cos(2\pi m \tau).$$

The $c_m$ with even index $m$ are zero with the exception of $c_0 = \frac{1}{2}$. For odd $m$ we have

$$c_m = c_{-m} = \frac{2}{\pi^2 m^2};$$

the estimate (20) is thus seen to hold for $K = 2\pi^{-2}$. We compute $\hat{c}_k$ numerically for $k = 0, 1, 3, 5, 7$ and for numbers $n$ of sampling points that are powers of 2. The following values result:

| $n$ | $\hat{c}_0$ | $\hat{c}_1$ | $\hat{c}_3$ | $\hat{c}_5$ | $\hat{c}_7$ |
|---|---|---|---|---|---|
| 4 | 0.50000000 | 0.25000000 | 0.25000000 | 0.25000000 | 0.25000000 |
| 8 | 0.50000000 | 0.21338835 | 0.03661165 | 0.03661165 | 0.21338835 |
| 16 | 0.50000000 | 0.20526674 | 0.02531116 | 0.01130049 | 0.00812161 |
| 32 | 0.50000000 | 0.20329467 | 0.02317831 | 0.00878935 | 0.00485303 |
| 64 | 0.50000000 | 0.20280520 | 0.02267929 | 0.00827043 | 0.00430224 |
| 128 | 0.50000000 | 0.20268306 | 0.02255655 | 0.00814651 | 0.00417649 |
| $\infty$ | 0.50000000 | 0.20264237 | 0.02251582 | 0.00810569 | 0.00413556 |

The numbers in the above table invite the following comments.

(i)   The coefficients $\hat{c}_k$, like their analytical counterparts $c_k$, are mathematically real. If they are computed by the Horner scheme as outlined in §7.1, the computed values of the imaginary parts will not automatically be zero. These values (not shown here) thus may be considered a measure of numerical stability. It turns out that, in the present example at any rate, the stability is excellent.

(ii)   The values of $\hat{c}_0$ are seen to be exact for all $n$. This is in accordance with the error formula (17), because in the present example all $c_{ln} = 0$ if $n$ is even and $l \neq 0$.

(iii)   The repeated values of $\hat{c}_k$ in the rows for $n = 4$ and $n = 8$ illustrate the fact that the sequence of the $\hat{c}_k$ is periodic with period $n$, which in view of $\hat{c}_{-k} = \hat{c}_k$ implies that $\hat{c}_k = \hat{c}_{n-k}$.

(iv)   A look at any column illustrates the fact that if $n$ is doubled, the error of $\hat{c}_k$ is approximately reduced by 75%, in accordance with the theory. Denoting by $\hat{c}_{k,m}$ the values of $\hat{c}_k$ computed with $n = 2^m$ sampling points, we thus may compute more accurate values $\hat{c}'_{k,m}, \hat{c}''_{k,m}, \ldots$ by the Romberg algorithm:

$$\hat{c}'_{k,m} = \tfrac{1}{3}(4\hat{c}_{k,m} - \hat{c}_{k,m-1}),$$

$$\hat{c}''_{k,m} = \tfrac{1}{15}(16\hat{c}_{k,m} - \hat{c}_{k,m-1}), \ldots.$$

Below are shown the values thus obtained for $k = 1$:

| $m$ | $\hat{c}_{1,m}$ | $\hat{c}'_{1,m}$ | $\hat{c}''_{1,m}$ |
|---|---|---|---|
| 2 | 0.25000000 | | |
| 3 | 0.21338835 | 0.20118447 | |
| 4 | 0.20526674 | 0.20255954 | 0.20265121 |
| 5 | 0.20329467 | 0.20263731 | 0.20264249 |
| 6 | 0.20280520 | 0.20264204 | 0.20264236 |
| 7 | 0.20268306 | 0.20264235 | 0.20264237 |

It is seen that $\hat{c}''_{1,5}$ has an error $< 10^{-7}$, while the value $\hat{c}_{1,7}$, which uses four times as many data points, has an error in excess of $4 \cdot 10^{-5}$.

(II)   If $f$ is analytic (that is, can be expanded in a Taylor series) at every real point $\tau$, then it is shown in complex analysis that constants $K > 0$ and $q$, $0 < q < 1$, exist so that

$$|c_m| \leq Kq^m \qquad \text{for all integers } m. \tag{21}$$

If this estimate is used in (17) and if it is assumed that $-n < k < n$, then there follows

$$|\hat{c}_k - c_k| \leq K \sum_{l=1}^{\infty} (q^{nl+k} + q^{nl-k}) = Kq^n \frac{q^k + q^{-k}}{1 - q^n}. \tag{22}$$

For large $n$ this error bound asymptotically behaves like const $\cdot q^n$. Thus, at each doubling of $n$, the dependence of the error bound on $n$ is squared. If the error bound correctly reflects the asymptotic behavior of the error, we find, denoting by $\hat{c}_{k,m}$ the value of $\hat{c}_k$ computed with $n = 2^m$ data points, that the limit of

$$\frac{|\hat{c}_{k,m+1} - c_k|}{|\hat{c}_{k,m} - c_k|^2}$$

exists, which is typical for quadratic convergence. Romberg acceleration in this situation will have no effect.

**DEMONSTRATION 7.3-2:**    The foregoing conditions are met by

$$f(\tau) := \frac{1}{\sqrt{1 + \varepsilon \cos 2\pi\tau}}$$

if $|\varepsilon| < 1$. The exact Fourier coefficients are real and satisfy $c_{-k} = c_k$. They can be expressed analytically as power series in $\varepsilon$, although not in closed form. The following numerical values result for $\varepsilon = 0.9$:

| $n$ | $\hat{c}_0$ | $\hat{c}_1$ | $\hat{c}_2$ | $\hat{c}_3$ | $\hat{c}_4$ |
|---|---|---|---|---|---|
| 4 | 1.47193848 | −0.60920035 | 0.47193848 | −0.60920035 | 1.47193848 |
| 8 | 1.34599790 | −0.45957292 | 0.23596924 | −0.14962743 | 0.12594058 |
| 16 | 1.33206002 | −0.44360501 | 0.21322178 | −0.11296730 | 0.06297029 |
| 32 | 1.33182165 | −0.44333628 | 0.21285397 | −0.11240530 | 0.06206682 |
| 64 | 1.33182155 | −0.44333618 | 0.21285382 | −0.11240509 | 0.06206648 |

The much more rapid convergence as $n \to \infty$ is clearly evident. The entries for $n = 64$ are correct in all cases. The comments (i) and (iii) made on Demonstration 7.3-1 here are likewise appropriate. The value of $\hat{c}_4$ given for $n = 8$ illustrates (18).

    (III)    If the function $f$ can be extended to complex values of $\tau$ so that the resulting function is analytic at all complex $\tau$, then the estimate (21) holds (with an appropriate constant $K = K(q)$) for *every* positive $q$, no matter how small. An even more rapid convergence of the $\hat{c}_k$ to the $c_k$ as $n \to \infty$ may then be expected.

**DEMONSTRATION 7.3-3:**    The function

$$f(\tau) := e^{ix \cos 2\pi\tau},$$

where $x$ is a parameter, meets these specifications. The exact Fourier coefficients $c_m$ for any integer $m$ are related to the Bessel functions $J_m(x)$ by means of

$$c_m = i^m J_m(x).$$

Because the $J_m$ cannot be expressed in elementary terms, Fourier analysis thus provides a method to compute these functions. The following are some values of $i^{-m}\hat{c}_m$, and thus approximations to $J_m(x)$, for $x = 4.5$:

| $n$ | $\hat{c}_0$ | $i^{-1}\hat{c}_1$ | $i^{-2}\hat{c}_2$ | $i^{-3}\hat{c}_3$ | $i^{-4}\hat{c}_4$ |
|---|---|---|---|---|---|
| 4 | 0.39460210 | −0.48876506 | 0.60539790 | 0.48876506 | 0.39460210 |
| 8 | −0.30229121 | −0.25865791 | 0.30269895 | 0.23010715 | 0.69689331 |
| 16 | −0.32054248 | −0.23106054 | 0.21784968 | 0.42469976 | 0.34844666 |
| 32 | −0.32054251 | −0.23106043 | 0.21784898 | 0.42470397 | 0.34842298 |

Here the entries for $n = 32$ are correct to all digits given. Thus, the fast convergence is clearly evident.

# PROBLEMS

**1** Let $f$ denote the function of period 1 whose graph is shown in Figure 7.3a.

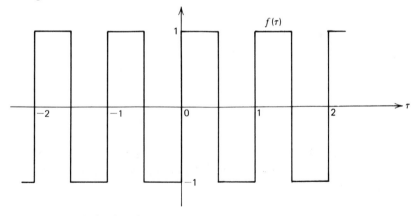

**Fig. 7.3a.** Periodic function.

  **(a)** Compute the exact Fourier coefficients of $f$, showing that $a_k = 0$ for all $k$,

$$b_k = \begin{cases} \dfrac{4}{\pi k}, & k \text{ odd} \\ 0, & k \text{ even} \end{cases}$$

and hence that

$$c_k = -\frac{2i}{\pi k}, \quad k \text{ odd}.$$

(b) Compute the discrete Fourier coefficients *analytically*, using $n = 2m$ data points. Result:

$$\hat{c}_k = \begin{cases} -\dfrac{2i}{n} \cot \dfrac{k\pi}{n}, & k \text{ odd}, \\ 0, & k \text{ even}. \end{cases}$$

(c) Verify that for each fixed $k$,

$$\hat{c}_k \to c_k \quad \text{as} \quad n \to \infty.$$

(d) Show theoretically that Romberg acceleration will work in this case.

(e) Verify experimentally that Romberg acceleration will work.

2 Let $w$ be a complex number, $|w| > 1$.

(a) Show that the complex Fourier coefficients of the periodic function

$$f(\tau) := \frac{1}{w - e^{2\pi i \tau}}$$

are given by

$$c_k = \begin{cases} w^{-k-1}, & k \geq 0, \\ 0, & k < 0. \end{cases}$$

[Write $f(\tau) = 1/w(1 - w^{-1}e^{2\pi i \tau})$ and use the geometric series.]

(b) Show that the discrete Fourier coefficients of $f$, using $n$ data points, are

$$\hat{c}_k = \begin{cases} w^{-k-1} \dfrac{1}{1 - w^{-n}}, & 0 \leq k < n; \\ w^{-k-n-1} \dfrac{1}{1 - w^{-n}}, & -n \leq k < 0, \end{cases}$$

and comment on the result.

(c) Compute some discrete coefficients numerically for $w = 2$, and verify the result of (b).

**3**  Let $f$ be defined, continuous, and of bounded variation on the interval $[0, 1]$, $f(0) = f(1) = 0$. By continuing $f$ as an odd function of period 2, show that $f$ can be expanded in a **Fourier sine series**,

$$f(\tau) = \sum_{k=1}^{\infty} b_k \sin k\pi\tau,$$

with the coefficients

$$b_k := 2 \int_0^1 f(\tau)\sin k\pi\tau \, d\tau, \qquad k = 1, 2, \dots .$$

**4**  Find the Fourier sine series of the function depicted in Figure 7.3b.

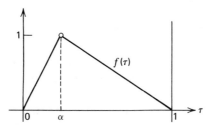

**Fig. 7.3***b*.  Plucked string.

**5  Motion of a Plucked String.**  Let a taut string of length 1 be fixed at its endpoints, and let $u(x, t)$ denote the displacement of the string at position $x$ ($0 \le x \le 1$) and time $t > 0$. It is shown in advanced calculus that if the string is released without initial velocity from its position at $t = 0$, then

$$u(x, t) = \sum_{k=1}^{\infty} b_k \sin \pi kx \cos \pi k\omega t, \qquad (23)$$

where

$$u(x, 0) = \sum_{k=1}^{\infty} b_k \sin \pi kx,$$

and where $\omega$ is a constant, depending on the physical properties of the string. The human ear will hear (23) as a superposition of vibrations of frequencies $f = k\omega/2$, $k = 1, 2, \dots$, called the $k$-th harmonics of the string. (See Figure 7.3c.) Assume now that the string is plucked. Its initial position then will resemble the triangle shape shown in Figure 7.3b, and the $b_k$ may be calculated as in Problem 4. Where should the string be plucked if the $k$-th harmonic is to be zero?

**Fig. 7.3c.**   Harmonics.

## §7.4   THE ERROR OF TRIGONOMETRIC INTERPOLATION

Two mathematical errors are committed in the numerical simulation of the Fourier series of a periodic function $f$:

(i)  the exact Fourier coefficients $c_k$ are replaced by their discretized forms $\hat{c}_k$;

(ii)  the Fourier series is truncated, that is, replaced by a finite partial sum.

The problem posed is to furnish a mathematically reliable appraisal of the cumulative effect of these two errors.

Suppose $n$ is even, $n = 2m$, and suppose that the $\hat{c}_k$ are computed by the trapezoidal rule as in (10), §7.3. Since, as we have seen, the discrete coefficients $\hat{c}_k$ are periodic with period $n$, they cannot be expected to furnish reasonable approximations to the exact coefficients $c_k$ unless $|k| < m$; moreover, $\hat{c}_m = \hat{c}_{-m}$ in a sense approximates $c_m + c_{-m}$. Thus, if in the exact Fourier series

$$f(\tau) = \sum_{k=-\infty}^{\infty} c_k e^{2\pi ik\tau} \tag{1}$$

the $c_k$ are replaced by the $\hat{c}_k$, there is no point in extending the summation beyond $k = \pm m$; moreover, the terms $k = \pm m$ need to be multiplied by $\frac{1}{2}$. Thus, the reasonable numerical equivalent to (1) is the function

$$t(\tau) := \sum_{k=-m}^{m}{}' \hat{c}_k e^{2\pi ik\tau}, \tag{2}$$

where the dash indicates that the terms with $k = \pm m$ bear the factor $\frac{1}{2}$.

The question as to the error committed in approximating $f(\tau)$ by $t(\tau)$ is very easily answered if $\tau$ is one of the points $\tau_j = j/n$ where $j$ is an integer. A glance at the formulas of §7.1 will show that the function (2) is identical with the trigonometric polynomial interpolating $f$ at the points $\tau_j$. Thus, the approximation of $f(\tau)$ by $t(\tau)$ is exact when $\tau = \tau_j$, and the errors (i) and (ii) mentioned above just cancel each other at these points.

The problem remains, however, to appraise $|f(\tau) - t(\tau)|$ for values $\tau \neq \tau_j$. We discuss two methods.

## First Method

The basic idea is to reduce the problem to a problem of estimating the error in ordinary, polynomial interpolation where it can be handled by Theorem 5.3. This can be done, in principle, by introducing the variable $x := \cos 2\pi\tau$. Two difficulties appear: (a) To each value of $x$ in $(-1, 1)$, there correspond two values of $\tau$ in $(0, 1)$; (b) while it is true that the functions $\cos 2\pi k\tau$ are polynomials (of degree $k$) in $x = \cos 2\pi\tau$, the same is not the case for the functions $\sin 2\pi k\tau$. The difficulties may be resolved by considering separately the even and the odd part of the trigonometric interpolating polynomial (2).

Let $t(\tau)$ be defined by (2). Then the even function

$$t^+(\tau) := \tfrac{1}{2}\{t(\tau) + t(-\tau)\}$$

$$= \tfrac{1}{2}a_0 + \sum_{k=1}^{m} a_k \cos 2\pi k\tau$$

$(a_k := c_k + c_{-k})$ interpolates the even function

$$f^+(\tau) := \tfrac{1}{2}\{f(\tau) + f(-\tau)\}$$

at the $m + 1$ points $\tau = \tau_j$ where $j = 0, 1, \ldots, m$, and hence automatically also at the points $\tau = -\tau_j$. It follows that the polynomial of degree $m$ in $x$,

$$t^{+*}(x) := t^+(\tau) \qquad (x = \cos 2\pi\tau)$$

$$= \tfrac{1}{2}a_0 + \sum_{k=1}^{m} a_k T_k(x)$$

interpolates

$$f^{+*}(x) := f^+(\tau) \qquad (x = \cos 2\pi\tau)$$

at the $m + 1$ points

$$x_j := \cos 2\pi\tau_j, \qquad j = 0, 1, \ldots, m,$$

and the error $f^{+*}(x) - t^{+*}(x)$ can be estimated in terms of the $(m + 1)$st derivative of $f^{+*}$ by Theorem 5.3. Similarly, the odd part of $t$,

$$t^-(\tau) := \tfrac{1}{2}\{t(\tau) - t(-\tau)\}$$

$$= \sum_{k=1}^{m-1} b_k \sin 2\pi k\tau$$

$(b_k := i(c_k - c_{-k}))$ interpolates the odd part of $f$,

$$f^-(\tau) := \tfrac{1}{2}\{f(\tau) - f(-\tau)\},$$

at the $m - 1$ points $\tau = \tau_j$, $j = 1, 2, \ldots, m - 1$. By symmetry, $t^-$ interpolates also at $\tau = -\tau_j$; furthermore, it trivially interpolates at $\tau = \tau_0$ and $\tau = \tau_m$. In order to obtain a polynomial in $x$, we consider

$$t^{-*}(x) := \frac{1}{\sin 2\pi\tau} t^-(\tau) \qquad (x = \cos 2\pi\tau)$$

$$= \sum_{k=1}^{m-1} b_k U_k(x),$$

where

$$U_k(x) := \frac{\sin 2\pi k\tau}{\sin 2\pi\tau}, \qquad (x = \cos 2\pi\tau)$$

the so-called Chebyshev polynomial of the second kind, in fact is a polynomial of degree $k - 1$. (See Problem 10, §3.1.) Thus, $t^{-*}$ is a polynomial of degree $m - 2$; it interpolates the function

$$f^{-*}(x) := \frac{1}{\sin 2\pi\tau} f^-(\tau) \qquad (x = \cos 2\pi\tau)$$

at the $m - 1$ points $x_j = \cos 2\pi\tau_j$, $j = 1, 2, \ldots, m - 1$, and the error committed in this interpolation can again be appraised, in terms of the derivative of order $m - 1$ of $f^{-*}$, by Theorem 5.3.

## Second Method

From the point of practicability, the method just discussed has two serious disadvantages: (i) Unless $n$ is very small or $f$ is especially contrived, the estimates obtained from it are clumsy due to complex expressions for the high derivatives involved; (ii) the method does not apply at all to functions that do not possess the required number of derivatives. These disadvantages are avoided in the following approach, which works directly with the Fourier series (1) of $f$, assumed to be absolutely convergent.

Let

$$\hat{f}(\tau) := \sum_{k=-m}^{m} {}' c_k e^{2\pi i k\tau} \tag{3}$$

be the $m$-th partial sum of the exact Fourier series of $f$, where the dash has the same meaning as in (2). We clearly have

$$|f(\tau) - t(\tau)| = |f(\tau) - \hat{f}(\tau)| + |\hat{f}(\tau) - t(\tau)|.$$

Concerning the first term on the right, there clearly holds

$$\left| f(\tau) - \hat{f}(\tau) \right| = \left| \sum_{|k| \geq m}' c_k e^{2\pi i k \tau} \right| \leq \sum_{|k| \geq m}' |c_k|; \tag{4}$$

Here again, the dash signifies that the terms $k = \pm m$ are taken with the factor $\frac{1}{2}$. To obtain an estimate for the second term, we use Theorem 7.3b to obtain

$$\left| \hat{f}(\tau) - t(\tau) \right| = \left| \sum_{k=-m}^{m}{}' (c_k - \hat{c}_k) e^{2\pi i k \tau} \right|$$

$$\leq \sum_{k=-m}^{m}{}' \left| \sum_{l=1}^{\infty} (c_{k+ln} + c_{k-ln}) \right|$$

$$\leq \sum_{k>m} |c_k| + \sum_{k<-m} |c_k| + \tfrac{1}{2}|c_m| + \tfrac{1}{2}|c_{-m}|$$

$$\leq \sum_{|k| \geq m}' |c_k|. \tag{5}$$

From (4) and (5) we now have the following very simple result:

**THEOREM 7.4:** *If $f$ is represented by the absolutely convergent Fourier series* (1) *and if $n = 2m$, then the balanced trigonometric polynomial* (2) *of degree $m$ that interpolates $f$ at the points $\tau = \tau_j = j/n$ ($j = 0, 1, \ldots, n - 1$) for all real $\tau$ satisfies*

$$| f(\tau) - t(\tau) | \leq 2 \sum_{|k| \geq {}_1 m}' |c_k|. \tag{6}$$

As in the case of Theorem 7.3b, the estimate (6) enables one to assess the quality of approximation by the interpolating trigonometric polynomial on the basis of qualitative information concerning the smoothness of $f$. For instance, if $f$ has simple jump discontinuities in the $p$-th derivative and consequently the $c_k$ satisfy Equation (19) of §7.3, then

$$| f(\tau) - t(\tau) | \leq \frac{2K}{m^p} \left( \frac{1}{m} + \frac{2}{p} \right) \tag{7}$$

implying a geometric convergence of $t(\tau)$ to $f(\tau)$ as the number of interpolating points is repeatedly doubled. (For an example, see Demonstration 7.2-3.) If $f$ is analytic, and consequently the $c_k$ satisfy Equation (21) of §7.3, then

$$| f(\tau) - t(\tau) | \leq 2Kq^m \frac{1+q}{1-q}, \tag{8}$$

and on doubling the number of interpolating points quadratic convergence is obtained.

**DEMONSTRATION 7.4-1:**   For the function

$$f(\tau) := \frac{1}{\sqrt{1 + \varepsilon \cos 2\pi\tau}},$$

we compute the trigonometric interpolating polynomials for $n = 2^l$ $(l = 0, 1, 2, \ldots)$ points, using the doubling algorithm described in Theorem 7.2b. The following values are obtained at $\tau = 0.2$ for $\varepsilon = 0.9$:

| $n$ | $t(0.2)$ |
|-----|----------|
| 1 | 0.72547625 |
| 2 | 1.56737043 |
| 4 | 0.71362571 |
| 8 | 0.96118001 |
| 16 | 0.88989875 |
| 32 | 0.88438477 |
| 64 | 0.88453488 |
| 128 | 0.88453492 |
| 256 | 0.88453492 |
| $\infty$ | 0.88453492 |

(Digits underlined are in error.) From about $n = 16$ onward, the approximate doubling of the numbers of the correct digits can be clearly observed.

## PROBLEMS

1   Make an experimental study of the convergence of the trigonometric interpolating polynomials for the function $f(0) = f(1) := 0$,

$$f(\tau) := \begin{cases} 1, & 0 < \tau < \frac{1}{2}, \\ -1, & \frac{1}{2} < \tau < 1, \end{cases}$$

periodically continued. (The Fourier series does not converge absolutely in this case.) What happens near the points of discontinuity?

2   Kepler's equation

$$M = E - \varepsilon \sin E$$

according to Demonstration 2.1-4 for each value of the mean anomaly $M$ and for each eccentricity $\varepsilon$ so that $|\varepsilon| < 1$ has a unique solution $E$ which we now call $E(M, \varepsilon)$.

(a)   Show that for each $\varepsilon$ so that $|\varepsilon| < 1$, the function

$$f(M) := E(M, \varepsilon) - M$$

is periodic with period $2\pi$.

   (b)   Show that the function $f$ is odd.

   (c)   The function $f$ being periodic, we may try to interpolate it by a trigonometric polynomial using $n$ data points. To this end, we calculate the values of $f$ at the points $M = 2\pi k/n$, $k = 1, 2, \ldots, n$, using the algorithm of Demonstration 2.1-4. The resulting trigonometric interpolating polynomial may then be evaluated by the barycentric formula. Assuming $\varepsilon = 0.2$, carry out this process for $n = 4, 8, 16$ and show that $n = 16$ points suffice to approximate $f$, and hence $E$, with an error not exceeding about $10^{-6}$.

---

# §7.5   THE EXISTENCE OF A "FAST" FOURIER TRANSFORM

The central operation in numerical Fourier analysis is the forming of the sums

$$y_k := \frac{1}{n} \sum_{j=0}^{n-1} x_j w_n^{-kj} \tag{1}$$

for $k = 0, 1, \ldots, n-1$, where $n$ is a positive integer, $x_0, x_1, \ldots, x_{n-1}$ are given (real or complex) numbers, and

$$w_n := \exp\left(\frac{2\pi i}{n}\right)$$

is a certain $n$-th root of unity. The sums (1) occur in the discrete computation of Fourier coefficients (see §7.3), in the construction of the interpolating trigonometric polynomial (see §7.1), as well as in certain applications to be discussed in subsequent sections.

   Generally, these operations may be thought of as various forms of Fourier *analysis*. The **spectrum**, that is, the sequence of coefficients $\{y_k\}$, is computed from the **data**, that is, from the sequence $\{x_j\}$. (In discrete Fourier analysis, both the data and the spectrum are periodic sequences.) The Fourier **synthesis**, the converse operation, also requires operations of the general form (1): If say, $n$ is even, $n = 2m$, and if the balanced trigonometric polynomial

$$t(\tau) = \sum_{k=-m}^{m}{}' c_k e^{2\pi i k \tau} \tag{2}$$

(terms $k = \pm m$ have factor $\frac{1}{2}$) is to be evaluated at $\tau = \tau_j = j/n$, this amounts to forming

$$t(\tau_j) = \sum_{k=-m}^{m}{}' c_k w_n^{kj},$$

which if $c_{k+n} := c_k$ in view of $w_n^{(k+n)j} = w_n^{kj}$ is identical with

$$t(\tau_j) = \sum_{k=0}^{n-1} c_k w_n^{kj},$$

and these sums are again of the form (1), except for the factor $1/n$ and the change of the sign in the exponent of $w_n$.

The operations (1) may be written in compact form by introducing the vectors

$$\mathbf{x} := \begin{pmatrix} x_0 \\ x_1 \\ \vdots \\ x_{n-1} \end{pmatrix}, \qquad \mathbf{y} := \begin{pmatrix} y_0 \\ y_1 \\ \vdots \\ y_{n-1} \end{pmatrix}$$

and the $n \times n$ matrix

$$\mathbf{F}_n := (f_{ij})$$

with the $(i, j)$-element

$$f_{ij} := \frac{1}{n} w_n^{-ij} \tag{3}$$

where $i, j = 0, 1, \ldots, n - 1$. Then obviously,

$$\mathbf{y} = \mathbf{F}_n \mathbf{x} \tag{4}$$

while the converse operation is given by

$$\mathbf{x} = n \overline{\mathbf{F}}_n \mathbf{y}, \tag{5}$$

where by $\overline{\mathbf{F}}_n$ we denote the matrix whose elements are the conjugates of those of $\mathbf{F}_n$. The matrix $\mathbf{F}_n$ will be called the (discrete) **Fourier operator** of order $n$, and the vector $\mathbf{y}$ defined by (4) is called the **discrete Fourier transform** of the vector $\mathbf{x}$.

By any conventional operations count, the construction of the vector $\mathbf{y}$ from a given data vector $\mathbf{x}$ requires $n^2$ complex multiplications $\mu$ and a similar number of additions $\alpha$. Certain modern applications of discrete Fourier analysis, such as time series analysis or digital image processing, require values of $n$ as large as $2^{10}$ or $2^{12}$. For such $n$ the work required to construct $\mathbf{y}$ from $\mathbf{x}$ is already quite substantial, and it becomes prohibitive if this operation is to be carried out many times. It is, therefore, a fact of paramount importance that there exists an algorithm which computes the discrete Fourier transform of a given vector $\mathbf{x}$ in a number of operations which is only a small multiple of (instead the square of) the number of components of $\mathbf{x}$. This algorithm, which was discovered in 1965 by J. W. Cooley and J. W. Tukey and which today is called the **Fast Fourier Transform**, is particularly effective if $n$ is a power of 2. The required number of multiplications then is only $\frac{1}{2}n \operatorname{Log}_2 n$; thus, for instance, if $n = 2^{10} = 1024$, only $5120\mu$ are required to compute $\mathbf{y} = \mathbf{F}_n \mathbf{x}$ in place of the more than $1,000,000\mu$ that are necessary when the formulas (1) are carried out literally.

The basic idea of the Fast Fourier Transform is to reduce the computation of the transform of a vector of size $n$ to the computation of the transforms of two vectors of size $\frac{1}{2}n$. Thus, to begin with, assume that $n$ is even, $n = 2m$. Let $\mathbf{x}$ be any vector of size $n$,

$$\mathbf{x} = \left(x_0, x_1, \ldots, x_{n-1}\right)^T.$$

Along with $\mathbf{x}$ we consider the two vectors of size $m = n/2$,

$$\mathbf{x}' := \left(x_0, x_2, \ldots, x_{2m-2}\right)^T,$$
$$\mathbf{x}'' := \left(x_1, x_3, \ldots, x_{2m-1}\right)^T.$$

We consider the problem of computing the $n$ components $y_0, y_1, \ldots, y_{n-1}$ of the vector

$$\mathbf{y} := \mathbf{F}_n \mathbf{x} \tag{6}$$

under the assumption that the discrete Fourier transforms

$$\begin{aligned} \mathbf{y}' &:= \mathbf{F}_m \mathbf{x}', \\ \mathbf{y}'' &:= \mathbf{F}_m \mathbf{x}'' \end{aligned} \tag{7}$$

are already known; that is, that the numbers

$$\begin{aligned} y_j' &= \frac{1}{m} \sum_{k=0}^{m-1} x_{2k} w_m^{-kj}, \\ y_j'' &= \frac{1}{m} \sum_{k=0}^{m-1} x_{2k+1} w_m^{-kj} \end{aligned} \tag{8}$$

are available for $j = 0, 1, \ldots, m-1$.

By the definition of the discrete Fourier transform,

$$y_j = \frac{1}{n}\sum_{k=0}^{n-1} x_k w_n^{-kj}.$$

Summing separately with respect to the even and to the odd indices,

$$y_j = \frac{1}{n}\left\{\sum_{k=0}^{m-1} x_{2k} w_n^{-2kj} + \sum_{k=0}^{m-1} x_{2k+1} w_n^{-(2k+1)j}\right\}.$$

In view of $w_m = w_n^2$, we have $w_n^{-2kj} = w_m^{-kj}$, hence

$$y_j = \frac{1}{n}\left\{\sum_{k=0}^{m-1} x_{2k} w_m^{-kj} + w_n^{-j}\sum_{k=0}^{m-1} x_{2k+1} w_m^{-kj}\right\}.$$

Using (8) this, in view of $n = 2m$, yields

$$y_j = \tfrac{1}{2}(y_j' + w_n^{-j}y_j'')$$

for $j = 0, 1, \ldots, m - 1$. If $j$ is one of the values $m, m + 1, \ldots, n - 1$, we express it in the form $m + j$ where $j = 0, 1, \ldots, m - 1$ and use the relations

$$w_m^{-k(m+j)} = w_m^{-kj}, \qquad w_n^{-(m+j)} = -w_n^{-j}$$

to obtain

$$y_{m+j} = \tfrac{1}{2}(y_j' - w_n^{-j}y_j'').$$

We thus obtain the following **duplication theorem**, which is fundamental for the Cooley-Tukey algorithm but which in essence was already known to Gauss:

**THEOREM 7.5a:** *With the notation introduced above, the components of* **y** *are expressed in terms of the components of the vectors* **y**′ *and* **y**″ *by means of the formulas:*

$$y_j = \tfrac{1}{2}(y_j' + w_n^{-j}y_j''),$$
$$y_{j+m} = \tfrac{1}{2}(y_j' - w_n^{-j}y_j''), \tag{9}$$

$j = 0, 1, \ldots, m - 1.$

We conclude that if the numbers $y_j'$ and $y_j''$ are already known, the computation of the $n$ numbers $y_j$ requires, in addition to $n\alpha$ and $n$ divisions by 2 (which are generally neglected in this kind of operations count), the forming of the $m$ nontrivial products

$$w_n^{-j}y_j'', \qquad j = 0, 1, \ldots, m - 1.$$

(One might even omit the multiplication by $w_n^0 = 1$ from the operations count by virtue of its triviality; however, to do so would seem pedantic because in a computer program to fuss about the special case $j = 0$ would, even for moderate $m$, cost at least as much time as to simply carry out the trivial product.)

If $n$ is a power of 2, $n = 2^l$, it is clear that the duplication theorem may be applied recursively until one arrives at Fourier transforms of vectors of size 1 which require no multiplications since $\mathbf{F}_1$ is the identity. Denoting by $\mu(l)$ the number of complex multiplications required to evaluate the Fourier transform of a vector of size $n = 2^l$ by the foregoing method, our considerations show that

$$\mu(l + 1) = 2\mu(l) + 2^l. \tag{10}$$

To solve this difference equation under the initial condition $\mu(0) = 0$, we try to construct a particular solution (see §3.1) of the form

$$\mu(l) = al2^l \tag{11}$$

with an unknown constant $a$. This will satisfy (10) for all $l$ if

$$a(l + 1)2^{l+1} = 2al2^l + 2^l;$$

that is, if $a = \frac{1}{2}$. We thus see that

$$\mu(l) = \tfrac{1}{2}l2^l$$

or, expressed in terms of $n = 2^l$,

$$\mu(l) = \tfrac{1}{2}n \, \mathrm{Log}_2 \, n.$$

This proves

**THEOREM 7.5b:**  *If $n$ is a power of 2, the discrete Fourier transform $\mathbf{y} = \mathbf{F}_n \mathbf{x}$ can be evaluated in merely $\tfrac{1}{2}n \, \mathrm{Log}_2 \, n$ complex multiplications.*

## PROBLEMS

1   Calculate the discrete Fourier transforms of the vectors $\mathbf{x} = (x_k)$ where
   (a)   $x_k = 1$, $k = 0, 1, \ldots, n - 1$;
   (b)   $x_0 = 1$, $x_k = 0$, $k = 1, 2, \ldots, n - 1$;
   (c)   $x_k = (-1)^k$, $k = 0, 1, \ldots, n - 1$ ($n$ even).

**2**  Show that the discrete Fourier transform of the vector $\mathbf{x} = (x_k)$, where $x_k = k \ (k = 0, 1, \ldots, n-1)$ is

$$y_m = \begin{cases} \dfrac{n-1}{2}, & m = 0; \\[2ex] -\dfrac{1}{2} + \dfrac{i}{2} \cot \dfrac{m}{n}, & m \neq 0. \end{cases}$$

**3**  Suppose $n = 3m$, and let

$$y_j' := \frac{1}{m} \sum_{k=0}^{m-1} x_{3k} w_m^{-kj},$$

$$y_j'' := \frac{1}{m} \sum_{k=0}^{m-1} x_{3k+1} w_m^{-kj},$$

$$y_j''' := \frac{1}{m} \sum_{k=0}^{m-1} x_{3k+2} w_m^{-kj}.$$

Show how to compute the discrete Fourier transform of the sequence $x_k$ from $y_j'$, $y_j''$, $y_j'''$, and estimate the number of complex multiplications required to compute the discrete Fourier transform of a sequence of length $n = 3^l$ by using these formulas recursively.

## §7.6  AN IMPLEMENTATION OF THE FAST FOURIER TRANSFORM*

Although the basic idea of the Fast Fourier Transform is simple, its implementation is no trivial matter, especially if attention is paid to the problem of keeping the storage requirements to a minimum. Here we describe an implementation that leads to an ultimately simple program and which requires essentially no storage beyond that required by the vector $\mathbf{x}$ itself.

We begin by taking a closer look at the hierarchy of vectors used in the repeated applications of the duplication theorem. If

$$\mathbf{x} = (x_k)$$

is a vector of size $n = 2^l$ whose elements $x_k$ are numbered from $k = 0$ to

---

* This section is of interest primarily to readers who wish to understand modern computer programs implementing the Fast Fourier transform.

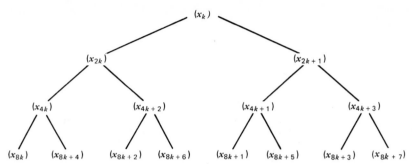

**Fig. 7.6a.**  Hierarchy of vectors in Fast Fourier Transform.

$k = n - 1$, then the array of vectors obtained by always combining the elements with even index into one vector and those with odd index into another begins as shown in Figure 7.6a.

Numbering the rows of this scheme from the top down, beginning with $j = 0$, the $j$-th row contains $2^j$ vectors, each of length $2^{l-j}$. We denote these vectors by $x^{(j, m)}$, where $m$ ranges from $m = 0$ to $m = 2^j - 1$, so that the scheme of Figure 7.6a appears as shown in Figure 7.6b.

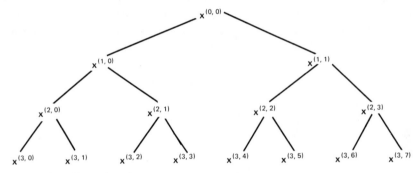

**Fig. 7.6b.**  Hierarchy of vectors, vector notation.

Denoting the $2^{-j}$ elements of $x^{(j, m)}$ by $x_k^{(j, m)}$ $(k = 0, 1, \ldots, 2^{-j} - 1)$, the vectors $x^{(j, m)}$ are defined by the recurrence relations

$$x_k^{(0, 0)} = x_k;\tag{1a}$$

$$x_k^{(j+1, 2m)} = x_{2k}^{(j, m)};\tag{1b}$$

$$x_k^{(j+1, 2m+1)} = x_{2k+1}^{(j, m)}.\tag{1c}$$

Our first task consists in identifying the elements $x_k^{(j, k)}$ directly in terms of the elements $x_q^{(0, 0)} = x_q$, which for convenience we also denote by $x[q]$. In other words, we try to determine the function $q = q(j, m, k)$ so that

$$x_k^{(j, m)} = x[q(j, m, k)]\tag{2}$$

for all triples $(j, m, k)$ that occur.

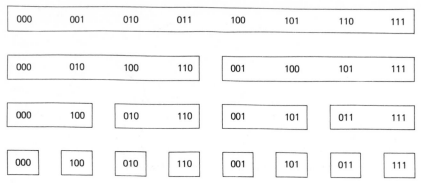

**Fig. 7.6c.** Indices of components of vectors in Figure 7.6b in binary notation. Each box represents a vector.

A clue is provided if we use binary notation for the indices of the components of the vectors involved. For instance, for $l = 3$ and $n = 2^3 = 8$ the indices of the components of the vectors occurring in Figure 7.6a are shown in Figure 7.6c.

A study of Figure 7.6c reveals that in the $j$-th row, the first $l - j$ binary digits number the components in increasing order, while the last $j$ binary digits number the vectors. The numbering of the vectors is not monotonic, however. For instance, in the second row, the vectors are numbered in the order

$$00 \quad 10 \quad 01 \quad 11,$$

while the natural numbering would be

$$00 \quad 01 \quad 10 \quad 11.$$

We see that in the actual numbering the bits of the natural numbering are reversed. This prompts us to introduce, for $j = 0, 1, 2, \ldots$, the **bit reversion function** $p_j$, which maps the set of integers $m = 0, 1, 2, \ldots, 2^j - 1$ onto itself by assigning to

$$m := m_0 + 2m_1 + 2^2 m_2 + \cdots + 2^{j-1} m_{j-1}$$

($m_i$ zero or one) the integer

$$p_j(m) := m_{j-1} + 2m_{j-2} + 2^2 m_{j-3} + \cdots + 2^{j-1} m_0. \tag{3}$$

For completeness, we set

$$p_0(0) = 0. \tag{4}$$

The bit reversion functions satisfy the recurrence relations

$$p_{j+1}(2m) = p_j(m),$$
$$p_{j+1}(2m + 1) = p_j(m) + 2^j, \tag{5}$$
$$j = 0, 1, 2, \ldots,$$
$$m = 0, 1, \ldots, 2^j - 1,$$

as follows immediately from the definition.

From Figure 7.6c, we now suspect that the function $q(j, m, k)$ in (2) is given by

$$q(j, m, k) = 2^j k + p_j(m).$$

In fact, we have

**THEOREM 7.6a:**  *For $l = 1, 2, \ldots$; $j = 0, 1, 2, \ldots, l$; $m = 0, 1, \ldots, 2^j - 1$; $k = 0, 1, \ldots, 2^{l-j} - 1$ there holds*

$$x_k^{(j, m)} = x[2^j k + p_j(m)]; \tag{6}$$

*in particular,*

$$x_0^{(l, m)} = x[p_l(m)]. \tag{7}$$

*Proof:*  The relation (6) is evidently true for $j = 0$. Assuming its truth for some $j \geq 0$, we have, using the relations (1) by means of which the vectors $\mathbf{x}^{(j, m)}$ were constructed,

$$x_k^{(j+1, 2m)} = x_{2k}^{(j, m)} = x[2^j 2k + p_j(m)]$$
$$x_k^{(j+1, 2m+1)} = x_{2k+1}^{(j, m)} = x[2^j(2k + 1) + p_j(m)].$$

In view of (5) we get

$$x_k^{(j+1, 2m)} = x[2^{j+1}k + p_{j+1}(2m)],$$
$$x_k^{(j+1, 2m+1)} = x[2^{j+1}k + p_{j+1}(2m + 1)];$$

that is, (6) holds with $j$ increased by 1.

We now apply to each vector of the scheme of Figure 7.6b the appropriate Fourier transform; that is, we define

$$\mathbf{y}^{(j, m)} := \mathbf{F}_{2^{l-j}} \mathbf{x}^{(j, m)}$$

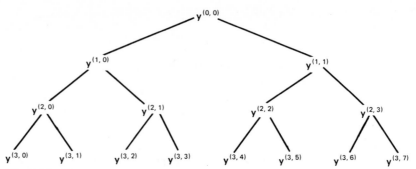

**Fig. 7.6d.** Fourier transforms of the vectors $\mathbf{x}^{(j, m)}$.

for $j = 0, 1, \ldots, l$ and $m = 0, 1, \ldots, 2^j - 1$. If the vectors $\mathbf{y}^{(j, m)}$ are arranged in a scheme similar to that of Figure 7.6b as shown in Figure 7.6d, then the top row contains the vector $\mathbf{y}^{(0, 0)} = \mathbf{F}_n \mathbf{x}^{(0, 0)}$ which we wish to construct, whereas the one-dimensional vectors in the bottom row are identical with the corresponding one-dimensional vectors $\mathbf{x}^{(l, m)}$ because $\mathbf{F}_1$ is the identity. The vectors in the intermediate rows are connected by the duplication theorem (Theorem 7.5a). Using (7), we have

$$y_0^{(l, m)} = x[p_l(m)], \qquad m = 0, 1, \ldots, 2^l - 1, \tag{8}$$

and for $j = l, l - 1, \ldots, 1$, using the abbreviation

$$r_j := w_{2^j} = \exp \frac{2\pi i}{2^j}, \tag{9}$$

the duplication theorem yields

$$
\begin{aligned}
y_k^{(j-1, m)} &= \tfrac{1}{2}\{y_k^{(j, 2m)} + r_{l-j+1}^{-k} y_k^{(j, 2m+1)}\} \\
y_{k+2^{l-j}}^{(j-1, m)} &= \tfrac{1}{2}\{y_k^{(j, 2m)} - r_{l-j+1}^{-k} y_k^{(j, 2m+1)}\}
\end{aligned}
\tag{10}
$$

where $k = 0, 1, \ldots, 2^{l-j} - 1$.

The formulas (10) may be simplified by combining the $2^j$ vectors $\mathbf{y}^{(j, m)}$ ($m = 0, 1, \ldots, 2^j - 1$) of size $2^{l-j}$ into a single vector

$$\mathbf{y}_j = (y_j[k])$$

of length $2^l$ by defining

$$y_j[2^{l-j}m + k] := y_k^{(j, m)} \tag{11}$$

($m = 0, 1, \ldots, 2^j - 1; k = 0, 1, \ldots, 2^{l-j} - 1$). From the point of view of programming this simply means that the vectors $\mathbf{y}^{(j, m)}$ for each fixed $j$ are stored consecutively in a single array of length $2^l$. In this new notation, the formulas (8) and (10) appear as

$$y_l[m] := x[p_l(m)],\tag{12}$$

$m = 0, 1, \ldots, 2^l - 1$, and

$$y_{j-1}[2^{l-j+1}m + k] \qquad = \tfrac{1}{2}\{y_j[2^{l-j+1}m + k] + r_{l-j+1}^{-k}y_j[2^{l-j+1}m + k + 2^{l-j}]\},$$

$$y_{j-1}[2^{l-j+1}m + k + 2^{l-j}] = \tfrac{1}{2}\{y_j[2^{l-j+1}m + k] - r_{l-j+1}^{-k}y_j[2^{l-j+1}m + k + 2^{l-j}]\}.$$

$$\tag{13}$$

Here we see that the same arguments occur in the brackets on the left and on the right. We let

$$h := 2^{l-j+1}m + k, \qquad s := 2^{l-j}$$

and observe that

$$r_{l-j+1}^{-h} = r_{l-j+1}^{2^{l-j+1}m - k} r_{l-j+1}^{-k} = r_{l-j+1}^{-k}$$

for all integers $k$. Thus, the formulas (13) may be written and executed as

$$\begin{aligned} y_{j-1}[h] &= \tfrac{1}{2}\{y_j[h] + r_{l-j+1}^{-h}y_j[h + s]\}, \\ y_{j-1}[h + s] &= \tfrac{1}{2}\{y_j[h] - r_{l-j+1}^{-h}y_j[h + s]\}, \end{aligned}\tag{14}$$

$j = l, l - 1, \ldots, 1$ where $s = 2^{l-j}$, and $h$ runs through the integers of the form $h = 2^{l-j+1}m + k$ with $k = 0, 1, \ldots, s - 1$ and $m = 0, 1, \ldots, 2^{j-1} - 1$.

**THEOREM 7.6b:** *If the vectors* $\mathbf{y}_j = (y_j[h])$ *are computed by (12) and (14), then the numbers* $y_0[m]$ *($m = 0, 1, \ldots, 2^l - 1$) are the* $n = 2^l$ *components of* $\mathbf{F}_n\mathbf{x}$.

The working of the algorithm is shown in the flow diagram of Figure 7.6e. We write

$$y := y_j$$
$$y' := y_{j-1}$$
$$e^{i\phi} := r_{l-j+1}$$
$$e^{i\psi} := r_{l-j+1}^h$$
$$s := 2^{l-j}$$

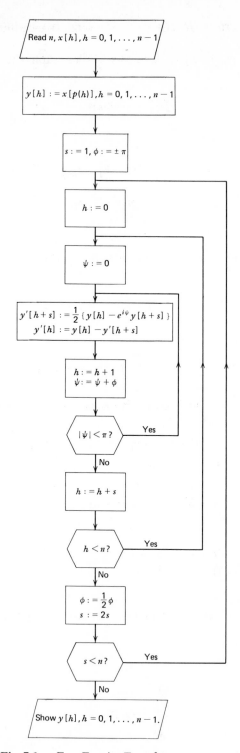

**Fig. 7.6e.** Fast Fourier Transform.

By choosing the correct initial value of $\phi$, we can compute both $\mathbf{F}_n \mathbf{x}$ ($\phi = -\pi$) and $\overline{\mathbf{F}_n} \mathbf{x}$ ($\phi = +\pi$).

**DEMONSTRATION 7.6-1:**  We use the Fast Fourier Transform to compute $\mathbf{F}_8 \mathbf{x}$ where $\mathbf{x} = (x_k)$,

$$x_k := f\left(\frac{k}{8}\right), \qquad k = 0, 1, \dots, 7;$$

$$f(\tau) := e^{ix \cos 2\pi\tau}, \qquad x = 4.5,$$

compare Demonstration 7.3-3. Each box contains the components of a vector $\mathbf{y}^{(j, m)}$.

| $h$ | $p(h)$ | $y_3[h]$ | $y_2[h]$ | $y_1[h]$ | $y_0[h]$ |
|---|---|---|---|---|---|
| 0 | 0 | $-0.210796$ $-0.977530i$ | $-0.210796$ $+0i$ | $0.394602$ $+0i$ | $-0.302291$ $+0i$ |
| 1 | 4 | $-0.210796$ $+0.977530i$ | $0$ $-0.977530i$ | $0$ $-0.488765i$ | $0$ $-0.258658i$ |
| 2 | 2 | $1.000000$ $+0i$ | $1.000000$ $+0i$ | $-0.605398$ $+0i$ | $-0.302699$ $+0i$ |
| 3 | 6 | $1.000000$ $+0i$ | $0$ $+0i$ | $0$ $-0.488765i$ | $0$ $-0.230107i$ |
| 4 | 1 | $-0.999185$ $-0.040377i$ | $-0.999185$ $+0i$ | $-0.999185$ $+0i$ | $0.696893$ $+0i$ |
| 5 | 5 | $-0.999185$ $+0.040377i$ | $0$ $-0.040377i$ | $0.020188$ $-0.020188i$ | $0$ $-0.230107i$ |
| 6 | 3 | $-0.999185$ $+0.040377i$ | $-0.999185$ $+0i$ | $0$ $+0i$ | $-0.302699$ $+0i$ |
| 7 | 7 | $-0.999185$ $-0.040377i$ | $0$ $+0.040377i$ | $-0.020188$ $-0.020188i$ | $0$ $0.258658$ |

# PROBLEMS

1   Write a program implementing the Fast Fourier Transform, and carry out the discrete Fourier transform required in earlier problems by this method.

2   Construct an efficient implementation of the Fast Fourier Transform for *real* sequences $\{x_k\}$.

3   How many fixed points does the function $p_j(h)$ have?

4   Show that the initial position of the elements $x[p_l(m)]$ can be achieved by interchanging pairs of suitable elements. How many such interchanges are required?

## §7.7   CONVOLUTION

For computational purposes, discrete Fourier analysis is best presented as taking place in the $n$-dimensional complex Euclidean space $\mathbb{C}^n$. In this context the Fourier operator $\mathbf{F}_n$ is represented by a certain matrix defined by Equation (3) of §7.5 which transforms vectors $\mathbf{x} \in \mathbb{C}^n$ into their discrete Fourier transforms $\mathbf{y} = \mathbf{F}_n \mathbf{x} \in \mathbb{C}^n$.

For mathematical purposes, as well as from the point of view of applications, however, it is better to think of discrete Fourier analysis as taking place in the space $\Pi_n$ of sequences $\{x_k\}$ that extend to infinity in both directions and that are periodic with period $n$; that is, whose elements $x_k$ are complex numbers satisfying

$$x_{k+n} = x_k \tag{1}$$

for all integers $k$. If $f$ is a function with period 1, we recall that the set of its sampling values $f_k := f(\tau_k)$, $\tau_k = k/n$, forms such a sequence; also, the discrete Fourier coefficients $\hat{c}_k$ of such a function form a sequence in $\Pi_n$.

Since no misunderstandings can arise, we shall use symbols such as $\mathbf{x}$, $\mathbf{y}$, ..., not only to denote vectors in $\mathbb{C}^n$, but also sequences in $\Pi_n$. If $\mathbf{x} = \{x_k\} \in \Pi_n$, we shall denote by $\mathbf{y} = \{y_m\} = \mathbf{F}_n \mathbf{x}$ the sequence in $\Pi_n$ with $m$-th element

$$y_m = \frac{1}{n} \sum_{k=0}^{n-1} x_k w_n^{-km}$$

where, as always, $w_n := \exp(2\pi i/n)$. The sequence $\mathbf{y}$ is again called the discrete Fourier transform of $\mathbf{x}$. The sequence $\mathbf{x}$ may be recovered from $\mathbf{y}$ by the formula

$$\mathbf{x} = \mathbf{F}_n^{-1} \mathbf{y},$$

where $\mathbf{F}_n^{-1} = n\overline{\mathbf{F}}_n$.

We now introduce several kinds of *multiplications* in $\Pi_n$ and discuss their interrelationships. It is clear how to define **multiplication by a scalar**. If $\mathbf{x} = \{x_k\}$ is a sequence in $\Pi_n$ and if $c$ is any complex number, we write

$$c\mathbf{x} := \{cx_k\}; \tag{2}$$

that is, each element of $\mathbf{x}$ is multiplied by the scalar $c$.

Another simple type of multiplication is the *element-by-element multiplication* or *Hadamard multiplication* of the two sequences. If $\mathbf{x} = \{x_k\}$ and $\mathbf{y} = \{y_k\}$ are two sequences in $\Pi_n$, their **Hadamard product** is obtained by multiplying for each $k$, the $k$-th element of $\mathbf{x}$ by the $k$-th element of $\mathbf{y}$. We denote the Hadamard product by

$$\mathbf{x} \cdot \mathbf{y} = \{x_k y_k\}; \tag{3}$$

we always write the dot for clarity. We note that the Hadamard product of two sequences in $\Pi_n$ is again a sequence in $\Pi_n$. Because only one period segment needs to be computed, the computation of the Hadamard product requires $n\mu$.

**EXAMPLE 7.7-1:** If $f$ and $g$ are two functions with period 1, and if $\mathbf{f} = \{f(\tau_k)\}$, $\mathbf{g} = \{g(\tau_k)\}(\tau_k := k/n)$ are the sequences of their sampling values, then the sequence of the sampling values of $fg$ is $\mathbf{f} \cdot \mathbf{g}$.

We finally introduce a third, somewhat more complicated kind of multiplication called *convolution*. If again $\mathbf{x} = \{x_k\}$ and $\mathbf{y} = \{y_k\}$ are sequences in $\Pi_n$, their **convolution product** or simply **convolution** is the sequence $\mathbf{z} = \{z_k\}$ with $k$-th element

$$z_k := \sum_{m=0}^{n-1} x_m y_{k-m}. \tag{4}$$

The $k$-th element of the convolution sequence thus is the sum of all products of the elements of the two factors where the indices add up to $k$. Here the sum is extended over a full period of each factor sequence. We use the notation

$$\mathbf{z} = \mathbf{x} * \mathbf{y}$$

to denote the convolution of the sequences $\mathbf{x}$ and $\mathbf{y}$. It is easy to see that the convolution of two sequences in $\Pi_n$ is again a sequence in $\Pi_n$. Since each element of the convoluted sequence in general contains $n$ products, the computation of a full period of $\mathbf{x} * \mathbf{y}$, if formed according to (4), now in general requires $n^2\mu$.

Although convolution as a concept may be less familiar than the two other kinds of multiplication which we have introduced, examples of it abound in computational analysis and in applied mathematics.

**EXAMPLE 7.7-2:** *Differencing.* Let $\mathbf{x} = \{x_k\}$ be a sequence in $\prod_n$. Then, the sequence of its differences

$$\Delta x_k := x_{k+1} - x_k$$

may be thought of as the convolution of $\mathbf{x}$ with the special sequence $\mathbf{d} = \{d_k\}$ which arises through periodic repetition of the segment

$$d_{-1} = 1, \qquad d_0 = -1, \qquad d_1 = d_2 = \cdots = d_{n-2} = 0.$$

Indeed, if $\mathbf{z} := \mathbf{d} * \mathbf{x}$, then according to (4)

$$z_k = \sum_{m=0}^{n-1} d_m x_{k-m} = -x_k + x_{k-n+1} = x_{k+1} - x_k,$$

because $d_{n-1} = 1$ and $x_{k-n+1} = x_{k+1}$ in view of the periodicity of both $\mathbf{x}$ and $\mathbf{d}$.

**EXAMPLE 7.7-3:** *Smoothing.* Often it is necessary to smooth a sequence $\mathbf{x}$, especially if the elements of $\mathbf{x}$ are experimental data. This is done by replacing each element $x_k$ of $\mathbf{x}$ by a weighted average of a certain number of neighboring elements $x_{k-d}, x_{k-d+1}, \ldots, x_{k+d}$, that is, by forming

$$\tilde{x}_k := c_d x_{k-d} + c_{d-1} x_{k-d+1} + \cdots + c_{-d} x_{k+d},$$

where $c_d, c_{d-1}, \ldots, c_{-d}$ are positive numbers independent of $k$ so that

$$c_{-d} + c_{-d+1} + \cdots + c_d = 1.$$

Thus, a simple smoothing formula might be

$$\tilde{x}_k := \tfrac{1}{4}x_{k-1} + \tfrac{1}{2}x_k + \tfrac{1}{4}x_{k+1}.$$

Clearly, the smoothed sequence $\tilde{\mathbf{x}} = \{\tilde{x}_k\}$ arises by convolution of $\mathbf{x}$ with a smoothing sequence $\mathbf{s} = \{s_k\} \in \Pi_n$ whose only non-zero elements in each period segment are the constants $s_k = c_k$. Thus, in the special case mentioned above, if $n = 8$, the smoothing sequence would be $\tfrac{1}{4}$ times

$$\cdots 1 \;\; 2 \;\; 1 \;\; 0 \;\; 0 \;\; 0 \;\; 0 \;\; 0 \;\; 1 \;\; 2 \;\; 1 \;\; 0 \;\; 0 \;\; 0 \;\; 0 \;\; 0 \cdots.$$
$$\uparrow$$
$$s_0$$

**EXAMPLE 7.7-4:** *Multiplication of polynomials.* Let

$$p(x) = p_0 + p_1 x + p_2 x^2 + \cdots + p_{m-1} x^{m-1},$$

$$q(x) = q_0 + q_1 x + q_2 x^2 + \cdots + q_{m-1} x^{m-1}$$

be two polynomials of degree $<m$. Their algebraic product

$$r(x) := p(x)q(x) = \sum_{i=0}^{2m-2} r_i x^i$$

then is a polynomial of degree $<2m-1$ whose coefficients $r_i$ are given by

$$r_0 = p_0 q_0$$
$$r_1 = p_0 q_1 + p_1 q_0$$
$$r_2 = p_0 q_2 + p_1 q_1 + p_2 q_0$$

and generally

$$r_i = \sum_{j=0}^{i} p_j q_{i-j} \qquad (5)$$

where, if $i \geq m$, coefficients $p_j$ and $q_j$ where $j \geq m$ are defined to be zero. Clearly, (5) resembles the rule for forming the convolution product. In order to perfect the resemblance, we let $n := 2m$, and define sequences **p** and **q** by periodic repetition of the segments of the $n$ elements

$$(p_0, p_1, \ldots, p_{m-1}, 0, 0, \ldots, 0)$$
$$(q_0, q_1, \ldots, q_{m-1}, \underbrace{0, 0, \ldots, 0}_{}).$$

$$m \text{ zeros}$$

Then evidently (5) may be replaced by

$$r_i = \sum_{j=0}^{n-1} p_j q_{i-j}, \qquad (6)$$

and we see that the $r_i$ are identical with the elements of the convolution

$$\mathbf{r} := \mathbf{p} * \mathbf{q}.$$

**EXAMPLE 7.7-5:** *Multiplication of large integers.* Let now $p$ and $q$ be integers, whose representations in the base $b$ are

$$p = p_0 + p_1 b + p_2 b^2 + \cdots + p_{m-1} b^{m-1},$$
$$q = q_0 + q_1 b + q_2 b^2 + \cdots + q_{m-1} b^{m-1},$$

where the $p_j$ and $q_j$ are integers, $0 \leq p_j < b$, $0 \leq q_j < b$. The integer $r := pq$ then equals

$$r = \sum_{i=0}^{2m-2} r_i b^i, \qquad (7)$$

where the $r_i$ are again given by (5) or, with the appropriate conventions, (6). In general, (7) is not the correct representation of $r$ in the base $b$, because the condition $0 \leq r_i < b$ does not necessarily hold. The correct representation is obtained by representing each $r_i$ in the base $b$,

$$r_i = r_{i0} + r_{i1}b + \cdots + r_{i,d-1}b^{d-1}, \tag{8}$$

substituting into (7) and performing the necessary adding and carrying operations. Because the $r_i$ satisfy

$$0 \leq r_i \leq m(b-1)^2 < mb^2,$$

the number $d$ of digits in the $b$-ary representation of the $r_i$ satisfies

$$d \leq \text{Log}_b(mb^2) = 2 + \text{Log}_b m,$$

where $\text{Log}_b$ denotes the logarithm to the base $b$. Thus, if $m$ is very large, the integers $r_i$ are very much shorter than either $p$ or $q$, and the main work in computing the product $pq$ is spent on computing the $r_i$; that is, the elements of the convolution $\mathbf{p} * \mathbf{q}$.

Since convolution is an expensive operation when $n$ is large, the incentive is substantial to search for ways to reduce the work. Because Fourier transforms are cheap, we try to find a formula for the Fourier transform of the convolution. Let $\mathbf{x} = \{x_k\}$ and $\mathbf{y} = \{y_k\}$ be sequences in $\Pi_n$, let

$$\mathbf{z} = \{z_k\} := \mathbf{x} * \mathbf{y}$$

be their convolution, and denote by $\hat{\mathbf{x}}, \hat{\mathbf{y}}, \hat{\mathbf{z}}$, respectively, the Fourier transforms of $\mathbf{x}, \mathbf{y}, \mathbf{z}$. We then have

$$\hat{z}_k = \frac{1}{n} \sum_{j=0}^{n-1} z_j w_n^{-jk}$$

$$= \frac{1}{n} \sum_{j=0}^{n-1} \left( \sum_{m=0}^{n-1} x_m y_{j-m} \right) w_n^{-jk}$$

$$= \frac{1}{n} \sum_{m=0}^{n-1} x_m \sum_{j=0}^{n-1} y_{j-m} w_n^{-jk}.$$

Because of periodicity, we can replace the summation with respect to $j$ by a summation where $p := j - m$ runs from $0$ to $n - 1$, and in view of $j = p + m$, we thus get

$$\hat{z}_k = \frac{1}{n} \sum_{m=0}^{n-1} x_m w_n^{-mk} \sum_{p=0}^{n-1} y_p w_n^{-pk}.$$

In the two sums on the right, we recognize the expressions

$$\hat{x}_k = \frac{1}{n} \sum_{m=0}^{n-1} x_m w_n^{-mk}$$

and

$$\hat{y}_k = \frac{1}{n} \sum_{p=0}^{n-1} y_p w_n^{-pk};$$

that is, the $k$-th elements of the Fourier transforms of $\mathbf{x}$ and $\mathbf{y}$. We thus have

$$\hat{z}_k = n\hat{x}_k \hat{y}_k;$$

that is, up to a trivial factor $n$, $\hat{\mathbf{z}}$ equals the *Hadamard product* of the Fourier transforms $\hat{\mathbf{x}}$ and $\hat{\mathbf{y}}$.

Thus, we see that for arbitrary sequences $\mathbf{x}$ and $\mathbf{y}$ in $\Pi_n$,

$$\mathbf{F}_n(\mathbf{x} * \mathbf{y}) = n\mathbf{F}_n \mathbf{x} \cdot \mathbf{F}_n \mathbf{y} \tag{9}$$

and thus, applying the operator $\mathbf{F}_n^{-1} = n\overline{\mathbf{F}}_n$ on both sides,

$$\mathbf{x} * \mathbf{y} = n^2 \overline{\mathbf{F}}_n(\mathbf{F}_n \mathbf{x} \cdot \mathbf{F}_n \mathbf{y}). \tag{10}$$

We thus have obtained:

**THEOREM 7.7a:** **(Convolution theorem).** *The convolution* $\mathbf{x} * \mathbf{y}$ *of any two sequences* $\mathbf{x}, \mathbf{y} \in \Pi_n$ *is given by* (10).

The convolution of two sequences thus may be computed by performing the following steps:
  (i)   Take the Fourier transforms of both sequences.
  (ii)  Form the Hadamard product of these Fourier transforms.
  (iii) Apply the conjugate Fourier transform.
  (iv)  Perform a scalar multiplication of the resulting sequence by $n^2$.
If $n$ is a power of 2, then by Theorem 7.5b each of these Fourier transforms costs only $\frac{1}{2}n \operatorname{Log}_2 n\mu$. To this must be added the $n\mu$ for forming the Hadamard product, while the scalar multiplication by $n^2$ (which in the binary number system amounts to a mere shift of the exponent) may be neglected. We thus have

**THEOREM 7.7b:** *If* $n$ *is a power of 2, the convolution of two sequences in* $\Pi_n$ *can be performed in no more than*

$$n(\tfrac{3}{2} \operatorname{Log}_2 n + 1) = n \operatorname{Log}_2(2n^{3/2})$$

*complex multiplications.*

Contrasted with the $n^2\mu$ necessary for forming the convolution in a naive fashion, order-of-magnitude savings are thus possible by using the Convolution Theorem. While the savings are most dramatic for large values of $n$, our demonstrations, in order to make them easily reproducible, work with small values of $n$.

**DEMONSTRATION 7.7-6:**    To smooth the sequence

$$\mathbf{x} := \|: \quad 7 \quad 4 \quad 8 \quad 3 \quad 2 \quad 9 \quad 6 \quad 5 \quad :\|$$

where the sign $\|: \; :\|$ indicates periodic repetition, by the smoothing sequence

$$\mathbf{s} := \tfrac{1}{4}\|: \quad 2 \quad 1 \quad 0 \quad 0 \quad 0 \quad 0 \quad 0 \quad 1 \quad :\|.$$

(See Example 7.7-3.) Naturally, the smoothed sequence is

$$\tilde{\mathbf{x}} = \|: \quad 5.75, \; 5.75, \; 5.75, \; 4.00, \; 4.50, \; 6.50, \; 6.50, \; 5.75 \quad :\|.$$

Let us check whether the same result is obtained by applying the Convolution Theorem. The Fast Fourier Transform yields for $\hat{\mathbf{x}} := \mathbf{F}_8\,\mathbf{x}$, the elements

| $k$ | $\hat{x}_k$ |
|-----|-------------|
| 0 | $5.500000 + 0i$ |
| 1 | $0.359835 + 0.368718i$ |
| 2 | $-0.625000 - 0.625000i$ |
| 3 | $0.890165 + 0.868718i$ |
| 4 | $0.250000 + 0i$ |
| 5 | $0.890165 - 0.868718i$ |
| 6 | $-0.625000 + 0.625000i$ |
| 7 | $0.359835 - 0.368718i$ |

The Fourier transform $\hat{\mathbf{s}}$ of $\mathbf{s}$ can be computed analytically. From

$$\hat{s}_k = \frac{1}{n} \sum_{m=0}^{n-1} s_m w_n^{-km}$$

we have for arbitrary $n$, using periodicity, since in each period segment only $s_0$, $s_1$, and $s_{n-1} = s_{-1}$ are different from zero,

$$\hat{s}_k = \frac{1}{4n}\left(2 + w_n^k + w_n^{-k}\right)$$

$$= \frac{1}{2n}\left(1 + \cos\frac{2\pi k}{n}\right),$$

which may also be written

$$\hat{s}_k = \frac{1}{n}\left(\cos\frac{\pi k}{n}\right)^2.$$

In order to take care of the final multiplication by $n^2$, we multiply $\hat{x}_k$ by

$$n^2\hat{s}_k = n\left(\cos\frac{\pi k}{n}\right)^2,$$

which for $n = 8$ yields

| $k$ | $n^2\hat{s}_k\hat{x}_k$ |
|---|---|
| 0 | $44.000000 + 0i$ |
| 1 | $2.457107 + 2.517764i$ |
| 2 | $-2.500000 - 2.500000i$ |
| 3 | $1.042893 + 1.017767i$ |
| 4 | $0 \quad . \quad + 0i$ |
| 5 | $1.042893 - 1.017767i$ |
| 6 | $-2.500000 + 2.500000i$ |
| 7 | $2.457107 - 2.517767i$ |

Applying $\mathbf{F}_8$ now yields

| $k$ | $\tilde{x}_k$ |
|---|---|
| 0 | $5.75000000 + 1 * 10^{-10}i$ |
| 1 | $5.75000000 - 5 * 10^{-10}i$ |
| 2 | $5.75000000 + 1 * 10^{-10}i$ |
| 3 | $4.00000000 - 3 * 10^{-10}i$ |
| 4 | $4.00000000 + 6 * 10^{-10}i$ |
| 5 | $6.50000000 + 5 * 10^{-10}i$ |
| 6 | $6.50000000 - 6 * 10^{-10}i$ |
| 7 | $5.75000000 + 3 * 10^{-10}i$ |

The results are, of course, correct; the small imaginary parts are indicative of the extraordinary numerical stability of the whole complicated process.

**DEMONSTRATION 7.7-7:** To compute the product

$$241 * 769$$

by the Convolution Theorem. To begin with, we require the Fourier transforms $\hat{\mathbf{p}}$ and $\hat{\mathbf{q}}$ of the sequences

$$\mathbf{p} := \| : \quad 1, 4, 2, 0, 0, 0, 0, 0 \quad : \|,$$

$$\mathbf{q} := \| : \quad 9, 6, 7, 0, 0, 0, 0, 0 \quad : \|.$$

We then form the products $64\hat{p}_k\hat{q}_k$ and apply $\overline{\mathbf{F}_8}$. This yields the numbers $r_i$ defined by (6). Only moderate accuracy is required since the $r_i$ are known to be integers.

| $k$ | $\hat{p}_k$ | $\hat{q}_k$ | $64\hat{p}_k\hat{q}_k$ | $r_k$ |
|---|---|---|---|---|
| 0 | $0.875 + 0i$ | $2.750 + 0i$ | $154.000 + 0i$ | $9 + 2 * 10^{-9}i$ |
| 1 | $0.479 - 0.604i$ | $1.655 - 1.405i$ | $-3.576 - 107.047i$ | $42 + 9 * 10^{-9}i$ |
| 2 | $-0.125 - 0.500i$ | $0.250 - 0.750i$ | $-26.000 - 2.000i$ | $49 - 4 * 10^{-9}i$ |
| 3 | $-0.229 - 0.104i$ | $0.595 + 0.345i$ | $-6.424 - 9.017i$ | $40 + 2 * 10^{-8}i$ |
| 4 | $-0.125 + 0i$ | $1.250 + 0i$ | $-10.000 + 0i$ | $14 + 4 * 10^{-9}i$ |
| 5 | $-0.229 + 0.104i$ | $0.595 - 0.345i$ | $-6.424 + 9.017i$ | $0 + 8 * 10^{-9}i$ |
| 6 | $-0.125 + 0.500i$ | $0.250 + 0.750i$ | $-26.000 + 2.000i$ | $0 - 4 * 10^{-9}i$ |
| 7 | $0.479 + 0.604i$ | $1.655 + 1.405i$ | $-3.576 + 107.047i$ | $0 - 2 * 10^{-8}i$ |

Indeed,

$$
\begin{array}{rl}
& 9 * 10^0 \\
+ & 42 * 10^1 \\
+ & 49 * 10^2 \\
+ & 40 * 10^3 \\
+ & 14 * 10^4 \\
\hline
= & 185329 \quad = 241 * 769.
\end{array}
$$

## PROBLEMS

**1** Does there exist a sequence **e** such that

$$\mathbf{e} * \mathbf{x} = \mathbf{x}$$

for all sequences $\mathbf{x} \in \Pi_n$?

**2** Let $\mathbf{a} = \{a_k\}$, $\mathbf{x} = \{x_k\}$, $\mathbf{b} = \{b_k\}$ be sequences in $\Pi_n$. Find a necessary and sufficient condition on **a** so that the equation

$$\mathbf{a} * \mathbf{x} = \mathbf{b} \tag{11}$$

has a solution **x** for every **b**. [Apply discrete Fourier transforms to (11).]

**3** A bar of length $L$ is simply supported at the endpoints. At the point $x$ $(0 \le x \le L)$ the bar is subject to a load $p(x)$. (See Figure 7.7a.) It is shown in elasticity that the vertical deflection of the bar satisfies the differential equation

$$y^{(4)} = \gamma p(x), \tag{12}$$

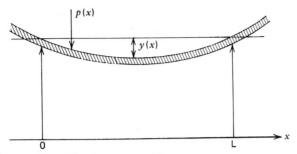

**Fig. 7.7a.** Simply supported bar.

where $\gamma$ is a constant, and the boundary conditions

$$y(0) = y''(0) = 0, \qquad y(L) = y''(L) = 0. \tag{13}$$

A discrete analog of (12) and (13) is obtained by letting $x_k := Lk/m$, $k = 0, 1, \ldots, m$ and by denoting by $y_k$ a value intended to approximate $y(x_k)$. Simulating derivatives by differences, (12) is modeled by

$$y_{k+2} - 4y_{k+1} + 6y_k - 4y_{k-1} + y_{k-2} = h^4\gamma p_k, \tag{14}$$

where $h = L/m$, $p_k := p(x_k)$, whereas the boundary conditions (13) are modeled by

$$y_0 = y_m = 0, \qquad y_1 - 2y_0 + y_{-1} = 0, \qquad y_{m+1} - 2y_m + y_{m-1} = 0. \tag{15}$$

It is proposed to solve the equations (14), (15) for given values $p_k$ by means of the discrete Fourier transform.

(a) Show that the boundary conditions (15) are taken care of by continuing $\{y_k\}$ and $\{p_k\}$ as *odd* sequences of period $n = 2m$.

(b) Defining the vectors **y** and **p** as in the main text, show that the vector on the left of (14) is the convolution of **y** with the sequence $\mathbf{d} = \{d_k\}$ where

$$d_0 = 1, \qquad d_1 = d_{-1} = -4, \qquad d_2 = d_{-2} = 6,$$

all other $d_k = 0$, and hence that (14), (15) is equivalent to

$$\mathbf{d} * \mathbf{y} = h^4\gamma\mathbf{p}. \tag{16}$$

(c) Denoting by $\hat{\mathbf{d}}, \hat{\mathbf{y}}, \hat{\mathbf{p}}$ the discrete Fourier transforms of **d**, **y**, **p**, show that by taking Fourier transforms and applying the Convolution Theorem, (16) becomes

$$\hat{\mathbf{d}} \cdot \hat{\mathbf{y}} = n^{-1}h^4\gamma\hat{\mathbf{p}}. \tag{17}$$

(d) Compute $\hat{\mathbf{d}} = \{\hat{d}_k\}$ analytically, and show that $\hat{d}_k \neq 0$ for $k \neq 0$. Hence, by looking at (17) componentwise, show that

$$\hat{y}_k = n^{-1}h^4\gamma\frac{\hat{p}_k}{\hat{d}_k}, \qquad k \neq 0. \tag{18}$$

Also show $\hat{y}_0 = 0$ by virtue of the fact that the sequence $\{y_k\}$ is odd.

(e) Show that $\mathbf{y}$ may be found as the inverse Fourier transform of the sequence defined by (18).

(f) Carry out the process sketched above for the special case where the load $p$ is concentrated at the midpoint of the bar, and possibly for other choices of $p$. (Choose a value of $n$ that is compatible with your Fast Fourier Transform program.)

**4 Integral Equations of the Convolution Type.** The functions $f(x)$ and $g(x)$ are defined on $[0, 1]$. The function $f$ is continuous; $g$ is continuous for $x > 0$ and absolutely integrable at $x = 0$. We consider the integral equation for an unknown function $u(x)$,

$$\int_0^x u(t)g(x - t)\,dt = f(x). \tag{19}$$

(a) Show how a discretized version of (19) may be solved quickly by means of the Convolution Theorem.

(b) Carry out the solution for the case where $g(x) := \sqrt{x}$, $f(x) := x$.

**5 Fast Fourier Transforms for Arbitrary $n$.** Let $n > 0$ be arbitrary, let $\mathbf{a} = \{a_k\}$ be a sequence with period $n$, and let $\hat{\mathbf{a}} = \{\hat{a}_m\}$ be the discrete Fourier transform of $\mathbf{a}$,

$$\hat{a}_m = \frac{1}{n}\sum_{k=0}^{n-1} a_k w^{-km}, \qquad m = 0, 1, \ldots, n - 1,$$

where

$$w := \exp\frac{2\pi i}{n}.$$

(a) Letting

$$v := \exp\frac{\pi i}{n},$$

so that $v^2 = w$, show that

$$\hat{a}_m = \frac{1}{n}v^{-m^2}c_m,$$

where

$$c_m := \sum_{k=0}^{n-1} (a_k v^{-k^2}) v^{(m-k)^2}, \qquad m = 0, 1, \ldots, n-1. \qquad (20)$$

**(b)**   Letting $N = 2n$ and defining sequences of period $N$ by

$$x_k := \begin{cases} a_k v^{-k^2}, & k = 0, 1, \ldots, n-1; \\ 0, & k = n, n+1, \ldots, N-1; \end{cases}$$

$$y_k := v^{k^2}, \qquad -\frac{N}{2} < k < \frac{N}{2},$$

show that the $c_m$ defined by (20) equal the first $n$ elements of

$$\mathbf{c} := \mathbf{x} * \mathbf{y}.$$

**(c)**   By selecting $N$ as the first power of 2 that exceeds $2n$, show that the quantities $n\hat{a}_m$ can be computed with a number of complex multiplications that does not exceed

$$6n \, \mathrm{Log}_2 \, n + 16n.$$

**(d)**   Compute the discrete Fourier transform of the sequence $\{a_k\}$ of period $n = 3$, where $a_k := k$, $k = 0, 1, 2$, by the above method.

6   Show that $\mathbf{x} \cdot \mathbf{y} = \mathbf{0}$ does not imply that either $\mathbf{x} = \mathbf{0}$ or $\mathbf{y} = \mathbf{0}$. How about the corresponding question for convolution?

---

## §7.8   TIME SERIES

A **time series** is a finite sequence of observations of a physical quantity $x$ at equidistant time intervals. The observation at time $\tau_k := \tau_0 + k\Delta\tau$ will be denoted by $x_k$, and we assume that $k$ ranges from $k = 0$ to $k = m - 1$. The quantity $x$ may be a brain current (as recorded in an encephalogram), the displacement of a seismograph during an earthquake, or the reading of a water gauge. In electroencephalograms, $\Delta\tau$ is of the order of $1/100$ of a second, and the length $m$ of a time series can be of the order of $2^{11}$ or $2^{12}$.

By analyzing a time series, important conclusions on the nature of the underlying physical process can often be drawn. For instance, in electroencephalography, epilepsy may be discovered. The analysis of a time series usually requires the calculation of the following new sequences from the given time series $\{x_k\}$:

(a)   The **covariance function**, that is, the sequence **r** with $t$-th element

$$r_t := \sum_{k=0}^{m-t-1} x_k x_{k+t}, \qquad t = 0, 1, \ldots; \tag{1}$$

(b)   the **power spectrum**, that is, the sequence **f** with $q$-th element

$$f_q := \frac{1}{n} \sum_{t=0}^{n-1} r_t \exp\left(-\frac{2\pi i}{n} tq\right), \qquad q = 0, 1, \ldots, \tag{2}$$

where $n$ is a fixed integer $> m$, usually $n = 2m$;
(c)   the **smoothed power spectrum g** defined by

$$g_q := \sum_{j=-h}^{h} c_j f_{q-j}, \qquad q = 0, 1, 2, \ldots, \tag{3}$$

where $h$ is a (small) positive integer, and the $c_j$ are constants satisfying

$$c_j \geq 0, \qquad |j| \leq h; \qquad \sum_{j=-h}^{h} c_j = 1. \tag{4}$$

It was recognized independently by Tukey and Bartlett around 1945 that smoothing of the power spectrum is essential if meaningful physical results are to be obtained.

It turns out that discrete Fourier analysis provides a common basis for describing operations with time series and for carrying them out in an efficient manner.

In addition to the Fourier operator $\mathbf{F}_n$ already introduced, we require the reversion operator **R** which replaces a sequence $\mathbf{x} = \{x_k\} \in \Pi_n$ by the sequence

$$\mathbf{Rx} := \{x_{-k}\}, \tag{5}$$

where the signs of the indices are reversed. If applied to a period segment of **x**, **R** is represented by the matrix

$$\mathbf{R} = \begin{pmatrix} 1 & 0 & 0 & \cdots & 0 & 0 & 0 \\ 0 & 0 & 0 & \cdots & 0 & 1 & 0 \\ 0 & 0 & 0 & \cdots & 1 & 0 & 0 \\ \multicolumn{7}{c}{\dotfill} \\ 0 & 1 & 0 & \cdots & 0 & 0 & 0 \end{pmatrix}$$

It is a direct consequence of the definitions that

$$\mathbf{F}_n \mathbf{R} = \mathbf{R} \mathbf{F}_n = \overline{\mathbf{F}_n}, \tag{6}$$

and consequently

$$\mathbf{F}_n^{-1} = n\mathbf{F}_n \mathbf{R} = n\mathbf{R}\mathbf{F}_n. \tag{7}$$

Let the given time series $\{x_0, x_1, \ldots, x_{m-1}\}$ be followed by $m$ zeros, and let $\mathbf{x}$ denote the sequence in $\Pi_n$ obtained by repeating these $n = 2m$ elements periodically. The other sequences $\mathbf{r}, \mathbf{f}, \mathbf{g}$ defined above are similarly embedded in $\Pi_n$. This enables us to express the operations (a), (b), (c) in the language of discrete Fourier analysis.

(a)   For $t < 0$, $t = -s$, we extend the definition (1) by setting

$$r_{-s} := \sum_{k=0}^{m-1+s} x_k x_{k-s}.$$

Then evidently,

$$r_{-s} = \sum_{k=s}^{m-1} x_k x_{k-s} = \sum_{j=0}^{m-1-s} x_{j+s} x_j = r_s,$$

and the sequence $\mathbf{r}$ becomes symmetric, $\mathbf{r} = \mathbf{R}\mathbf{r}$. Setting $\mathbf{y} := \mathbf{R}\mathbf{x}$, we have for $t \geq 0$

$$r_t = r_{-t} = \sum_{k=0}^{m-1} x_k x_{k-t} = \sum_{k=0}^{m-1} x_k y_{t-k} = (\mathbf{x} * \mathbf{y})_t,$$

and we see that

$$\mathbf{r} = \mathbf{x} * \mathbf{R}\mathbf{x}. \tag{8}$$

(b)   If $\mathbf{F} := \mathbf{F}_n$, the power spectrum is

$$\mathbf{f} = \mathbf{F}\mathbf{r} = \mathbf{F}(\mathbf{x} * \mathbf{R}\mathbf{x}). \tag{9}$$

(c)   As already noted in §7.7, the smoothing operation also can be expressed by a convolution. Denoting by $\mathbf{c}$ the sequence with elements $c_j$ in the positions $\equiv j \pmod{n}$ and with zeros elsewhere, we have

$$\mathbf{g} = \mathbf{c} * \mathbf{f}. \tag{10}$$

It was observed before the advent of the Fast Fourier Transform that the calculation of $\mathbf{g}$ could be simplified by performing the smoothing in the time domain. Let

$$\mathbf{t} := \mathbf{F}^{-1}\mathbf{c}. \tag{11}$$

Because $\mathbf{f} = \mathbf{F}\mathbf{r}$, we have by the Convolution Theorem (Theorem 7.7a)

$$\mathbf{c} * \mathbf{f} = \mathbf{F}(\mathbf{F}^{-1}\mathbf{c} \cdot \mathbf{F}^{-1}\mathbf{f}) = \mathbf{F}(\mathbf{t} \cdot \mathbf{r}).$$

Thus, (10) may be replaced by

$$g = F(t \cdot r).\tag{12}$$

The sequence $t = F^{-1}c$ is called the **time window** of the smoothing process defined by the constants $c_j$. If the smoothing formula is simple enough, its time window may be calculated analytically. For instance, for the smoothing $c_0 = \frac{1}{2}$, $c_1 = c_{-1} = \frac{1}{4}$ used in Example 7.7-3 and in Demonstration 7.7-6, the time window is $t = \{t_k\}$ where

$$t_k = \left(\cos\frac{k\pi}{n}\right)^2.\tag{13}$$

Smoothing by the time window amounts to one Hadamard multiplication, and thus to a mere $n\mu$, as opposed to the $(2h + 1)n\mu$ required by the formula (3). The Fourier transform which is needed in both cases requires the same amount of work in either case.

Much greater savings are possible by using the Fast Fourier Transform to calculate $r$. Let $a := F^{-1}x$. Then, $y = Rx = FRa$, and the determination of $r$ requires forming

$$x * y = Fa * FRa.$$

By the Convolution Theorem,

$$x * y = F(a \cdot Ra).$$

Because the sequence $x$ is real, $Ra = \bar{a}$, the complex conjugate sequence, hence

$$r = x * y = F(a \cdot \bar{a}).\tag{14}$$

In view of

$$F^2 = RFRF = \frac{1}{n}R,$$

it further follows that

$$f = \frac{1}{n}R(a \cdot \bar{a}) = \frac{1}{n}a \cdot \bar{a}.\tag{15}$$

Thus, in order to find $r$ and $f$, it is best to first compute $a = F^{-1}x$ by one application of the Fast Fourier Transform. $f$ is then obtained trivially from (15), and $r$ from (14) by one further application of the Fast Fourier Transform. The smoothed power spectrum can now be obtained from (10) (if $h$ is small enough) or, if the time window is known, from (12).

**DEMONSTRATION 7.8-1:** We consider the time series

$$\{0.2, 0.9, -0.4, -0.1\},$$

where $m = 4$. (This is merely intended to show the workings of the various algorithms. In any real-life application, $m$ would be much larger.) The sequence $\mathbf{x}$ thus is

$$\mathbf{x} := \|: \quad 0.2, 0.9, -0.4, -0.1, 0, 0, 0, 0 \quad :\|.$$

By (1) and (8), the covariance function is

$$\mathbf{r} = \|: \quad 1.02, -0.14, -0.17, -0.02, 0.00, -0.02, -0.17, -0.14 \quad :\|.$$

An application of the Fourier transform yields the power spectrum $\mathbf{f} = \mathbf{Fr} = \{f_q\}$ where

| $q$ | $f_q$ |
|---|---|
| 0 | 0.0450 |
| 1 | 0.1063 |
| 2 | 0.1700 |
| 3 | 0.1487 |
| 4 | 0.1250 |
| 5 | 0.1487 |
| 6 | 0.1700 |
| 7 | 0.1063 |

Our analysis shows that the numbers $f_q$ must be real. In actual computation the imaginary parts of the $f_q$ will usually not be exactly zero. In the implementation of the Fast Fourier Transform discussed in §7.6, working with a ten-digit mantissa, these imaginary parts will be $\leq 5 * 10^{-10}$, which again show the basic stability of the process. Smoothing with the weights $\frac{1}{4}, \frac{1}{2}, \frac{1}{4}$ now yields the smoothed power spectrum $\mathbf{g} = \{g_q\}$ where

| $q$ | $g_q$ |
|---|---|
| 0 | 0.0757 |
| 1 | 0.1069 |
| 2 | 0.1487 |
| 3 | 0.1481 |
| 4 | 0.1369 |
| 5 | 0.1481 |
| 6 | 0.1487 |
| 7 | 0.1069 |

If the power spectrum is computed by the Convolution Theorem, we first have to form $\mathbf{a} := \mathbf{F}^{-1}\mathbf{x} = \{a_k\}$. Then, from (15),

$$f_q = \frac{1}{n} |a_q|^2.$$

The results are

| $q$ | $a_q$ | $f_q$ |
|---|---|---|
| 0 | $0.6000 + 0i$ | 0.0450 |
| 1 | $0.9071 + 0.1657i$ | 0.1063 |
| 2 | $0.6000 + 1.0000i$ | 0.1700 |
| 3 | $-0.5071 + 0.9657i$ | 0.1487 |
| 4 | $-1.0000 + 0i$ | 0.1250 |
| 5 | $-0.5071 - 0.9657i$ | 0.1487 |
| 6 | $0.6000 - 1.0000i$ | 0.1700 |
| 7 | $0.9071 - 0.1657i$ | 0.1063 |

in agreement with the results obtained previously. The computation of the smoothed power spectrum via time window and (12) yields, using the time window (13),

| $q$ | $(t \cdot r)_q$ | $(F(t \cdot r))_q$ |
|---|---|---|
| 0 | 1.0200 | 0.0757 |
| 1 | $-0.1195$ | 0.1069 |
| 2 | $-0.0850$ | 0.1488 |
| 3 | $-0.0029$ | 0.1481 |
| 4 | 0 | 0.1369 |
| 5 | $-0.0029$ | 0.1481 |
| 6 | $-0.0850$ | 0.1488 |
| 7 | $-0.1195$ | 0.1069 |

# PROBLEMS

1  Show that $\mathbf{Rx} * \mathbf{Ry} = \mathbf{R}(\mathbf{x} * \mathbf{y})$.
2  Denoting by $E$ the shift operator,

$$(E\mathbf{x})_k := x_{k+1},$$

show that for arbitrary sequences with period $n$,

$$FE^k\mathbf{x} = \mathbf{w}^k \cdot F\mathbf{x},$$

where the sequence $\mathbf{w}$ is defined by

$$\mathbf{w} := \{w^k\}, \qquad w := \exp\frac{2\pi i}{n}.$$

3  For the time series $\{4, 3, 2, 1\}$ compute the covariance function, the power spectrum, and the power spectrum smoothed by $\{\cdots 0\ \frac{1}{4}\ \frac{2}{4}\ \frac{1}{4}\ 0 \cdots\}$.

4  Two simple smoothing sequences $\mathbf{c} = \{c_j\}$ are

(a)  $c_j = \begin{cases} \dfrac{1}{2h+1}, & j = -h, -h+1, \ldots, h; \\ 0 & \text{otherwise,} \end{cases}$

(rectangular smoothing, see Figure 7.8a)

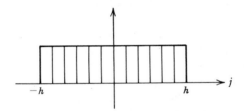

**Fig. 7.8a.** Rectangular smoothing.

(b)  $c_j = \begin{cases} \dfrac{1}{h}\left(1 - \dfrac{j}{h}\right), & j = -h, -h+1, \ldots, h \\ 0 & \text{otherwise,} \end{cases}$

(triangular smoothing, see Figure 7.8b)

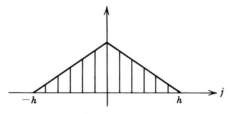

**Fig. 7.8b.** Triangular smoothing.

Show that the respective time windows

$$\mathbf{t} := \mathbf{F}_n^{-1}\mathbf{c} = \{t_m\}$$

are

$$
\text{(a)} \qquad t_m = \frac{1}{2h+1} \frac{\sin \dfrac{(2h+1)m\pi}{n}}{\sin \dfrac{m\pi}{n}},
$$

$$
\text{(b)} \qquad t_m = \frac{1}{h^2} \left[ \frac{\sin \dfrac{hm\pi}{n}}{\sin \dfrac{m\pi}{n}} \right]^2.
$$

**5**   Under what conditions on $\mathbf{x}$ is the sequence $\mathbf{F}_n \mathbf{x}$ real?

---

### Bibliographical Notes and Recommended Reading

On discrete Fourier analysis in general, see the survey articles by Henrici [1979a] and, in a more theoretical vein, Auslander and Tolmieri [1979]. On applications of discrete Fourier analysis to digital signal processing, see Oppenheim and Shafer [1975], and to digital image processing, Gonzalez and Wintz [1977]. The barycentric formula for trigonometric interpolation is given by Henrici [1979b]. The original reference on the Fast Fourier Transform is Cooley and Tukey [1965], although Goldstine [1977] found the root of the idea in a posthumous paper by Gauss. The Fast Fourier Transform for arbitrary $n$ dealt with in Problem 5 of §7.7 is based on an oral communication by A. Schönhage. Blackman and Tukey [1959], Tukey [1967] and Bloomfield [1976] are standard references on time series. Theorem 7.4 is due to Gaier [1974].

# BIBLIOGRAPHY

**H. Anton** [1980]   Calculus with analytic geometry. Wiley, New York.

**L. Auslander** and **R. Tolmieri** [1979]   Is computing with the finite Fourier transform pure or applied mathematics? Bull. Amer. Math. Soc. (New Series) **1**, 847–897.

**E. Batschelet** [1975]   Introduction to mathematics for life scientists. Springer, New York.

**R. B. Blackman** and **J. W. Tukey** [1959]   The measurement of power spectra. Dover, New York.

**P. Bloomfield** [1976]   Fourier analysis of time series. Wiley, New York.

**W. E. Boyce** and **R. C. DiPrima** [1977]   Elementary differential equations, 3rd ed. Wiley, New York.

**R. C. Buck** [1956]   Advanced calculus. McGraw-Hill, New York.

**T. F. Chan, G. H. Golub,** and **R. J. LeVeque** [1979]   Updating formulae and a pairwise algorithm for computing sample variances. STAN-CS-79-773, Department of Computer Science, Stanford University.

**E. W. Cheney** [1966]   Introduction to approximation theory. McGraw-Hill, New York.

**J. W. Cooley** and **J. W. Tukey** [1965]   An algorithm for the machine calculation of complex Fourier series. Math. Comp. **19**, 297–301.

**I. W. Cotton** [1975]   Remark on stably updating mean and standard deviation of data. Comm. ACM **18**, 458. Corrigendum, same Comm. **18**, 591.

**G. Dahlquist** [1963]   A special stability problem for linear multistep methods. BIT **3**, 27–43.

**G. Dahlquist** and **A. Björck** [1974]   Numerical methods. Prentice-Hall, Englewood Cliffs.

**P. Davis** [1961]   Interpolation and approximation. Blaisdell, Waltham.

**P. Davis** and **P. Rabinowitz** [1967]   Numerical integration. Blaisdell, London.

**J. J. Dongarra, J. R. Bunch, C. B. Moler,** and **G. W. Stewart** [1979]   LINPACK user's guide. Soc. Indust. Appl. Math., Philadelphia.

**M. Fisz** [1963]   Probability and mathematical statistics. Wiley, New York.

**G. E. Forsythe, M. A. Malcolm,** and **C. B. Moler** [1977]   Computer methods for mathematical problems. Prentice-Hall, Englewood Cliffs.

**G. E. Forsythe** and **C. B. Moler** [1967]   Computer solutions of linear algebraic systems. Prentice-Hall, Englewood Cliffs.

**G. E. Forsythe** and **W. Wasow** [1960]   Finite difference methods for partial differential equations. Wiley, New York.

**399**

**D. Gaier** [1974] Ableitungsfreie Abschätzunger bei trigonometrischer Interpolation und Konjugiertenbestimmung. Computing **12**, 145–148.

**S. Gass** [1969] Linear programming, methods and applications, 3rd ed. McGraw-Hill, New York.

**H. H. Goldstine** [1977] A history of numerical analysis from the 16th through the 19th century. Springer, New York.

**R. C. Gonzalez** and **P. Wintz** [1977] Digital image processing. Addison-Wesley, Reading.

**R. J. Hanson** [1975] Stably updating mean and standard deviations of data. Comm. ACM **18**, 57–58.

**P. Henrici** [1962] Discrete variable methods in ordinary differential equations. Wiley, New York.

—— [1964] Elements of numerical analysis. Wiley, New York.

—— [1974, 1977] Applied and computational complex analysis, vols I and II. Wiley, New York.

—— [1979a] Fast Fourier methods in computational complex analysis. SIAM Review **21**, 481–527.

—— [1979b] Barycentric formulas for interpolating trigonometric polynomials and their conjugates. Numer. Math. **33**, 225–234.

**A. S. Householder** [1970] The numerical treatment of a single non-linear equation. McGraw-Hill, New York.

**J. D. Lambert** [1971] Computational methods in ordinary differential equations. Wiley, New York.

**L. Lapidus** and **J. H. Seinfield** [1971] Numerical solution of ordinary differential equations. Academic Press, New York.

**C. L. Lawson** and **R. J. Hanson** [1974] Solving least square problems. Prentice-Hall, Englewood Cliffs.

**P. M. Neely** [1966] Comparison of several algorithms for computation of means, standard deviations, and correlation coefficients. Comm. ACM **9**, 496–499.

**A. V. Oppenheim** and **R. W. Schafer** [1975] Digital signal processing. Prentice-Hall, Englewood Cliffs.

**J. M. Ortega** and **W. C. Rheinboldt** [1970] Iterative solution of nonlinear equations in several variables. Academic Press, New York.

**A. M. Ostrowski** [1966] Solution of equations and systems of equations, 2nd ed. Academic Press, New York.

**R. D. Richtmyer** and **K. W. Morton** [1967] Difference methods for initial value problems, 2nd ed. Interscience, New York.

**W. Rudin** [1964] Principles of mathematical analysis. McGraw-Hill, New York.

**H. Rutishauser** [1952] Ueber die Instabilität von Methoden zur Integration gewöhnlicher Differentialgleichungen. Z. angew. Math. Physik **3**, 65–74.

—— [1954] Der Quotienten-Differenzen-Algorithmus. Z. angew. Math. Physik **5**, 233–251.

—— [1976] Vorlesungen über numerische Mathematik, 2 vols. Birkhäuser, Basel.

**S. L. Salas** and **Einar Hille** [1978] Calculus: One and several variables, 3rd ed. Wiley, New York.

**H. E. Salzer** [1972] Lagrangian interpolation at the Chebyshev points $x_{n,\nu} = \cos(\nu\pi/n)$, $\nu = 0(1)n$; some unnoted advantages. The Computer J. **15**, 156–159.

**A. Sard** and **S. Weintraub** [1971] A book of splines. Wiley, New York.

**M. H. Schultz** [1973] Spline analysis. Prentice-Hall, Englewood Cliffs.

**L. F. Shampine** and **M. K. Gordon** [1975] Computer solutions of ordinary differential equations: The initial value problem. Freeman.

**O. D. Smith** [1978] Numerical solution of partial differential equations: Finite difference methods. Clarendon Press, Oxford.

**G. W. Stewart** [1973] Introduction to matrix computations. Academic Press, New York.

**E. L. Stiefel** [1963] Introduction to numerical mathematics, transl. by Werner Rheinboldt. Academic Press, New York.

**G. Strang** [1976] Linear algebra and its applications. Academic Press, New York.

**W. G. Strang** and **G. Fix** [1973] An analysis of the finite element method. Prentice-Hall, Englewood Cliffs.

**F. Stummel** [1980] Rounding error analysis of elementary numerical algorithms. Computing, Suppl. 2, 169–195.

**W. J. Taylor** [1945] Method of Lagrangian curvilinear interpolation. J. Research National Bureau of Standards **35**, 151–155.

**J. W. Tukey** [1967] An introduction to the calculation of numerical spectrum analysis. The spectral analysis of time series, B. Harris, ed. John Wiley, New York.

**J. V. Uspensky** [1948] Theory of equations. McGraw-Hill, New York.

**A. J. van Reeken** [1968] Dealing with Neely's algorithms. Comm. ACM **11**, 149–150.

**R. S. Varga** [1962] Matrix iterative analysis. Prentice-Hall, Englewood Cliffs.

**B. P. Welford** [1962] Note on a method for calculating corrected sums of squares and products. Technometrics **4**, 419–420.

**J. H. Wilkinson** [1963] Rounding errors in algebraic processes. H'M's Stationary Office, London.

——— [1965]   The algebraic eigenvalue problem. Oxford University Press, Oxford.

**J. H. Wilkinson** and **C. Reinsch** [1971]   Linear algebra. Handbook for automatic computation, vol. II. Springer, Berlin.

**D. M. Young** [1971]   Iterative solution of large linear systems. Academic Press, New York.

# Index

Principal references (definitions) are printed in **bold face**.